REF
QH
651
.P47
Vol.8 Cop.1

PHOTOPHYSIOLOGY

Current Topics in Photobiology
and Photochemistry

Volume VIII

Editorial Board

David Fork	*Carnegie Institution*
John Jagger	*University of Texas*
Cyril Ponnamperuma	*University of Maryland*
Kendric Smith	*Medical School, Stanford University*
Frederick Urbach	*Medical School, Temple University*
R. A. Weale	*Institute of Ophthalmology, London*

PHOTOPHYSIOLOGY

CURRENT TOPICS IN PHOTOBIOLOGY AND PHOTOCHEMISTRY

Edited by

Arthur C. Giese
Department of Biological Sciences
Stanford University, Stanford, California

Volume VIII

1973

ACADEMIC PRESS · NEW YORK and LONDON
A Subsidiary of Harcourt Brace Jovanovich, Publishers

Copyright © 1973, by Academic Press, Inc.
ALL RIGHTS RESERVED.
NO PART OF THIS PUBLICATION MAY BE REPRODUCED OR
TRANSMITTED IN ANY FORM OR BY ANY MEANS, ELECTRONIC
OR MECHANICAL, INCLUDING PHOTOCOPY, RECORDING, OR ANY
INFORMATION STORAGE AND RETRIEVAL SYSTEM, WITHOUT
PERMISSION IN WRITING FROM THE PUBLISHER.

ACADEMIC PRESS, INC.
111 Fifth Avenue, New York, New York 10003

United Kingdom Edition published by
ACADEMIC PRESS, INC. (LONDON) LTD.
24/28 Oval Road, London NW1

LIBRARY OF CONGRESS CATALOG CARD NUMBER: 63-16961

PRINTED IN THE UNITED STATES OF AMERICA

CONTENTS

List of Contributors ix
Preface xi
Contents of Other Volumes xiii

CHAPTER 1. COMPARATIVE STUDIES ON PHOTOSYNTHESIS IN HIGHER PLANTS
Olle Björkman

1. Introduction 1
2. The Photosynthetic Process 4
3. Comparative Studies of Component Photosynthetic Reactions 9
4. Photosynthetic Adaptation 37
   References 59

CHAPTER 2. ANALYSIS OF PHOTOSYNTHESIS IN GREEN ALGAE THROUGH MUTATION STUDIES
Norman I. Bishop

1. Introduction 65
2. Methodology 66
3. Types of Mutations Directly Influencing the Photosynthetic Mechanism 67
   References 93

CHAPTER 3. SEPARATION OF PHOTOSYNTHETIC SYSTEMS I AND II
Jeanette S. Brown

1. Introduction 97
2. Differentiation of the Photosystems 99
3. Morphological Separation of the Photosystems in Plants 101
4. Physical Separation of the Photosystems *in Vitro* . . . 105
   References 110

CHAPTER 4. THE ROLE OF CATION FLUXES IN CHLOROPLAST ACTIVITY

GEOFFREY HIND AND RICHARD E. MCCARTY

1. General Introduction 114
2. The Measurement of Ion Fluxes in Organelle Suspensions . 114
3. Relation of Cation Fluxes to Energy Transduction . . 123
4. Cation Functions in Catalysis and Control 142
5. Mechanisms for Possible Control of Photosynthesis by Mg Ion 148
   References 153

CHAPTER 5. NITROGEN FIXATION IN PHOTOSYNTHETIC BACTERIA

DONALD L. KEISTER AND DARRELL E. FLEISCHMAN

1. Introduction 157
2. Relationship of Hydrogenase and Nitrogenase . . . 159
3. Role of Nitrogen Fixation in Metabolism 162
4. Control of Nitrogenase 164
5. Enzymology of Nitrogen Fixation 168
6. Ecological Significance of Nitrogen Fixation in Photosynthetic Bacteria 178
7. Concluding Remarks 179
   References 179

CHAPTER 6. PARALLEL AND SEQUENTIAL PROCESSING OF VISUAL INFORMATION IN MAN: INVESTIGATION BY EVOKED POTENTIAL RECORDING

D. REGAN

1. Introduction 185
2. Methods of Recording Evoked Potentials in Man . . . 188
3. Examples of Evoked Potential Studies of Visual Information Processing 192
4. Summary 207
   References 207

CHAPTER 7. INHIBITION OF GROWTH AND RESPIRATION BY VISIBLE AND NEAR-VISIBLE LIGHT

B. L. Epel

1. Introduction	209
2. Light and Growth: General Studies	210
3. Mechanisms of Photoinhibition	215
4. Concluding Remarks	228
References	228

CHAPTER 8. DENATURATION IN ULTRAVIOLET-IRRADIATED DNA

Ronald O. Rahn

1. Introduction	231
2. Types of Photodamage to DNA	232
3. Qualitative Detection of Defects	233
4. Quantitative Detection of Defects	244
5. Nature of Defect Region	249
6. Summary and Future Experiments	254
References	254

Author Index	257
Subject Index	267

LIST OF CONTRIBUTORS

Numbers in parentheses indicate the pages on which the authors' contributions begin.

Norman I. Bishop, *Department of Botany, Oregon State University, Corvallis, Oregon* (65)

Olle Björkman, *Department of Plant Biology, Carnegie Institution of Washington, Stanford, California* (1)

Jeanette S. Brown, *Department of Plant Biology, Carnegie Institution of Washington, Stanford, California* (97)

B. L. Epel, *Department of Botany, Tel Aviv University, Tel Aviv, Israel* (209)

Darrell E. Fleischman, *Charles F. Kettering Research Laboratory, Yellow Springs, Ohio* (157)

Geoffrey Hind, *Biology Department, Brookhaven National Laboratory, Upton, New York* (113)

Donald L. Keister, *Charles F. Kettering Research Laboratory, Yellow Springs, Ohio* (157)

Richard E. McCarty, *Section of Biochemistry, Molecular and Cell Biology, Cornell University, Ithaca, New York* (113)

Ronald O. Rahn, *Biology Division, Oak Ridge National Laboratory, Oak Ridge, Tennessee* (231)

D. Regan, *Department of Communication, University of Keele, Keele, Staffordshire, England* (185)

PREFACE

Volume VIII of *Photophysiology* is largely concerned with photobiology of plants. It was not so planned but is the result of default since the manuscripts dealing with other subjects did not meet the deadline. However, this has advantages inasmuch as it brings in review in one volume a number of interrelated aspects of plant photobiology. Chapters 1 to 4 on photosynthesis include (1) comparative studies on photosynthesis blending discussion at the plant, cellular, and molecular levels and relating these to the environment of higher plants; (2) an analysis of photosynthesis in the green algae by genetic methods; (3) studies on the physical separation of photosynthetic system I from photosynthetic system II; (4) the role of ion fluxes in chloroplast activity. Chapter 5 considers the interrelations between photosynthesis and nitrogen fixation in the photosynthetic bacteria. Chapter 6 is an account of photically induced potentials in man, and Chapter 7 reviews the inhibition of growth and respiration in a variety of cells by visible and near-visible light. Chapter 8 deals with denaturation of DNA by ultraviolet light. Volume VIII is thus primarily concerned with the photobiology of visible light.

Photophysiology, since its inception in 1964, has attempted to get interplay between photobiologists in widely separated areas, by sampling in each volume progress in a variety of areas relevant to photobiology. Thus it serves as a clearing house at an advanced level. This continues to be one of its major objectives. Because, at an advanced level the treatment is also often quite specialized, *Photophysiology* cannot in any one volume fully demonstrate the extensive interaction of light with organisms and cells and the need to consider the effects of light whenever organisms are in an environment in which light reaches them. The series can, however, serve as a repository of reviews and thus make possible quick access to original work in many areas of photobiology of general interest. In this way *Photophysiology* relates specialized findings to general biological thinking.

Photobiology does not appear in the curriculum of many colleges and universities and although individuals on the staff of many of these institutions are studying the effects of light on organisms in one way or another, some biologists have little contact with photobiology and its problems which to them may even appear esoteric.

ARTHUR C. GIESE

CONTENTS OF OTHER VOLUMES

VOLUME I

HISTORICAL INTRODUCTION
  Arthur C. Giese

PRINCIPLES OF PHOTOCHEMISTRY AND PHOTOCHEMICAL METHODS
  Stig Claesson

ELECTRON SPIN RESONANCE AND ITS APPLICATION TO PHOTOPHYSIOLOGY
  M. H. Blois, Jr., and E. C. Weaver

PHOTOCHEMICAL ACTION OF LIGHT ON MACROMOLECULES
  A. D. McLaren

ABSORPTION SPECTRA, SPECTROPHOTOMETRY, AND ACTION SPECTRA
  Mary Belle Allen

THE PHOTOCHEMICAL REACTIONS OF PHOTOSYNTHESIS
  F. R. Whatley and M. Losada

PHYSICAL ASPECTS OF THE LIGHT REACTION IN PHOTOSYNTHESIS
  Roderick K. Clayton

ACCESSORY PIGMENTS AND PHOTOSYNTHESIS
  L. R. Blinks

PHOTOTROPISM IN HIGHER PLANTS
  Winslow R. Briggs

SOME EFFECTS ON CHLOROPLASTS AND PLANT PHOTOPLASM
  Hemming I. Virgin

PHOTOCHEMICAL ASPECTS OF PLANT PHOTOPERIODICITY
  Sterling B. Hendricks

THE ROLE OF LIGHT IN PERSISTENT DAILY RHYTHMS
  J. Woodland Hastings

Author Index—Subject Index

VOLUME II

ANIMAL PHOTOPERIODISM
Albert Wolfson

PHOTOTAXIS IN MICROORGANISMS
Roderick K. Clayton

THE PHOTORECEPTOR PROCESS IN LOWER ANIMALS
Donald Kennedy

VISION AS A PHOTIC PROCESS
W. A. H. Rushton

THE PHYSICAL LIMITS OF VISUAL DISCRIMINATION
H. B. Barlow

STUDIES ON ULTRAVIOLET RADIATION ACTION UPON ANIMAL CELLS
Arthur Giese

MUTAGENIC EFFECTS OF ULTRAVIOLET AND VISIBLE LIGHT
G. Zetterberg

PHOTOREACTIVATION OF ULTRAVIOLET DAMAGE
Claud S. Rupert

PHOTOCHEMISTRY OF THE NUCLEIC ACIDS
Kendric C. Smith

BIOLUMINESCENCE—PRODUCTION OF LIGHT BY ORGANISMS
Aurin M. Chase

Author Index—Subject Index

VOLUME III

PHOTOCHEMICAL METHODS
Gilbert R. Seely

PHOTODYNAMIC ACTION
John D. Spikes

PHOTOTROPISM IN FUNGI
  Robert M. Page

STUDIES ON THE EFFECT OF LIGHT ON CHLOROPLAST STRUCTURE
  Lester Packer and David W. Deamer

THE PROTECTIVE FUNCTION OF CAROTENOID PIGMENTS
  Norman I. Krinsky

STRUCTURE OF THE PHOTOSYNTHETIC APPARATUS
  Daniel Branton

PRIMARY PROCESSES IN PHOTOSYNTHESIS
  G. Hoch and R. S. Knox

ULTRAVIOLET RADIATION AND THE ORIGIN OF LIFE
  Cyril Ponnamperuma

Author Index—Subject Index

VOLUME IV

PHOTOCHEMISTRY AND VISION
  R. A. Weale

PHOTOPERIODISM IN INSECTS
  Anthony D. Lees

RESPONSE OF HUMAN SKIN TO ULTRAVIOLET LIGHT
  Brian E. Johnson, Farrington Daniels, Jr.,
    and Ian A. Magnus

CELLULAR RECOVERY FROM PHOTOCHEMICAL DAMAGE
  Philip C. Hanawalt

A PHYSICAL APPROACH TO BIOLUMINESCENCE
  H. H. Seliger and Richard A. Morton

BIOLUMINESCENCE: ENZYMIC ASPECTS
  Milton J. Cromier and John R. Totter

Author Index—Subject Index

VOLUME V

FLASH PHOTOLYSIS RESEARCH IN PHOTOBIOLOGY
L. I. Grossweiner

PHYSICAL PROPERTIES OF SINGLET OXYGEN
E. A. Ogryzlo

CHEMICAL AND BIOLOGICAL ASPECTS OF SINGLET EXCITED MOLECULAR OXYGEN
Thérèse Wilson and J. Woodland Hastings

SPECTROPHOTOMETRIC STUDIES OF THE MECHANISM OF PHOTOSYNTHESIS
David C. Fork and Jan Amesz

THE PHOTOPHYSIOLOGY OF VERTEBRATE COLOR VISION
H. Ripps and R. A. Weale

THE ACTION OF ULTRAVIOLET LIGHT ON MAMMALIAN CELLS
Robert B. Painter

PHOTOREACTIVATION IN ANIMAL CELLS
John S. Cook

ULTRAVIOLET CARCINOGENESIS
John H. Epstein

Author Index—Subject Index

VOLUME VI

CHLOROPHYLL FLUORESCENCE AND PHOTOSYNTHESIS: FLUORESCENCE TRANSIENTS
Govindjee and George Papageorgiou

MECHANISM OF LIGHT-INDUCED CAROTENOID SYNTHESIS IN NONPHOTOSYNTHETIC PLANTS
Prem P. Batra

PHOTOSENSITIZATION BY NATURAL PIGMENTS
Arthur C. Giese

SOLAR UV IRRADIATION AND THE GROWTH AND DEVELOPMENT OF HIGHER PLANTS
  Martyn M. Caldwell

PHOTOBIOLOGY OF PLANT VIRUSES
  A. Kleczkowski

THE ROLES OF GENETIC RECOMBINATION AND DNA POLYMERASE IN THE REPAIR OF DAMAGED DNA
  Kendric C. Smith

THE STUDY OF PHOTOENZYMATIC REPAIR OF UV LESIONS IN DNA BY FLASH PHOTOLYSIS
  Walter Harm, Claud S. Rupert, and Helga Harm

PHOTOPERIODISM AND REPRODUCTIVE CYCLES IN BIRDS
  Donald S. Farner and Robert A. Lewis

Author Index—Subject Index

VOLUME VII

ELECTRON RESONANCE STUDIES OF PHOTOSYNTHETIC SYSTEMS
  Ellen C. Weaver and Harry E. Weaver

PHYTOCHROME, A PHOTOCHROMIC SENSOR
  W. Shropshire, Jr.

PHOTOSYNTHETIC NITROGEN FIXATION BY BLUE-GREEN ALGAE
  Mary Belle Allen

THE ROLE OF LIGHT IN NITRATE METABOLISM IN HIGHER PLANTS
  Leonard Beevers and R. H. Hageman

MODE OF PHOTOSENSITIZING ACTION OF FUROCOUMARINS
  Luigi Musajo and Giovanni Rodighiero

PHOTOSENSITIVITY IN PORPHYRIA
  Walter J. Runge

ULTRAVIOLET ACTION AND PHOTOREACTIVATION IN ALGAE
  Per Halldal and Örn Taube

CHROMOSOME ABERRATIONS INDUCED BY ULTRAVIOLET RADIATION
Sheldon Wolff

PHOTOCHEMISTRY OF NUCLEIC ACIDS AND THEIR CONSTITUENTS
A. J. Varghese

ENZYMATIC AND NONENZYMATIC BIOLUMINESCENCE
Frank H. Johnson and Osamu Shimomura

Author Index—Subject Index

Chapter 1

COMPARATIVE STUDIES ON PHOTOSYNTHESIS IN HIGHER PLANTS*

Olle Björkman

Department of Plant Biology, Carnegie Institution of Washington, Stanford, California

1. Introduction 1
2. The Photosynthetic Process 4
   2.1 The Overall Reaction 4
   2.2 Component Reactions and Their Relation to Environmental Factors . 5
3. Comparative Studies of Component Photosynthetic Reactions . . . 9
   3.1 Photochemical Events 9
   3.2 Photosynthetic Electron Transport 18
   3.3 Carbon Fixation and Reduction 22
   3.4 Oxygen Inhibition of Photosynthesis and Photorespiration . . . 31
   3.5 Diffusion Paths and the Stomatal Mechanism 34
4. Photosynthetic Adaptation 37
   4.1 Adaptations Involving Quantitative Differences 38
   4.2 Adaptive and Evolutionary Aspects of C_4 Photosynthesis and Crassulacean Acid Metabolism 51
   References 59

1. Introduction

Since photosynthesis is the source of nearly all chemical energy and organic carbon that enter into any terrestrial ecosystem, it is obvious that the efficiency and the capacity with which the constituent plants are able to carry out this process ultimately determine the potential productivity. An estimation of its limits in any one of the widely diverse natural environments that exist on earth therefore requires knowledge of the relationship between the photosynthetic process and the environment. This includes not only the dependence of photosynthesis on the immediate environment to which the plant is exposed and the ability of a given species, or genotype, to adjust its photosynthetic characteristics in re-

* CIW DPB Publication No. 492.

sponse to an environmental change, but also the kind and extent of genetically determined differences—quantitative and qualitative—that have evolved among plants occupying contrasting natural habitats as a result of natural selection.

Information on these fundamental relationships is also of great importance to research on plant evolution and distribution, and it is basic to the applied plant sciences, particularly research concerned with improving the productivity and extending the range of cultivation of economically important plants.

It is evident that such information can most effectively be obtained through *comparative* investigations of photosynthetic performance and characteristics of plants from ecologically diverse environments. Many early workers in photosynthesis research realized this, and much information on the response of the overall rate of photosynthesis to changes in external factors was gathered during the first three decades of this century. To a considerable degree the main purpose of these intensive studies of the kinetics of photosynthesis was to obtain insight into its biochemical mechanism, and information on the relationship between photosynthesis and the environment was often merely a by-product, albeit an important one. However, experiments were also specifically designed to elucidate environmental and adaptive aspects of photosynthesis, and several comparisons of the photosynthetic performance of species occupying ecologically diverse habitats were made. Among the pioneers in this area of research were Harder in Germany, Lundegårdh and Stålfelt in Sweden, Boysen-Jensen and Müller in Denmark, and Livingston and Shreve in the United States. These workers were no doubt aware of the great possibilities of comparative studies of photosynthesis, but the crude techniques available at that time and lack of knowledge of the basic mechanisms of photosynthesis severely limited progress in this field.

During the past three decades tremendous progress has been made in uncovering the mechanisms of photosynthesis. It is now possible to take apart the highly complex overall process and to study its component steps. This rapid progress necessitated a high degree of specialization in photosynthesis research, and for most investigators it was no longer possible to study the process as a whole. As a result, environmental and comparative aspects of photosynthesis ceased to be in the mainstream of photosynthesis research, and meaningful communication between the photosynthesis specialist and the ecologically oriented plant biologist came almost to a halt. In the quest for understanding the fundamental mechanisms of photosynthesis it also became necessary to use limited plant materials, and it was generally believed that variability among higher plants, occupying ecologically diverse habitats, would be restricted to small quantitative differences that were mainly attributable to leaf morphology and stomatal behavior.

This was the situation until some six years ago, when interest in comparative studies of photosynthesis in higher plants was revived. Undoubtedly, much of the stimulation of new research in this area came from the recent discovery of the C_4 dicarboxylic acid pathway of photosynthetic CO_2 fixation which characterizes large groups of higher plants and from the realization that the high oxygen content of the present atmosphere causes a marked inhibition of net photosynthesis and primary productivity in all plants that possess the conventional Calvin-Benson pathway of CO_2 fixation. This resurgence of interest in comparative studies is not restricted to the biochemistry of plants possessing qualitative differences in their photosynthetic pathways, but has carried over to the study of environmental, evolutionary, and adaptive aspects of photosynthesis in general.

Investigations in these study areas are much more likely to meet with considerably greater success now than some thirty years ago. Not only does the investigator have at his disposal the necessary techniques for precise quantitative measurements of photosynthesis and of external parameters in the field and the laboratory as well as the facilities for growing the experimental plants under a series of controlled environments, but he also has available to him a mass of information on the basic mechanisms of the photosynthetic process. I believe that the renaissance in the area of physiological ecology has just begun, and it is probably not too optimistic to predict that we will be witnessing intensive activity and major advances in this area within this decade.

In this chapter I will attempt to review developments in some areas of comparative studies of higher plant photosynthesis that have taken place in the past ten-year period as seen from the viewpoint of a plant physiologist who is interested in ecological and adaptive aspects of the process. The approach is nevertheless "mechanism-oriented," and no attempt is made to cover such important study areas as those concerned with comparisons of photosynthesis and productivity of whole natural biomes or of different crop stands. Neither does this article deal with studies of the influence of plant canopy architecture on photosynthesis or the relationship between photosynthesis and yield. General aspects of the mechanisms of the photosynthetic process and its dependence on the various components of the physical environment are treated in Section 2, and comparative studies of the capacities and pathways of its various constituent processes among plants, occupying ecologically diverse environments, are discussed in Section 3. Throughout the article the main emphasis is on functional differentiation of the photosynthetic apparatus and its adaptive significance; therefore, differences rather than similarities are stressed. Photosynthetic adaptation to different habitats, the possible mechanisms underlying these adaptations, and evolutionary aspects of photosynthetic differentiation are treated in Section 4. The treatment undoubtedly reflects the particular interests of the author, and often examples are taken from the work carried out in the author's laboratory. This in no way implies that these are the best or only examples. Throughout the article citations should be regarded as illustrative rather than inclusive.

2. The Photosynthetic Process

2.1 The Overall Reaction

Photosynthesis may be defined as the conversion of light energy to chemically bound energy in plant constituents. In all higher plants and most algae, water serves as the final electron donor and other inorganic compounds, predominantly CO_2 but also nitrate and sulfate serve as the final electron acceptors. As a result, water is oxidized to molecular oxygen and inorganic compounds are reduced to organic compounds of higher free energy content. Water and carbon dioxide are required for every final product of photosynthesis. When carbohydrates are the sole final organic products the overall reaction may be expressed by the equation:

$$H_2O + CO_2 + light \rightarrow O_2 + (CH_2O)$$

Consequently, for each mole of CO_2 that is reduced to the level of carbohydrate, 1 mole of oxygen is evolved and the net gain in free energy by the plant is 112,000 cal. Under these conditions, the molar ratio CO_2 consumed:O_2 evolved is equal to unity and it makes no difference whether the rate of photosynthesis is measured as the rate of CO_2 uptake, O_2 release, carbohydrate formation, or energy gain, since any one of these can be calculated from the other. It should be noted, however, that recent research has shown (Bassham, 1971) that although they are usually the compounds formed in the largest amount, carbohydrates are not the sole organic products of photosynthesis but, in addition, amino acids (from which proteins are synthesized), glycerol phosphate (from which fats are made), and organic acids are produced in substantial amounts. It is obvious that whenever compounds which are less reduced than carbohydrates (such as organic acids) are formed to a significant extent, the ratio CO_2 consumed:O_2 evolved is greater than unity and the energy gain per CO_2 consumed becomes less than 112,000 cal. On the other hand, when a significant amount of compounds that are more reduced than carbohydrate are formed, or when nitrate and sulfate are reduced for the synthesis of nitrogen- or sulfur-containing compounds, the ratio CO_2

consumed:O_2 evolved is less than unity and the energy gain exceeds 112,000 cal per mole of CO_2 fixed. Fock *et al.* (1969, 1972) reported that for higher plants photosynthesizing in normal air the ratio net CO_2 consumed:net O_2 evolved is as high as 1.3 and exceeds 2.0 at elevated O_2 concentrations even though the ratio is equal to, or slightly less than unity in an atmosphere with 2% O_2 concentration. In view of these remarkable results the net energy gain may be considerably lower than predicted if one assumes that 112,000 cal are gained per mole of CO_2. These observations, if proved to be correct, are of particular importance in view of the fact that rates of higher plant photosynthesis are almost exclusively measured as CO_2 uptake alone. The ratio CO_2 consumed:O_2 evolved may vary not only among different species and in different environments, but may also be expected to vary with the age of a leaf, as there is a shift from protein synthesis in the young leaf to carbohydrate synthesis in the mature leaf and perhaps a considerable synthesis of organic acids in the old leaf.

2.2 Component Reactions and Their Relation to Environmental Factors

The equation for the overall process gives no indication of the complex mechanism of photosynthesis. In the first quarter of this century, it was considered as a reversal of the respiratory process in which combustion of carbohydrate yields CO_2 and H_2O. It was thought that CO_2 in association with chlorophyll participates in a photochemical reaction in which the oxygen of CO_2 is split off and replaced with H_2O. A drastic change in the conception of the process stemmed from Van Niel's comparative studies in the 1930's of higher plant and algal photosynthesis with that of photosynthetic bacteria, which are able to use inorganic substrates such as hydrogen sulfide and where the assimilation of CO_2 is associated with the production of sulfur rather than oxygen. Van Niel proposed that higher plant and algal photosynthesis is a special case, where water serves as an electron donor, and that the evolved oxygen comes from water, not from CO_2, a hypothesis whose correctness has now been well established. The reduction of CO_2 has therefore nothing to do with the photochemical part of photosynthesis, except that the latter process provides the necessary reductant. Van Niel's studies amply demonstrate the great potential of the comparative approach.

As currently conceived, the overall process is composed of four major stages (Fig. 1). These component stages are interdependent and generally form a linear sequence. The rate of the overall process may be limited by the capacity of any one of these stages. The first stage is the absorption of light quanta by the light-harvesting pigment molecules; as a result the energy of these molecules is raised, and the molecules are said to be in an excited state. The light-harvesting pigment molecules then

Formation of reducing power and ATP

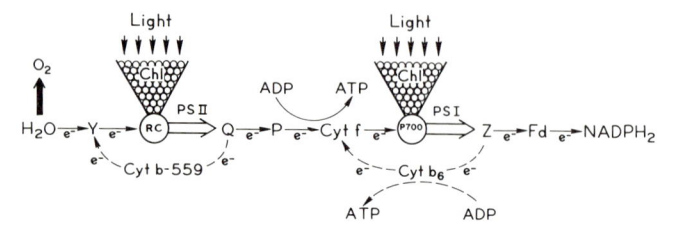

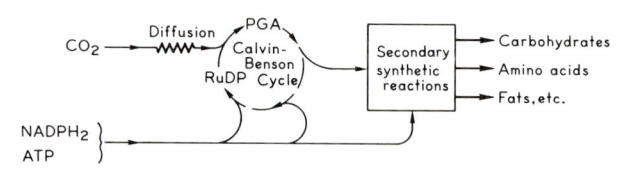

Fig. 1. A current simplified scheme of higher plant photosynthesis. Abbreviations and symbols: PS II, photosystem II; chl, light-harvesting chlorophyll; RC, reaction center of PS II; Y, oxidant and Q, reductant of photoact II; P, electron carrier, probably identical to a plastoquinone; cyt, cytochrome; PS I, photosystem I; P700, reaction center chlorophyll of PS I; Z, primary reductant of photoact I; Fd, ferredoxin; RuDP, ribulose 1,5-diphosphate; PGA, 3-phosphoglyceric acid. Several electron carriers and many other intermediates are not shown. The C₄ dicarboxylic acid pathway of CO_2 fixation is depicted in Fig. 6.

transfer their absorbed energy to reaction centers at which oxidizing and reducing entities are formed. These are the primary photochemical events in photosynthesis. In higher plant chloroplasts (and algae) there are two separate photosystems, each with its own reaction center and light-harvesting pigment system (see Chapter 3 by J. S. Brown). The rate at which these light reactions proceed is dependent on the quality and intensity of light alone and is totally unaffected by other external factors such as CO_2 concentration and temperature as long as the system remains intact.

The oxidant and reductant produced in the photochemical reactions are used in sequence to drive electron transport. Electrons are taken from water in reactions which are closely linked with the oxidant of photosystem II, and as a result, gaseous oxygen is evolved.* The re-

---

* The zigzag formulation of the mechanism of photosynthesis which was suggested by Hill and figured by Duysens is conceived as consisting of two photosynthetic systems (PS) connected by electron carriers. Some workers prefer to designate these PS 2 and PS 1, and other workers prefer to designate them PS II and PS I. The choice is arbitrary, nothing specific being implied by a choice between the two. (Ed.)

ductant of photosystem II is apparently linked via electron carriers to the oxidant of photosystem I. The strong reductant of this photosystem in turn reduces NADP, a diffusible electron carrier which transports the reducing power, generated in the photosynthetic membranes, to other sites within the chloroplast. Concomitantly with electron transport, ATP is synthesized in photophosphorylation. The electron transport reactions, while they are quite rapid, are still much slower than the photochemical reactions.

This second stage of photosynthetic electron transport and phosphorylation, is driven by light only indirectly, and the rate by which it proceeds is affected primarily by temperature.

A third stage is the biochemical fixation of CO_2. Until recently it was thought that in the light all plants utilized the Calvin-Benson (the reductive pentose phosphate) pathway for CO_2 fixation. In this very important pathway, one molecule of CO_2 reacts in the chloroplast with the 5-carbon sugar phosphate ribulose 1,5-diphosphate (RuDP) to form two molecules of the 3-carbon compound 3-phosphoglyceric acid (C_3 plants). This key reaction is catalyzed by the enzyme RuDP carboxylase or carboxydismutase. Recent comparative studies have shown that in many higher plants the initial fixation of CO_2 in the light instead takes place via the Hatch-Slack-Kortschak, or the C_4-dicarboxylic acid pathway of photosynthesis. In this pathway, CO_2 reacts with the 3-carbon phospho-(enol)pyruvic acid (PEP) to form the 4-carbon dicarboxylic acid, oxaloacetate, which is rapidly reduced to malic acid or aminated to aspartic acid (C_4 plants). This reaction is catalyzed by the enzyme PEP carboxylase. The CO_2 fixation step in this pathway is similar to that used in dark fixation of CO_2 by certain succulent plants exhibiting crassulacean acid metabolism (CAM-plants). Regardless of the pathway used, the rate at which the CO_2 fixation step can proceed is only indirectly dependent on light, but it is directly and strongly dependent on temperature and particularly on the concentration of CO_2 in the leaf. The CO_2 fixation by RuDP carboxylase may also be markedly affected by oxygen concentration.

The reduction of CO_2 to carbohydrates, lipids, etc., involves a great number of reactions that utilize reduced NADP and ATP formed in the previous stages. In addition, NADP and ATP are required for the regeneration of the CO_2 acceptor. These various reactions involve a multitude of enzyme-catalyzed reactions whose rates are directly and strongly dependent on temperature but are only indirectly related to factors such as light and CO_2 concentration.

The diffusive flow of CO_2 from the external air to its site of reaction in the leaf must also be considered as a component of the photosynthetic process. During steady-state photosynthesis a concentration gradient develops along which CO_2 will diffuse. In this diffusion path CO_2 encounters several physical barriers, or resistances, among which are the boundary layer between the ambient air and the leaf surface, the stomatal pore, and the mesophyll cell walls. The rate of diffusive transport is determined by the concentration gradient between the atomsphere and the mesophyll cells. Thus, it is directly dependent on the CO_2 concentration external to the leaf. It is directly affected by temperature, but much less so than most enzyme-catalyzed reactions.

It is clear from what has been said above that the different component processes of photosynthesis are directly and strongly dependent on one or more of the environmental factors. As a result, the rate of the overall process of photosynthesis exhibits a great dependence on all major physical variables of the external environment and is more sensitive to environmental changes than any other growth process.

The fact that the different component steps of photosynthesis are affected by different environmental factors has very important implications as to the possible mechanisms of photosynthetic adaptation to different environments since in one environment the capacity of certain component steps may be expected to exert the major limitation to the overall photosynthetic rate while in a different environment other component steps are potentially limiting. For example, at low light intensities, such as those prevailing on shaded forest floors, the overall rate can be expected to be governed primarily by the capacity of the light-harvesting network of pigment molecules and by the efficiency of conversion of the excitation energy at the reaction centers to oxidizing and reducing entities. An increase in the rate of overall photosynthesis under these conditions can therefore only be achieved by an increased capacity of these component steps. In an environment with prevailing high light intensities, on the other hand, the overall rate may be expected to be determined by one or several of the subsequent steps. If temperature is low and CO_2 supply is adequate, enzyme-catalyzed steps such as those involved in the fixation and reduction of CO_2 and in the electron transport chain would be expected to exert the main limitation, and only an increase in the capacity of those components would lead to an increase in overall rate. In environments with high light and temperatures, and particularly when water supply is restricted, the resistance of the stomata to gaseous diffusion often is necessarily high in order to prevent excessive water loss. Under these conditions, the diffusion of CO_2 can be expected to be a major limiting factor and the diffusion rate can be increased only if the efficiency of the CO_2 fixation step is increased so that a greater concentration gradient can develop.

On the basis of the differential dependence of the various component steps of photosynthesis, one can thus infer that the overall photosynthetic performance of plants in different environments can be increased by an adjustment of the *relative* capacities of the component steps without necessarily altering the total investment in constituents of the photosynthetic apparatus. It follows that these relative capacities might be expected to differ among plants from ecologically diverse habitats so as to enable each plant to perform efficiently under the conditions prevailing in its particular environment. Experimental evidence for the existence of such adaptative quantitative differences among higher plants will be discussed in more detail in later sections. While this kind of photosynthetic adaptation appears to be a common and powerful adaptive mechanism, it is not the only kind. Recent work on the function of the C_4 dicarboxylic acid pathway provides strong evidence that it represents a special adaptation that has evolved in response to the selective forces operating in particular kinds of ecological habitats. Qualitative differentiation in biochemical pathways may therefore be a more frequent occurrence than previously anticipated.

3. Comparative Studies of Component Photosynthetic Reactions

3.1 Photochemical Events

3.1.1 Light-Harvesting Pigments

The first step in the conversion of light energy is its absorption by photosynthetic pigments. Clearly only that fraction of the light which is absorbed by these pigments can be utilized for photosynthesis. All photosynthetic organisms contain some form of chlorophyll and carotenoids. Many algae also contain other photosynthetic pigments, such as the phycoerythrin and phycocyanin of red algae. Photosynthetic bacteria contain a special type of chlorophyll not found in green plants. In higher plant chloroplasts various *in vivo* forms of chlorophyll a, chlorophyll b, and carotenoids are the only pigments responsible for the absorption of light energy used for photosynthesis. The relative proportions of these pigments as well as their total concentration may, however, show considerable variation among, as well as within, species, resulting in somewhat different efficiencies in light absorption by the leaves.

The effect of increasing chlorophyll concentration on the light-harvesting efficiency of a leaf is illustrated in Fig. 2. The absorptance,

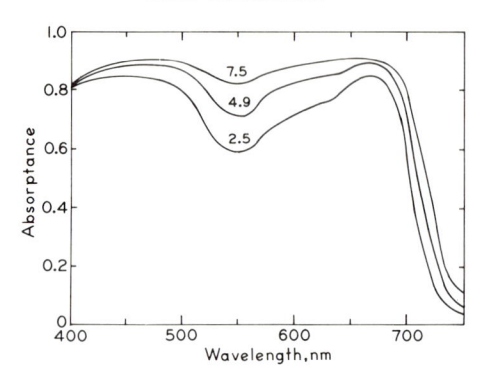

FIG. 2. Spectral leaf absorptance. Numbers on curves indicate the content of chlorophyll (a + b) in milligrams per square decimeter of leaf area. (Redrawn from Björkman, 1968b.)

i.e., that fraction of the incident light which is absorbed by the leaf, increases with increasing chlorophyll concentration, but the absorptance increase is less than directly proportional to increase in pigment. The relationship can roughly be expressed as

$$A_\lambda = 1 - \exp(k_\lambda \cdot l_\lambda \cdot c)$$

where A_λ is the absorptance (not to be confused with absorbance or optical density) of the leaf, k_λ is constant at wavelength λ, l_λ the effective path length of the light in the leaf, and c, the amount of pigment per leaf volume. This expression is similar to that given by Beer's law for light absorption in homogeneous pigment solutions. However, the leaf is a highly heterogeneous system and the *in vivo* forms of chlorophyll have different absorption characteristics than extracted chlorophyll in solution (French, 1971). As a result, the values of k_λ are considerably different from the extinction coefficient for chlorophyll in solution. Moreover, owing to multiple reflections and scattering within a leaf, the effective length of the light path is greater than that given by the thickness of the leaf and it varies with wavelength as well as with leaf anatomy. A leaf is therefore a more effective absorber than a solution of equal pigment content. An average green leaf absorbs as much as 80% of the incident light in the wavelength region 400–700 nm. Only a small fraction of this absorption is caused by nonphotosynthetic pigments.

As shown in Fig. 2 an increase in chlorophyll concentration, or amount per unit leaf area, from 2.5 to 7.5 mg dm^{-2} (most higher plant leaves fall within this range) has only a small effect in the blue and red, where chlorophyll has a high specific absorption, and the effect is greatest in the green and far red, where it has a low absorption. Thus, a high content of chloroplast pigments and the multiple reflection of light within the leaf tends to compensate for the low specific absorption of these pigments in certain parts of the spectrum. It is obvious that an increase in pigment concentration in the range found in leaves of higher plants invariably leads to an increase in the overall effectiveness by which the leaf is capable of absorbing light in any natural habitat, even though the increase in absorption is less than directly proportional to the increased pigment content, and the increase may be negligible at the absorption peaks of the pigments. Consequently, as far as light-harvesting capacity is concerned the widely held concept of "excessive chlorophyll" is untenable.

The phycoerythrin and phycocyanin of red algae fill much of the gap in the absorption spectrum of the chlorophylls and carotenoids and probably represent adaptations that enable the red algae to absorb a greater fraction of the light energy available at the great ocean depths where they often occur. The bacteriochlorophyll of photosynthetic bacteria extends to wavelengths >800 nm. Hence, these organisms are capable of utilizing light of considerably longer wavelengths than are higher plants and algae.

It might be conjectured that also in green plants the presence of photosynthetic pigments which make possible the utilization of energy of longer wavelengths would be advantageous, particularly in habitats where light intensity is low and a great proportion of the total energy is in the far red. The existence of such an adaptive differentiation among green plants is indicated by Halldal's studies (1968) of the green algae *Ostreobium reinecky* which lives within the surface of the brain coral *Favia* in the Great Barrier Reef. Since the coral is covered by a dense canopy of epizoic dinoflagellates, the light penetrating to the green algae is extremely low and most of it is in the wavelength region between 700 and 750 nm. The action spectrum of photosynthesis for this algae differs from the spectra of other species of green algae and higher plants primarily in being capable of absorbing and utilizing light of long wavelengths (700–750 nm). Halldal concluded that a new form of chlorophyll a occurs in *Ostreobium* with maximum absorption around 720 nm. It might be expected that, since this pigment absorbs light at wavelengths greater than the absorption range of P700, the latter could not serve as an effective reaction center and hence the reaction center of photosystem I must also be different. There is as yet no evidence indicating that adaptive differentiation similar to that of *Ostreobium* has occurred among higher plants even though the selective advantage of being able to utilize far red must be considerable in habitats such as the floor of densely shaded rain forests where the available energy

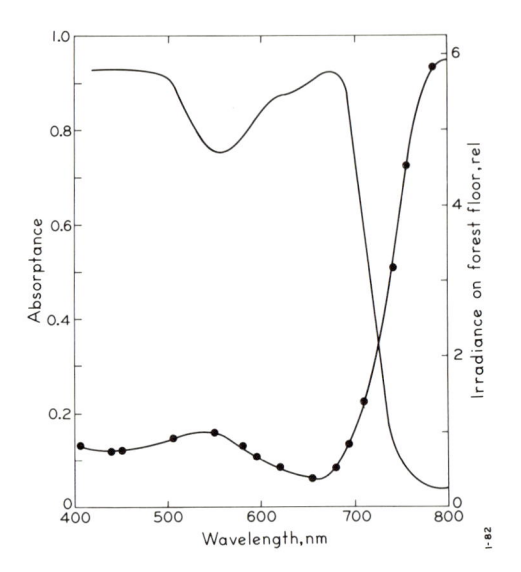

Fig. 3. Spectral distribution of diffuse radiant energy (●—●) on the floor of a dense rain forest and absorptance (—) of an *Alocasia* leaf occupying this site. (From Björkman and Ludlow, 1972.)

is predominantly at wavelengths greater than 700 nm (Fig. 3). Although the absorptance of some extreme shade plants does appear to be somewhat greater in the 700–710 nm range than is usually the case (O. Björkman, unpublished), this can probably be explained simply by an increase in the quantity of chlorophyll and its remarkably dense packing in the grana stacks (Fig. 10, Goodchild *et al.*, 1972). Systematic studies concerning the pigment composition *in vivo* of such plants however, are notably lacking.

3.1.2 Photosynthetic Unit Size

One cornerstone in the current views of photosynthesis is the concept of the photosynthetic unit. This concept originated from Emerson and Arnold's experiments of the maximum yield of photosynthesis per flash of light. It was originally defined as the minimum number of chlorophyll molecules required for the evolution of one molecule of O_2 and the consumption of one molecule of CO_2. These experiments inferred that rather than acting independently, a great number of chlorophyll molecules act in unison to collect the necessary light quanta required to drive the photosynthetic machinery once. In the current view of photosynthesis the photosynthetic unit is conceived of as a functional entity consisting of a photochemical reaction center together with its "antenna" of light-harvesting pigments.

The energy of the light quanta absorbed by the carotenoids and chlorophylls absorbing at shorter wavelengths are considered to be transferred to chlorophylls absorbing at longer wavelengths. Ultimately all the energy collected by the various light-harvesting pigments of a photosynthetic unit is transferred to the reaction center. The reaction center of photosystem I in green plants has been identified as a special form of chlorophyll a, with peak absorption at approximately 700 nm; it has been termed P700. The identity of the reaction center of photosystem II has not yet been established. Photosynthetic bacteria which cannot use water as an electron donor contain only a single photosystem; bacteriochlorophyll serves as the light-harvesting pigment, and P800 as the reaction center.

Photosynthetic unit size may be defined as the number of light-harvesting pigment molecules that serve one reaction center. It is now commonly expressed as the molar ratio of chlorophyll to reaction centers. In higher plant chloroplasts the ratio of chlorophyll to P700, the reaction center of photosystem I, is generally considered to be in the order of 400. If it is assumed that the light-harvesting pigment is equally divided between the two photosystems there would be about 200 chlorophyll molecules for each reaction center. Clayton (1965) considered the possibility that the most primitive photosynthetic organisms had no light-harvesting system at all, but that the reaction center chlorophyll itself served this function. With a chlorophyll content, and hence a light absorption efficiency equal to that of a normal leaf, it would, however, take several minutes for a single chlorophyll molecule in a dimly illuminated plant to collect 1 light quantum, and at least 2 quanta are required for the transport of one electron from water to NADP. For the collection of the 8–10 quanta needed for the evolution of 1 molecule of O_2 and fixation of one molecule of CO_2, an hour would be required. Obviously, such a system, where only one chlorophyll molecule is used for collecting the quanta required for every unit reaction would be an extremely inefficient mechanism. The development of an antenna of a great number of chlorophyll molecules which harvest the light quanta so that the sum of their energy could be used to drive photosynthetic electron flow, must have had a tremendous selective advantage, particularly under conditions of low light intensities. The cost to the plant of a system where at least 1 molecule of each electron carrier must be produced for every chlorophyll molecule would certainly be very great and would be of little use if the rate of absorption of quanta is low in comparison with the capacity for electron flow of the electron transport chain. On the other hand, under conditions of bright light the electron transport capacity must be commensurate with the rate of quantum absorption. If it is not, a fraction of the quanta will be wasted.

It would appear that the average value of some 400 chlorophyll molecules per photosynthetic unit (photosystem I + II) determined for higher plant chloroplasts, is a compromise between the demand of reasonably efficient performance at both low and high light intensities in terms of a given investment in photosystem constituents.

Recent comparative studies have indicated that the photosynthetic unit size might vary both among higher plants and green algae and that it may be determined by nuclear genes. Results obtained with certain chlorophyll-deficient mutants of tobacco (Schmid and Gaffron, 1971), pea (Highkin et al., 1969), and *Chlorella* (Wild et al., 1971) suggested that the average size of the photosynthetic unit is considerably smaller in these mutants than in the normal plants.

In this context the definition of photosynthetic unit size and the criteria on which it is based are important. Because of its close association with photosystem I and because it occurs in roughly equimolar concentrations with P700 in normal spinach chloroplasts, cytochrome f has often been used as a basis for photosynthetic unit size, as have also other electron carriers, such as plastoquinone and cytochromes b_6 and b-559. The use of the ratio of chlorophyll to these electron carriers rather than to P700 as a criterion of photosynthetic unit size is questionable in the light of recent results. Keck et al. (1970) comparing various functions of chloroplasts isolated from "wild type" and a chlorophyll-deficient mutant of soybean, found considerable differences in the ratio of chlorophyll to plastoquinones and to cytochromes but not to P700. Other studies indicate (see Table I) that the ratio between several electron carriers including cytochrome f on the one hand, and P700 on the other, is severalfold greater in plants occupying habitats with high light intensities than in plants growing in densely shaded habitats (Boardman et al., 1972). Moreover, comparisons of higher plant individuals, having the same genetic constitution but grown under a series of controlled light intensities, showed that the ratio of these electron carriers to P700 can be directly modified by the light intensity during growth, but the ratio between chlorophyll and P700 remained essentially constant at about 400–500 chlorophyll molecules per reaction center (Björkman et al., 1972b). Hence, whereas there is little doubt that photosynthetic adaptation to contrasting light climates involves large changes in ratio between light-harvesting pigments and electron carriers, associated with both photosystem I and photosystem II, modifications in the ratio of light-harvesting pigments to the reaction centers themselves appear to be small or insignificant in the higher plants and algae investigated to date. Recent estimates of photosynthetic unit size by the method of Emerson and Arnold, rather than by determinations of chlorophyll:P700 ratios, indicate that a 10-fold reduction in the light intensity for growth results in an increase in photosynthetic unit size by a factor of approximately 1.5 in the green alga *Chlorella* (Myers and Graham, 1971; Sheridan, 1972). Comparisons of chlorophyll:P700 ratios between sun species and extreme shade species of higher plants gave even smaller differences (Boardman et al., 1972). These results are in sharp contrast with those obtained with the photosynthetic bacterium *Rhodopseudomonas spheroides* in which reduction in light intensity for growth resulted in a 10-fold increase in photosynthetic unit size in terms of light-harvesting bacteriochlorophyll per reaction center bacteriochlorophyll (Aagaard and Sistrom, 1972).

Comparing several higher plant species possessing the C_4 dicarboxylic pathway for CO_2 fixation (C_4 plants) with species lacking this pathway (C_3 plants), Black and Mayne (1970) found that the chlorophyll:P700 ratio in extracts of whole leaves was, on the average, about 1.5 times higher in the C_4 species. At the time these authors interpreted the results as an indication of a smaller photosynthetic unit size in C_4 as compared with C_3 plants. However, subsequent studies have provided a different interpretation. C_4 plants, in contrast to C_3 plants, have a highly specialized leaf anatomy, characterized by two different types of chloroplast-containing cell layers, commonly referred to as mesophyll and bundle-sheath. It is now widely accepted that these cells serve specific functions in C_4 photosynthesis (see Section 3.3.2). Results indicate that the lower chlorophyll:P700 ratio of *Sorghum* and certain other C_4 species is not due to a different photosynthetic unit size but is rather related to a deficiency of photosystem II in the bundle-sheath chloroplasts (Section 3.2.1). It should be noted that in many other C_4 species the bundle-sheath chloroplasts have well-developed grana and normal photosystem II activity.

3.1.3 QUANTUM YIELD

In consideration of evolutionary unity and economy, it is not surprising that the basic mechanism of the primary photochemical events appears to be the same in all green plants, although the quality and quantity of light-harvesting pigments may show considerable variation. As a corollary, the intrinsic efficiency of the conversion of absorbed light quanta into primary photoproducts should be the same, as should also the efficiency of O_2 evolution and CO_2 fixation under conditions where the rate of quantum absorption is the sole limiting factor of photosynthesis (maximum quantum yield). Measurements of actual quantum yields provide experimental evidence that this is so. Accurate quantum yield determinations for a large number of higher plant species are of recent date. Measurements made by McCree (1971–1972) of 22 different species of crop plants, grown in the field as well as in artificial light, gave absolute values ranging from 0.07 to 0.08 mole CO_2 per absorbed Einstein (or molecules of CO_2 per absorbed quantum). The relative quantum yields at different wavelengths were also similar in the various species. Figure 4 shows the mean spectral quantum yield for all the species. Both the absolute values and the essential features of the curve for spectral quantum yield are similar to those determined by Björkman (1968b) for leaves of certain wild species. The curve shape differs little from the classical quantum yield spectrum for the green algae *Chlorella pyrenoidosa*, determined by Emerson and Lewis (1943). The main difference in spectral quantum yield among higher plants is in the region 400–500 nm, where carotenoids are responsible for a considerable fraction of light absorption. It is a widely held view that the transfer to the reaction centers of energy absorbed by these pigments is less efficient than the energy absorbed by chlorophyll and that this lower efficiency is responsible for the lower quantum yield in this spectral region. If this view is correct it is not surprising if species differences in quantum yield occur mainly in this region, since there exist marked variations in the carotenoid:chlorophyll ratio among higher plants.

The evidence that the efficiency of quantum conversion to primary photoproducts and of maximum quantum yield for O_2 evolution and CO_2 reduction shows no marked variations should not, however, be taken to mean that the actual rate of net photosynthesis at low light intensities does not vary among higher plants. There exist marked differences in pigment composition and concentration as well as other

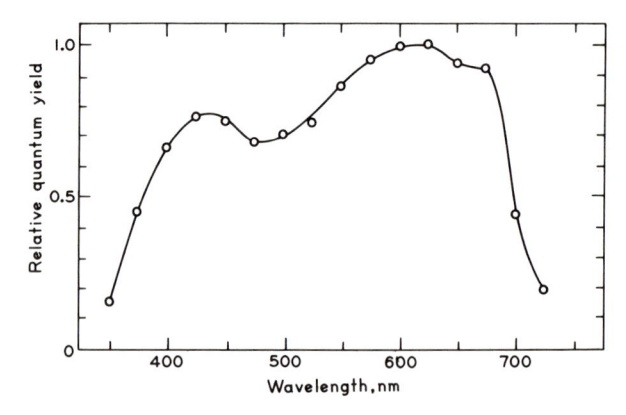

FIG. 4. Mean quantum yield of net CO_2 uptake as a function of wavelength of the absorbed light for leaves of 22 higher plant species. (Redrawn from McCree, 1971–1972.)

optical characteristics of leaves which influence the ratio of absorbed incident light, and therefore affect the photosynthetic yield per quantum of incident light. There are also several other factors. For one, in air of normal oxygen and CO_2 concentrations the actual quantum yield of net CO_2 uptake in C_3 plants is markedly inhibited by oxygen but it is unaffected in C_4 plants (Björkman, 1966a, 1967; Bulley et al., 1969). For another, because of the much lower rate of dark respiration of extreme shade plants, their rate of net CO_2 fixation in light of low intensities is often many times greater than in plants occupying open sunny habitats. A striking example of this is provided by a recent study of photosynthetic gas exchange by plants growing on the floor of an extremely dense rain forest in Queensland (Björkman et al., 1972a,b). This is illustrated in Fig. 5, where the rate of CO_2 exchange as a function of the rate of quantum absorption for a rain forest species, *Alocasia macrorrhiza*, grown in its native habitat is compared with that for a sun species *Atriplex patula* ssp. *hastata*, grown under a light regime simulating that prevailing in its natural habitat. At a quantum flux of about 1.0 nanoeinstein cm^{-2} sec^{-1}, which is similar to the highest quantum flux that the rain forest plant receives during a clear day in the absence of sunflecks, this plant is capable of a net CO_2 uptake of 0.3 μmole dm^{-2} min^{-1}. The sun species, grown at high light intensity, would lose carbon at a rate of 1.7 μmoles of CO_2 dm^{-2} min^{-1}.

The differences in the efficiency with which plants are capable of

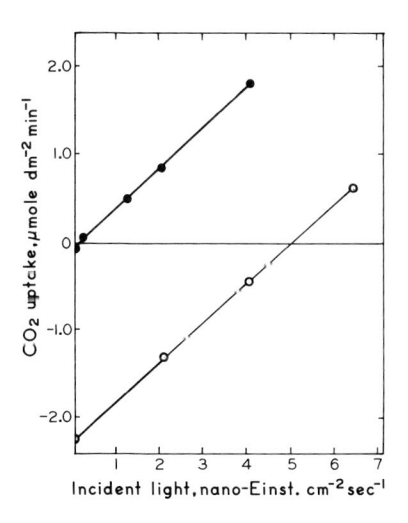

Fig. 5. Rate of net CO_2 exchange as a function of incident quantum flux for leaves of a "sun" plant (*Atriplex*) (○—○) and an extreme shade plant (*Alocasia*) (●—●). (Drawn from data of Björkman et al., 1972a,b.)

utilizing light of low intensities in photosynthesis become particularly striking if one considers the investment in leaf constituents. Shade leaves, such as those of *Alocasia*, although they have more light-harvesting pigments on the basis of unit leaf area, contain less than one-third of the protein and one-half of the total dry matter of sun leaves of *Atriplex*.

A special case of unusually low intrinsic quantum yields of photosynthesis is that of shade plants which happen to be exposed to light intensities much greater than usually occur in their native habitat and which are beyond the potential range of phenotypic adjustment by the genotype. In such plants the low quantum yield is due to an inactivation of the reaction centers themselves (see Section 4.1.1).

3.2 Photosynthetic Electron Transport

The net transport of an electron from water to the reductant that drives the reduction of CO_2 is carried out via the electron transport chain. In the current series formulation of photosynthesis, an electron is transferred from water via one or several electron carriers to the initial photoreductant of photosystem II. This reaction is driven by excitation of the reaction center of this photosystem. An electron from the primary photoreductant is then transferred to the reaction center of photosystem II via several electron carriers (designated Q, P, and cyt f in Fig. 1) to the primary photoreductant of photosystem I. This reaction is driven by excitation of the photosystem I reaction center. Finally, an electron from the primary photoreductant is transferred to NADP via several electron carriers including ferredoxin (Fd in Fig. 1). In contrast to many other electron carriers which are believed to be membrane-bound, NADP is a relatively small molecule which is free to diffuse within the chloroplast to the various sites of carbon reduction.

The path of electron transport and the identity of the various electron carriers are not well established. Alternative pathways may exist and the nature of the various electron carriers may differ among different plant species. Although photosynthetic electron transport is the subject of intensive investigation, the plant materials used are limited, and no truly comparative studies have been carried out so far.

Two recent findings are of particular interest in this context. As shown by Smillie and Entsch (1971), phytoflavin, a flavoprotein found in the blue-green algae *Anacystis nidulans* and *Anabaena cylindrica*, can replace ferredoxins (nonheme iron-containing protein) in the transfer of electrons to NADP, both in these algae and in higher plant chloroplasts. Also, Ben-Amotz and Avron (1972) have shown that a reduced photoproduct preceding ferredoxin in the electron transport chain can serve as the reductant of CO_2, and that neither ferredoxin nor NADP may be necessary as photosynthetic electron carriers in the halophytic green alga *Dunaliella parva*.

3.2.1 ELECTRON-TRANSPORT CAPACITY AND CONTENT OF ELECTRON CARRIERS

At low rates of quantum absorption by the light-harvesting pigments the rate of electron flow is low, and a small electron-carrying capacity is thus sufficient to support this flow. At high light intensities, on the other hand, the rate of quantum conversion may exceed the electron-carrying capacity of the electron transport chain. The rate of electron flow will then not respond to a further increase in light intensity, i.e., it is light saturated. The light intensity at which light saturation occurs will depend on the ratio between the electron-carrying capacity of the electron transport chain and the absorbing capacity of the light-harvesting pigment network. The light-saturated rate of electron flow will depend on the electron-carrying capacity alone. It can therefore be expected that the light-saturated rate of which a plant is capable will be related to its content of electron carriers.

By supplying natural or artificial compounds that can donate, and others that can accept, electrons (Hill reagents) to illuminated suspensions of chloroplast and algae, it is possible to determine the maximum light-saturated capacity for electron flow of the chain in that portion which lies between the donor and the acceptor. Such methods are widely used for studying localization and interactions of intermediates of the electron transport chain. They are, of course, also useful in comparative studies of quantitative aspects of the capacity for electron flow, but relatively few such studies have been carried out on plants from ecologically diverse habitats. Comparative studies of Hill reaction rates with ferricyanide as the electron acceptor on chloroplasts isolated from the leaves of *Teucrium scorodonia* (Mousseau et al., 1967) suggested that the light-saturated capacity of electron flow increases with light intensity for growth. Other observations suggested that this capacity is much lower in species limited to shaded habitats than in those occupying open locations (Björkman, 1968a).

Comparisons of the response of the plastoquinone:chlorophyll ratio to light intensity for growth have recently been carried out with both algae and higher plants. Some form of plastoquinone is believed to serve as an electron carrier, and the carrier P in Fig. 1 may be identical to plastoquinone A. Lichtenthaler (1971) found that fully expanded sun leaves of *Fagus sylvatica* had a much higher plastoquinone concentration both per unit cell volume and per unit chlorophyll than did shade leaves of the same species. However, only a small fraction of the total plastoquinone content is likely to serve in photosynthetic electron transport, and much of it was indeed found to accumulate outside the grana-lamellae regions of the chloroplasts. Hence it is difficult to relate the observed differences in plastoquinone content to capacity for electron transport.

As a part of a recent study of leaf factors underlying photosynthetic adaptation to light intensity, Boardman *et al.* (1972) and Björkman *et al.* (1972b) compared the capacities for electron flow in species from densely shaded rain forests and from sunny habitats. They also investigated the influence of light intensity during growth on the electron-transport capacity of the same plant grown at a series of light intensities. Electron flows driven by photosystem I and photosystem II were determined separately. At rate-limiting light intensities the rate of electron flow on the basis of equal quantum absorption was similar in the different species and it was unaffected by the light intensity at which the plants were grown. This is to be expected if the photosynthetic unit size and the intrinsic efficiency of quantum conversion are the same. The light intensity (rate of quantum absorption) at which light-saturation of electron flow occurred was, however, many times greater in the sun plant. So was also the light-saturated rate of electron flow, both that driven by photosystem I and that driven by photosystem II (see Fig. 11). Growing the sun plant at the lowest light intensity at which it is capable of a significant growth rate resulted in similar but less striking differences in the capacity for electron flow. The absolute capacities for electron flow on the basis of chlorophyll are similar to those obtained for photosynthetic CO_2 uptake by the intact leaves, even when limitations to CO_2 uptake by gaseous diffusion have been accounted for. The correlation between photosynthetic CO_2 uptake of whole leaves among the different species and treatments and the electron transport capacity is particularly striking for that driven by photosystem II.

Subsequent determinations showed that the ratios of certain electron carriers to light-harvesting chlorophyll were also several times greater in the sun than in the shade species and that these ratios decreased with decreasing light intensity for growth (Table I). Thus, on a chlorophyll basis, the contents of cytochrome f and carrier P were about 3.5 times greater in the sun than in the shade species and the amounts of these carriers were reduced to approximately one-half when the former species was grown at a low light intensity. Similar results were obtained for other cytochromes. However, as was mentioned in Section 3.1.2,

TABLE I

QUANTITY OF SEVERAL PHOTOSYNTHETIC ELECTRON CARRIERS, P700, AND SOLUBLE PROTEIN IN RELATION TO CHLOROPHYLL IN TWO RAIN FOREST SPECIES AND IN A "SUN" SPECIES GROWN AT THREE DIFFERENT LIGHT INTENSITIES[a,b]

Plant	Light for growth	P	Cyt f	Cyt b₆	P700[c]	Q	Soluble protein
Alocasia macrorrhiza	Very low	1.0	1.0 (1120)	1.0 (640)	1.0 (570)	1.0	1.0
Cordyline rubra	Very low	1.0	1.2 (950)	1.2 (530)	0.9 (645)	1.0	0.7
Atriplex patula	High	3.6	3.4 (330)	2.8 (225)	1.3 (430)	1.0	4.9
Atriplex patula	Medium	3.0	2.2 (509)	2.6 (250)	1.3 (434)	1.0	4.2
Atriplex patula	Low	2.1	1.6 (689)	1.9 (342)	1.2 (468)	1.0	2.5

[a] Based on data from Boardman *et al.* (1972), Björkman *et al.* (1972b), and Goodchild *et al.* (1972).

[b] The quantity of each chloroplast component in relation to chlorophyll is set equal to unity for *Alocasia*. Numbers in parentheses are the measured molar ratios of light-harvesting chlorophyll to electron carrier or P700.

[c] Means based on both chemical and photochemical determinations of P700.

the content of P700, the reaction center of photosystem I, showed only slight differences among the different plants. Moreover, no differences were detected in Q, the quencher of fluorescence from photosystem II, and believed to be identical to or very closely associated with the primary reductant, formed upon excitation of the reaction center of this photosystem.

The above results clearly indicate that the capacity for electron transport and the content of electron carriers relative to the photochemical reaction centers and light-harvesting pigments can vary among higher plants and that these factors are subject to regulation by the light climate in which the plants grow. The results further suggest that such regulation is an important mechanism (although not the only one) in the adaptation of the photosynthetic apparatus to contrasting natural light regimes (Section 4.1.1).

The capacity for the Hill reaction with dichlorophenolindophenol as the electron acceptor may also vary among different populations of the same species (McNaughton, 1967; Tieszen and Helgager, 1968; Williams, 1971) and is influenced by both the temperature regime and the photoperiod under which the plant is grown. The possible adaptive significance of these intraspecific differences is unclear. As shown by Keck and Boyer (1972), the capacity for electron transport and photophosphorylation is also affected by the water status of the leaves from which the chloroplasts are isolated. Increased water stress in sunflower leaves resulted in a decrease in whole-leaf photosynthesis (corrected for stomatal responses) which was correlated with a decreased capacity for electron transport and photophosphorylation in chloroplast isolated from these leaves.

As mentioned in Section 3.1.2, chloroplasts from bundle-sheath cells of certain C_4 plants possess unusually low rates of electron flow driven by photosystem II whereas they exhibit high rates of electron flow and cyclic photophosphorylation driven by photosystem I (Anderson *et al.*, 1971a; Bazzaz and Govindjee, 1971; Downton *et al.*, 1970; Mayne *et al.*, 1971; Woo *et al.*, 1970). These activities are probably related to a reduced level of photosystem II itself, since absorption spectra as well as fluorescence emission and action spectra of isolated bundle-sheath chloroplasts with poorly developed grana resemble those of photosystem I fractions from spinach chloroplasts and are unlike those of photosystem II fractions (Anderson *et al.*, 1971a; Bazzaz and Govindjee, 1971; French and Berry, 1971). It appears likely that this unusual imbalance between the levels of the two photosystems represents an adjustment to special requirements for reduced NADP and ATP in the carbon metabolism of these bundle-sheath cells (Berry, 1970; Mayne *et al.*, 1971).

3.3 Carbon Fixation and Reduction

3.3.1 THE CALVIN-BENSON CYCLE

During the ten-year period following the Second World War, the group led by Calvin, Benson, and co-workers in a very elegant series of experiments were able to map the fundamental process by which carbon dioxide is assimilated and reduced to final carbon products of photosynthesis. Since this time Bassham and co-workers (and also other workers in laboratories in various parts of the world) have refined our knowledge of this process, and its mechanism is probably better known than any other aspect of photosynthesis. One key reaction in the Calvin-Benson, or the reductive pentose phosphate cycle, is the reaction between CO_2 and ribulose 1,5-diphosphate to form 3-phosphoglyceric acid. In a number of subsequent reactions, some of which require $NADPH_2$ and ATP, carbon is reduced to the level of carbohydrate while the CO_2 acceptor, ribulose 1,5-diphosphate, is regenerated. For the reduction of one CO_2, the equivalent of at least 2 $NADPH_2$ and 3 ATP would be required. A recent account for the detailed operation of the cycle has been given by Bassham (1971).

The CO_2 fixation reaction in the Calvin-Benson cycle is catalyzed by the enzyme ribulose 1,5-diphosphate (RuDP) carboxylase, also called carboxydismutase. RuDP carboxylase is the principal component of fraction 1 protein of leaves. This protein constitutes a large proportion of the total leaf protein and perhaps more than one-half of the protein in the chloroplast stroma. The fact that plants put such a high proportion of their total investment in protein into one single enzyme strongly suggests that this enzyme is a potential rate-limiting step in photosynthesis. Also, the light-saturated rate of photosynthesis is strongly CO_2 dependent at CO_2 concentrations of normal air; hence the carboxylation reaction, catalyzed by RuDP carboxylase, and barriers to the diffusive transport of CO_2, are implicated as important rate-limiting reactions. In a recent review, Raven (1970) has discussed the relative importance of carboxylation and diffusion limitation in photosynthesis.

There are many studies showing that conditions that affect the level of RuDP carboxylase affect the capacity for light-saturated photosynthesis in a similar manner. The light intensity under which a given plant is grown has a striking and parallel influence on both (Björkman, 1966b, 1968b; Björkman et al., 1969a, 1972b; Bowes et al., 1972; Gauhl, 1969; Medina, 1970, 1971). Partial defoliation of plants in several species resulted in a parallel increase in the capacity for light-saturated photosynthesis and RuDP carboxylase activity in the remaining leaves (Neales et al., 1971; Wareing et al., 1968). Woolhouse (1967) showed that the decline in photosynthetic capacity with increasing leaf age in *Perilla* was accompanied by a similar decline in the fraction 1 protein content of the leaves. Reduction in the nitrogen nutrient in the root medium caused a similar decline in photosynthetic capacity and RuDP carboxylase activity, and these were more affected than the total protein content and the capacity for the Hill reaction (Medina, 1970, 1971; Wojcieska et al., 1972). Virus infection of sugar beet leaves caused a parallel reduction in photosynthesis and RuDP carboxylase activity (Hall, 1970). In many of these studies a decrease in photosynthetic rate was also accompanied by an increased diffusive resistance of the stomata, but in all cases the changes in this resistance were insufficient to account for the change in photosynthetic capacity.

The level of RuDP carboxylase also differs among species from ecologically diverse habitats. Björkman (1966a, 1968a,b) and Gauhl (1969) have shown that the capacity of photosynthesis, when light is saturating and CO_2 limiting, is proportional to RuDP carboxylase activity both in a range of sun and shade species and in a range of genotypically adapted sun and shade ecotypes of the same species (see Section 4.1.1). The role of RuDP carboxylase in plants possessing the C_4 dicarboxylic acid pathway is discussed in Section 3.3.2.

3.3.2 The C_4 Dicarboxylic Acid Pathway

Although other CO_2 fixation pathways were known in biological systems, the one catalyzed by RuDP carboxylase was until recently generally considered to be the only quantitatively important carboxylation pathway in photosynthesis. Plants in which 3-phosphoglyceric acid is the first carboxylation product in the light include all algae, pteridophytes, and most higher plants tested so far. In 1965 Kortschak and co-workers reported that the first stable carbon compounds formed in photosynthesis by sugar cane leaves are 4-carbon dicarboxylic acids, malic and aspartic acid, rather than the 3-carbon 3-phosphoglyceric acid. Hatch and Slack (1966; Hatch et al., 1967) confirmed these observations and extended their findings to include a number of other grasses and also dicotyledonous plants, mostly of suspected tropical origin. On the basis of comparative studies of the radioactive labeling kinetics of intermediate compounds and of enzyme activities, Hatch, Slack, and their co-workers proposed that these plants utilized a pathway for CO_2 fixation which is catalyzed by the enzyme phospho(enol)pyruvate (PEP) carboxylase rather than RuDP carboxylase and that the first product of this fixation is oxaloacetic acid (OAA), which is rapidly reduced to malic acid or aminated to aspartic acid. The carbon atom in number 4 position of these acids subsequently appeared in 3-phosphoglyceric acid, hexose phosphates, and finally in sugars and other end products of photosynthesis. Mainly on the basis of the observation that the activity of PEP carboxylase was very high but that of RuDP carboxylase was very low, and insufficient to play an integral role in photosynthesis in plants which incorporated CO_2 into C_4 acids, Hatch and Slack originally proposed that the new C_4 pathway replaced the Calvin-Benson pathway of CO_2 fixation. A change in this view was initiated by the finding (Björkman and Gauhl, 1969) that, contrary to earlier observations, leaves of C_4 plants contain RuDP carboxylase levels comparable to those of C_3 plants. The literature on the biochemistry of C_4 plants is now voluminous, and several recent reviews have been published (see articles in *Photosynthesis and Photorespiration*, Hatch et al., 1971; Coombs, 1971; Kanai and Black, 1972). The current views on the basic features of its mechanism therefore will be treated only briefly here.

In what is probably the most widely accepted current model of C_4 photosynthesis (Fig. 6) CO_2 is initially fixed in the mesophyll cells by PEP carboxylase (which is predominantly located in these cells) into C_4 dicarboxylic acids. These C_4 acids are then transported to the bundle-

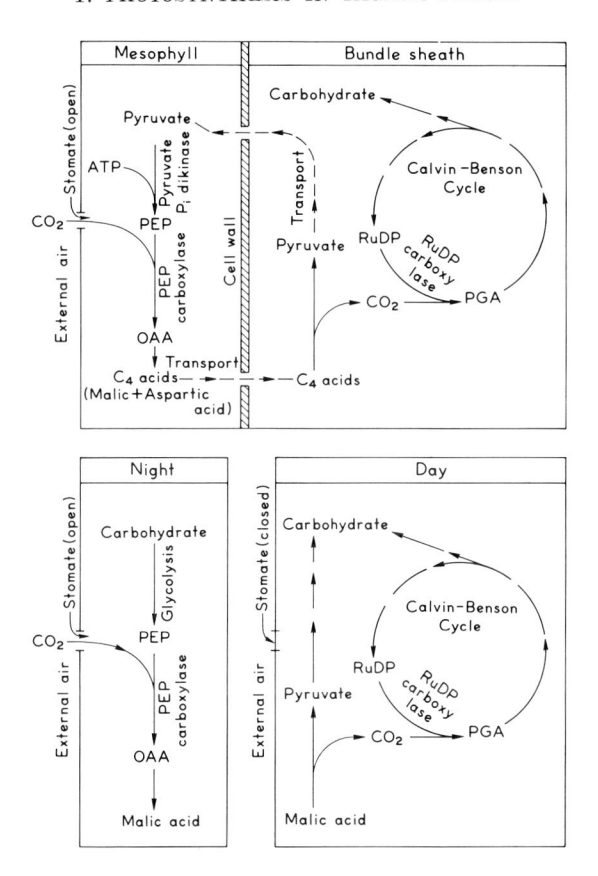

Fig. 6. Schemes for C₄ photosynthesis (upper half) and crassulacean acid metabolism (CAM) (lower half). In C₄ plants all the reactions occur simultaneously in the light, but they are spatially separated. In CAM plants, all the reactions take place within the same cell, but those resulting in the fixation of CO_2 from the atmosphere into C₄ acids occur in the dark whereas those leading to the reduction of carbon to the level of carbohydrate occur in the light. Abbreviations: PEP, phospho(enol)pyruvic acid; OAA, oxaloacetic acid; RuDP, ribulose 1,5-diphosphate; PGA, 3-phosphoglyceric acid.

sheath cells, where they are decarboxylated to CO_2 and pyruvate. This CO_2 is refixed in the bundle-sheath cells by the RuDP carboxylase of the Calvin-Benson cycle in a conventional way. The pyruvate is phosphorylated to PEP in a reaction catalyzed by pyruvate P_i dikinase, an enzyme which is present in C₄ but not in C₃ plants. The required ATP is supplied by photophosphorylation and the initial CO_2 acceptor, PEP, is regenerated. Thus, there are two sequential carboxylations. The first, utilizing atmospheric CO_2, is catalyzed by PEP carboxylase and takes place in the mesophyll cells. The second, utilizing internally generated CO_2, is catalyzed by RuDP carboxylase and occurs in the bundle-sheath cells.

There probably exist a number of variations on this theme among different C_4 species. Depending on the particular species, either malic or aspartic acid is the main product of the initial CO_2 fixation step, although both are invariably formed, and either of these C_4 acids may be predominantly transported to and decarboxylated in the bundle sheath (Downton, 1971). The enzyme responsible for this decarboxylation may also vary. In some species, such as maize, where malic acid is the predominant product, the bundle-sheath cells contain high activities of malic enzyme and this enzyme presumably catalyzes the decarboxylation of malic acid in these cells. In certain other C_4 species, such as *Panicum maximum*, the activity of malic enzyme is very low or absent, but instead, PEP carboxykinase, another enzyme capable of decarboxylating oxaloacetic acid, is present (Edwards *et al.*, 1971). Oxaloacetic acid can readily be formed from malic or aspartic acid since the activity of the enzymes catalyzing these conversions, malate dehydrogenase and aspartate aminotransferase, respectively, are high. A third group of C_4 species, such as *Amaranthus retroflexus* (or *Atriplex spongiosa* and *A. rosea*), in which aspartic acid is a major product, lack substantial levels of both malic enzyme and PEP carboxykinase. Attempts to detect an appropriate decarboxylation enzyme in these plants have been unsuccessful so far (Kanai and Black, 1972).

As mentioned previously (Section 3.2.1) bundle-sheath chloroplasts of certain C_4 plants possess unusual capacities for electron transport and photophosphorylation, and this may reflect a photochemical adjustment to differential demands for $NADPH_2$ in the carbon metabolism of the bundle-sheath cells. The variation among various C_4 species may well be related to the kind of decarboxylating system operative in those cells. If, for example, malic acid is formed in the mesophyll cells and is decarboxylated in the bundle-sheath cells by malic enzyme, this decarboxylation would result in the formation of one $NADPH_2$ in the latter cells and consequently the demand for $NADPH_2$, generated by photosynthetic electron transport, would be less than normal. If, on the other hand, aspartic or oxaloacetic acid is transported to the bundle-sheath cells and the latter acid is decarboxylated by PEP carboxykinase there would be no net movement of reducing equivalents from the mesophyll to the bundle sheath.

The minimum total energy requirement for the fixation of one CO_2 and its reduction to the level of carbohydrate in C_4 photosynthesis would be 5 ATP and 2 $NADPH_2$, i.e., 2 ATP more than in C_3 photosynthesis.

The extra ATP's are required for the phosphorylation of pyruvate to PEP. However, these are intrinsic energy requirements and may not be valid for *net* photosynthesis under conditions of normal O_2 and CO_2 concentrations.

In the outline of C_4 photosynthesis given above, the specialized leaf anatomy, characterized by distinct mesophyll and bundle-sheath cell layers, serves an essential role in providing spatial compartmentation of the two sequential CO_2 fixation reactions (Fig. 7). The CO_2 fixation in the mesophyll cells, catalyzed by the PEP carboxylase, functions as a very efficient trap for fixation of CO_2 from the external air. This

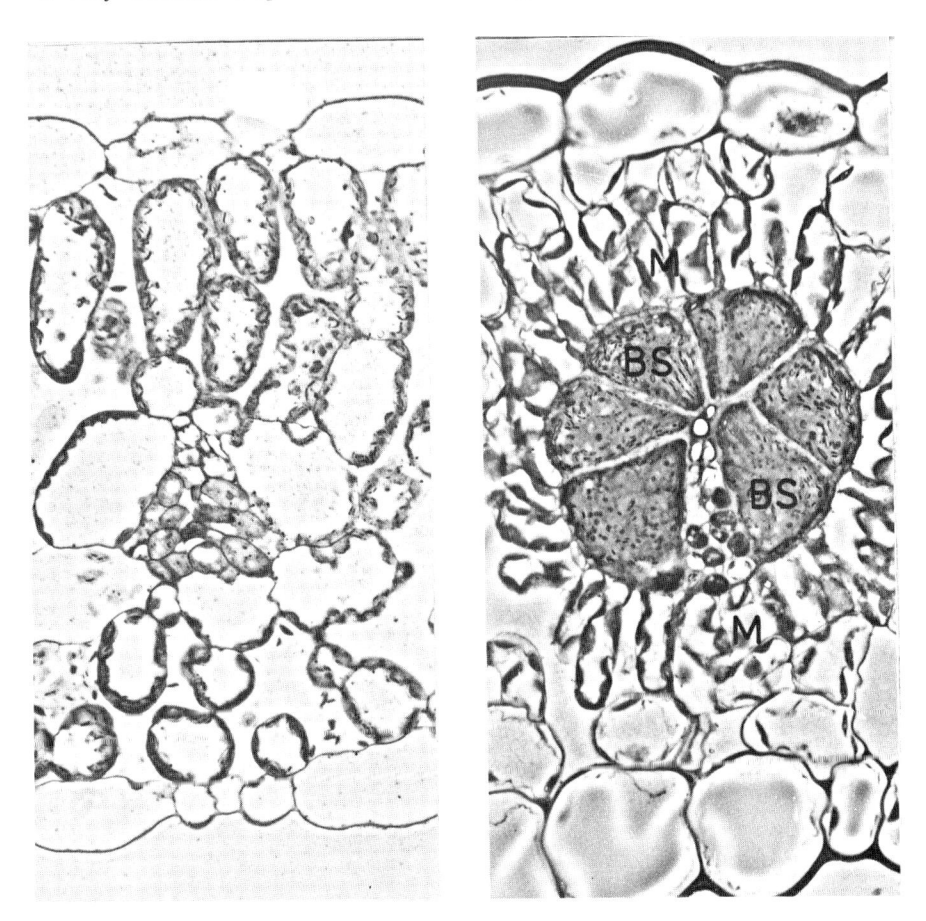

FIG. 7. Phase contrast light micrographs of cross sections of leaves of the C_3 species *Atriplex patula* (left) and the C_4 species *Atriplex rosea* (right). The leaves had been fixed and embedded in plastic for electron microscopy. BS, bundle-sheath cells. (Micrographs by courtesy of Dr. J. Boynton.)

carbon is transported in the form of C_4 acids to the bundle-sheath cells, where it is released as CO_2. Thus the C_4 pathway serves as a metabolic "CO_2-concentrating mechanism" that permits the carboxylation sites in the Calvin-Benson cycle to operate at an effectively higher CO_2 concentration than in C_3 plants where the CO_2 concentration at the carboxylation site is in a more direct equilibrium with the external atmosphere. Experimental evidence for the existence of a larger CO_2 pool in leaves of C_4 than of C_3 species has been provided by Hatch (1971). An increased CO_2 concentration may have several important beneficial effects, all of which are presumably related. It increases the rate of carboxylation by RuDP carboxylase when this reaction is CO_2 limited, and it suppresses the inhibitory effect of oxygen on the carboxylation reaction (cf. Section 3.4). In the author's opinion the concept of the C_4 pathway as serving primarily as a CO_2 concentrating mechanism is very attractive and is consistent with several independent lines of evidence, which include photosynthetic gas exchange characteristics of C_4 plants *in vivo* as well as with leaf anatomical traits and most enzyme distribution and ^{14}C-labeling patterns. This concept is a unifying one.

The specialized leaf anatomy (Fig. 7) is present in all naturally occurring plants with C_4 photosynthesis regardless of their taxonomic and phylogenetic relationships, although the detailed structure may show considerable variation among different C_4 species (Laetsch, 1971). Studies of segregating F_2 and F_3 populations of hybrids between C_4 and C_3 species provide evidence that this type of anatomy is in itself not sufficient to give the photosynthetic gas exchange characteristics of C_4 plants and neither is the presence of high PEP-carboxylase activity or even the capacity to initially fix a high proportion of CO_2 into C_4 acids (Björkman et al., 1970, 1971). Efficient C_4 photosynthesis apparently can occur only if the component biochemical steps are coordinated and spatially compartmented. The leaf anatomy of C_4 species was known over fifty years before the discovery of C_4 photosynthesis, and there has been considerable speculation as to its functional significance by plant anatomists in the past. The compartmentation function now attributed to this anatomy, if proved to be correct, provides a particularly illustrative example of interaction between biochemical and anatomical differentiation in the evolution of functional adaptations to specific natural habitats. Adaptive and evolutionary aspects of C_4 photosynthesis are discussed in Section 4.2.

Some recent criticism has been raised against the role of leaf anatomy in providing spatial compartmentation in C_4 photosynthesis and should be mentioned here. One objection is based on results indicating that mesophyll cells of maize contain RuDP carboxylase activities similar to those of the bundle-sheath cells (Baldry et al., 1971; Bucke and Long, 1971; Poincelot, 1972). This contrast with results obtained with many C₄ species (e.g., Berry et al., 1970; Björkman and Gauhl, 1969; Edwards et al., 1970; Huang and Beevers, 1972; Osmond and Harris, 1971; Slack, 1969). Another objection stems from the discovery that cells of the C₄ species *Froelichia gracilis*, grown photoheterotrophically in tissue culture, are capable of fixing CO_2 into C₄ acids even though only one type of cell is present (Laetsch and Kortschak, 1972). On this basis the latter authors contend that structural specialization of cells is not causally related to the C₄ pathway. It is, however, well to remember that although certain hybrids between C₄ and C₃ species are capable of fixing CO_2 into C₄ acids they do not possess complete and functional C₄ photosynthesis. This may also be true of the cells grown in tissue culture, and the results therefore do not necessarily prove that compartmentation is not a requisite in complete and efficient C₄ photosynthesis. Thus, while doubts have been raised, the author is yet to be convinced that C₄ photosynthesis occurs without the spatial compartmentation of reactions discussed above.

3.3.3 Crassulacean Acid Metabolism (CAM)

It has long been known that in many succulent plants the diurnal course of CO_2 and water vapor exchange is inverted, i.e., carbon dioxide is consumed and water vapor is released at nighttime. The uptake of CO_2 during the night is accompanied by a massive increase in the malic acid content and decrease in the carbohydrate content of the leaves. Conversely, during the day the malic acid content falls and the carbohydrate content increases. Thus, in contrast to the situation in other plants, including C₄ plants, CO_2 fixation is temporally separated from photosynthesis with CO_2 uptake occurring in the dark and the reduction of this carbon to the level of carbohydrates occurring in the light. Since this unusual behavior was first intensively studied in members of the Crassulaceae, it is known as crassulacean acid metabolism (CAM). Excellent reviews of various aspects of CAM have recently been published (Ting, 1971; Ting et al., 1972).

Until very recently, many workers in this field thought that CO_2 fixation in the *dark* by CAM plants involves a double carboxylation: CO_2 from the atmosphere reacts with RuDP in the *dark* to form 3-phosphoglyceric acid (3-PGA), as it does in C₃ plants in the *light*. The 3-PGA is then converted to PEP, which reacts with atmospheric CO_2 to form oxaloacetate and subsequently malic acid, which is stored. This CO_2 fixation step is catalyzed by PEP carboxylase, as is also the case in C_4 plants. The CO_2 acceptors RuDP and PEP are both derived from stored carbohydrate. In the light the stomata are closed and the stored malic acid is decarboxylated to CO_2 and pyruvate; this internally generated CO_2 is refixed by RuDP carboxylase and reduced to carbohydrate via the Calvin–Benson cycle.

The involvement of RuDP in the *dark* fixation of CO_2 was based on the observed distribution of ^{14}C between carbon atoms 4 and 1 within malic acid-^{14}C, isolated from CAM plants after exposure to $^{14}CO_2$ in the dark. Earlier work had indicated that the distribution of radioactivity between carbon in the C-4 and C-1 carboxyl was approximately in a ratio of 2:1. However, Sutton and Osmond (1972) were able to demonstrate that nearly all the label is initially fixed into the C-4 carboxyl and that the previously determined ratio can probably be attributed to artifacts. The recent results remove the basis for a double carboxylation reaction in the dark. Thus, only one CO_2 fixation step, the β-carboxylation of PEP, is probably involved in the principal dark fixation of CO_2 by CAM plants. This makes the similarity in the CO_2 fixation pathways between CAM and C_4 photosynthesis even greater than was previously thought. Figure 6 (lower half) shows a simplified scheme of CAM. The energetic requirement for the fixation of one molecule of CO_2 is estimated to be approximately 6 ATP and 2 $NADPH_2$, which is greater than both C_3 and C_4 photosynthesis.

It should be emphasized that there are also important differences between CAM and C_4 photosynthesis. In the latter, regeneration of the CO_2 acceptor, PEP, is continuous and driven by photochemically produced energy, whereas in CAM this acceptor is derived from stored carbohydrate. Moreover, in C_4 photosynthesis the initial CO_2 fixation step is spatially separated from the decarboxylation and refixation steps, and the C_4 decarboxylic acids never accumulate, whereas in CAM these processes are separated in time and malic acid is stored in great quantities.

Another comparative aspect is also important: C_4 photosynthesis is genetically fixed, and manipulation of the environment during growth and leaf development does not change the photosynthetic pathway. While the ability to carry out CAM is presumably also genetically determined, the extent to which, if any, the phenotype will carry out CAM is remarkably dependent on environmental factors. Thus, a given plant may fix all of the CO_2 in the dark by CAM when grown under appropriate conditions (in general, hot days with high light intensities and cool nights) whereas under other growing conditions it may fix most or all of its CO_2 during the day, probably by direct fixation by the Calvin-Benson cycle. There are numerous examples where the situation is intermediate, i.e., the plant fixes CO_2 by CAM at nighttime and by C_3 photosynthesis during at least a part of the daylight hours.

3.4 Oxygen Inhibition of Photosynthesis and Photorespiration

The present high concentration of oxygen and low concentration of carbon dioxide in the atmosphere are generally considered to have arisen from photosynthetic activity by plants. The effect of these factors in limiting present photosynthesis and productivity and the possible existence of adaptations to overcome these limitations are therefore problems of great interest in several branches of biology and is particularly important from an ecophysiological viewpoint.

It was discovered many years ago by Warburg that elevated concentrations of O_2 inhibit the rate of net photosynthesis by green algae, provided the light intensity was high. Considerable work was subsequently done by other workers, but until the mid 1960's there seems to have been no general suspicion that the oxygen concentration of normal air had a significant adverse effect on higher plant photosynthesis. At this time comparative studies of a number of plant species from diverse environments demonstrated that atmospheric oxygen causes a 30% to 40% inhibition of net photosynthetic uptake (Björkman, 1966a; Fock and Egle, 1966; Forrester et al., 1966a; Tregunna et al., 1966; Hesketh, 1967) and that this inhibition is present not only at high, but also at low, light intensities (Björkman, 1966a). Stomatal responses were not involved, and different growing conditions, although strongly affecting the photosynthetic rate, did not markedly affect the degree on inhibition by oxygen (Björkman, 1966a; Gauhl and Björkman, 1969; Troughton, 1969). While inhibition by normal O_2 concentrations proved to be a very widespread phenomenon among higher plants, there were some notable exceptions (Forrester et al., 1966b; Hesketh, 1967; Downes and Hesketh, 1968). It is now well established that those higher plants whose photosynthetic CO_2 uptake is not sensitive to O_2 all possess the C_4 pathway. These plants were also known to have other unusual characteristics of photosynthetic gas exchange (Section 4.2).

At about the same time workers studying the effect of light on plant respiration arrived at the conclusion that the rate of respiratory CO_2 release was greatly enhanced by light and this rate could be suppressed by a low O_2 concentration. There is much evidence that glycolate formation in the light and its subsequent metabolism is responsible for this photorespiratory CO_2 release (see recent reviews by Gibbs, 1969; Jackson and Volk, 1970; Goldsworthy, 1970; Tolbert and Yamazaki, 1970; Wolf, 1970; Zelitch, 1971; also articles in Hatch et al., 1971). Until recently, it was generally considered that the inhibitory effect of oxygen on the rate of net CO_2 uptake was mainly caused by a high rate of photorespiratory CO_2 release via the glycolate pathway, counteracting photosynthetic CO_2 uptake. The absence of an apparent O_2 inhibition of net photosynthesis in C_4 plants was attributed to a low capacity of glycolate formation and metabolism in these plants or to an efficient refixation of photorespiratory CO_2 before it can escape from the leaf. Some recent studies indicate, however, that there are alternative explanations to the presence of an O_2 inhibition of net photosynthesis in C_3 but not in C_4 plants. Oxygen has been shown to inhibit CO_2 fixation by RuDP carboxylase isolated from higher plants, probably by competing with CO_2 for binding to the same site on the enzyme (Ogren and Bowes, 1971; Bowes and Ogren, 1972; Berry, 1971). A similar competitive inhibition is obtained *in vivo* with leaves of C_3 plants (Björkman, 1971), and indeed also with isolated bundle-sheath cells of maize, a C_4 plant (Chollet and Ogren, 1972), although photosynthetic CO_2 uptake by *intact* maize leaves is unaffected by O_2. Studies with RuDP carboxylase *in vitro* also indicate that this enzyme catalyzes an alternative reaction which results in an oxidative cleavage of RuDP to phosphoglycerate and phosphoglycolate (Bowes et al., 1971; Lorimer et al., 1972). This reaction thus may be responsible for the *in vivo* production of the glycolate metabolized in photorespiration, although there are also other possible mechanisms for glycolate formation (Gibbs, 1969).

It thus appears that O_2 may compete with CO_2 for RuDP, and therefore inhibits the carboxylation reaction. This may also lead to the formation of glycolate and subsequent photorespiration. The amount of the total O_2 inhibition that is attributable to the direct inhibition of the carboxylation reaction and that attributable to photorespiration *in vivo* is difficult to estimate. However, recent studies by Osmond and Björkman (1972) with closely related C_3 and C_4 species of *Atriplex* suggest that production of glycolate and the subsequent release of CO_2 in photorespiration is probably the minor component of the O_2 effect on net CO_2 uptake in normal air when stomata are fully open and CO_2 supply is not severely restricted. Whereas the inhibition of net CO_2 uptake by 21% O_2 in the C_3 species could be suppressed completely by raising the CO_2 concentration in the intercellular spaces, the production of glycolate pathway intermediates was stimulated by 21% O_2 at both normal and high CO_2 concentrations. This was true of the C_3 as well as the C_4 species. These results are also consistent with those of Ludwig and Canvin (1971), who found that the rate of CO_2 release in photorespiration is not markedly suppressed by elevated CO_2 concentrations, and give indirect support to the view that O_2 has a direct inhibiting effect on photosynthesis, perhaps by inhibiting RuDP carboxylase. If the C_4 pathway serves as a mechanism for concentrating CO_2 at the site of CO_2 fixation by RuDP carboxylase in the bundle-sheath cells, one would predict that photosynthesis of C_4 plants at *normal* CO_2 concentrations would be unaffected by CO_2 concentration and that it would respond to other external factors in a similar manner as do C_3 plants at *high* external CO_2 concentrations. These predictions are substantially in agreement with experimental results (Section 4.2).

The physiological function of the complex metabolic machinery of the glycolate pathway, which is associated with the peroxisomes and which is specifically developed to metabolize photosynthetically produced glycolate (Tolbert, 1971), has been subject to much speculation. It certainly is highly unlikely than an energy-wasting process such as photorespiration would have evolved unless it could provide a selective advantage to the plant. Goldsworthy (1969) suggested that it might represent an adaptation to some past environmental factor and that it may now be regarded as an evolutionary hangover. Although in the light of recent studies photorespiration probably does not result in such massive loss of carbon during photosynthesis as previously thought, a significant loss appears inevitable. In the absence of a useful function, it seems likely that it would have been eliminated.

Osmond and Björkman (1972) proposed that photorespiration represents a mechanism that protects the photochemical reaction centers from photoinactivation under conditions of severely restricted CO_2 supply. There is much evidence that inhibition of photosynthesis by high light intensities is primarily caused by an inactivation of the reaction centers themselves as a result of an excess of excitation energy being transferred to them from the light-harvesting pigments (cf. Section 4.1.1). Under conditions of adequate CO_2 supply this excess may be small even in bright light since energy is being consumed in photosynthesis. However, in natural habitats with high irradiance and water stress, stomata are often closed in order to prevent excessive water loss through transpiration. As a result, CO_2 diffusion becomes severely restricted and the rate of photosynthesis and the consumption of photochemically generated energy fall to a very low level. Light absorption nevertheless continues at an undiminished rate, and there would be a vast excess of excitation energy at the reaction centers, causing their destruction. The production of glycolate and its subsequent metabolism provides a possible mechanism whereby much of the energy normally utilized in net CO_2 fixation can be consumed without supply of CO_2 from the external air (Fig. 8). The glycolate pathway enables the Calvin-Benson cycle to turn over without net carbon gain with normal

Adequate CO₂ supply

```
              3X CO₂
3X RuDP  ─────┬────→  6X 3-PGA
    ↑         │          │
    │   Carbon reduction cycle
    │                    │
    └────────────────────┴──→ 1X 3-PGA
              ↑
            9 ATP
            6 NADPH₂
```

No CO₂ supply

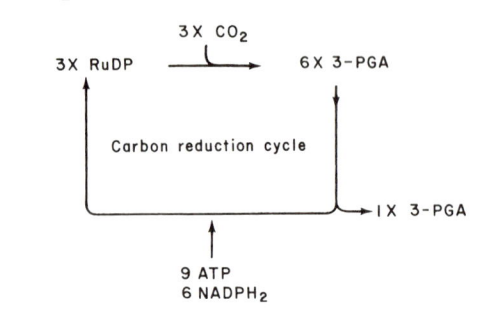

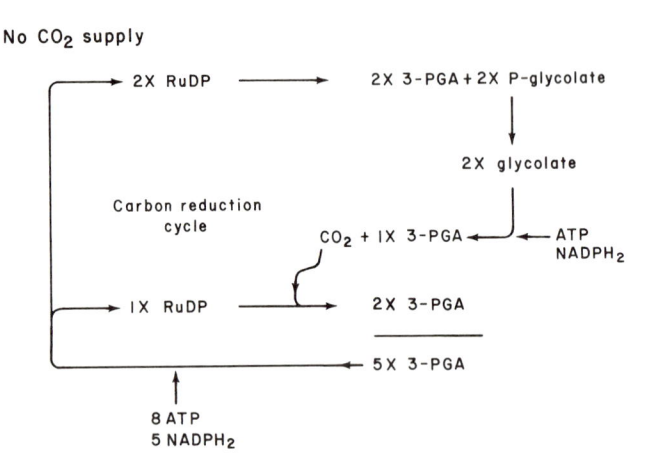

```
  ┌──→ 2X RuDP  ──────→  2X 3-PGA + 2X P-glycolate
  │                                    │
  │                                    ↓
  │                              2X glycolate
  │     Carbon reduction                │
  │         cycle       CO₂ + 1X 3-PGA ←┴── ATP
  │                                         NADPH₂
  ├──→ 1X RuDP  ──────→  2X 3-PGA
  │                      ──────────
  └──────────────────── 5X 3-PGA
              ↑
            8 ATP
            5 NADPH₂
```

FIG. 8. Schemes depicting the operation of the carbon reduction cycle and the use of NADPH₂ and ATP in a brightly illuminated leaf under conditions of fully open stomata (adequate CO₂ supply) and closed stomata (no CO₂ supply). The latter scheme is tentative. RuDP, ribulose 1,5-diphosphate; PGA, 3-phosphoglyceric acid. (From Osmond and Björkman, 1972.)

consumption of NADPH₂ and ATP per turn of the cycle. Since both C₃ and C₄ plants may frequently encounter such conditions, it seems very likely that both types of plants would make use of this proposed function of the glycolate pathway.

3.5 Diffusion Paths and the Stomatal Mechanism

In the colonization of terrestrial habitats, plants had to face a new problem with regard to photosynthesis. For CO_2 uptake to occur, CO_2 must be permitted to diffuse into the leaf. In order to maintain a high water content in the plant, essential for all metabolic activity, excessive water loss by diffusion of water vapor from the leaf must be prevented.

This is obviously a dilemma since there seems to be no way in which the resistance to the diffusion of CO_2 can be kept low while the resistance to the diffusion of water vapor is kept high. (A semipermeable membrane with these properties, if it exists, yet remains to be discovered.) The stomatal apparatus, which is present in all higher land plants, can be considered as a mechanism which was the best possible solution to this problem. It enables the plant to regulate the resistance to gaseous diffusion. As long as the supply of water to the leaves through the vascular system keeps up with the loss of water vapor from them, the diffusion resistance will decrease in response to an increased demand for CO_2. The biochemical and biophysical mechanisms by which control of stomatal opening and closing is achieved has been subject to intensive investigation for nearly a century and much recent progress has been made. However, space does not permit even a brief account of this work here. For the present purpose it may be sufficient to say that stomata open in response to increased light intensity and decreased CO_2 concentration in the leaf. However, even in bright light and low CO_2 concentration the stomata close in response to a decreased leaf water potential. With the notable exception of CAM plants, which often keep their stomata open during the night when they fix CO_2 and keep them closed during the day, stomata are generally closed in the dark.

When the stomata are fully open the resistance to gaseous diffusion is obviously at its minimum. This minimum diffusion resistance is inversely proportional to the number of stomata per unit leaf area and increases with increasing stomatal size although not necessarily in direct proportion. A detailed treatment of this subject has been given by Parlange and Waggoner (1970). Stomatal number and size vary greatly among different species and are also markedly affected by growing conditions, particularly light intensity. Hence, the minimum stomatal diffusion resistance also shows great variation (Meidner and Mansfield, 1968). In addition to the diffusion resistance imposed by the stomata that imposed by the boundary layer at the leaf/air interface also affects the diffusion of both water vapor and CO_2. This resistance is usually small in comparison with the total diffusion resistance to CO_2 at wind speeds usually encountered under conditions where photosynthesis is markedly influenced by diffusion. This boundary layer resistance increases with increased leaf size.

Under conditions of high light intensities and high temperatures the maximum intrinsic rate by which the photosynthetic machinery can operate is dependent on the concentration of CO_2 in the intercellular spaces. Whenever photosynthesis is CO_2 dependent it will inevitably also be partially limited by the resistance of the stomata to diffusion since the CO_2 concentration gradient between the ambient air and the intercellular spaces must increase with increasing diffusion resistance. When this resistance is low a further decrease will have only a small influence on photosynthesis, but the stomata will always impose some limitation on light-saturated photosynthetic rate in habitats with bright light and high temperatures. When light intensity or temperature are low photosynthesis is not appreciably CO_2 dependent and is therefore not influenced by stomatal diffusion resistance.

When comparing *intrinsic* photosynthetic characteristics of higher plants by gas exchange measurements under conditions where CO_2 is a limiting factor, it is obviously necessary that the intercellular CO_2 concentration be known. This can be obtained if water vapor release is measured simultaneously with CO_2 uptake. Since the pathways for the two gases between the ambient air and the intercellular spaces are the same, the resistance to CO_2 diffusion can be calculated from the measured values for the resistance to water vapor transfer, R_{H_2O}, by multiplying the latter with the ratio between the diffusivity of CO_2 and that of water vapor, D_{CO_2}/D_{H_2O}. The intercellular CO_2 concentration, C_i, is equal to that of the ambient air, C_a, minus the measured CO_2 flux, P, multiplied by the diffusion resistance for CO_2:

$$C_i = C_a - P \times R_{H_2O} \times (D_{CO_2}/D_{H_2O})$$

In addition to these resistances to *gaseous* diffusion of CO_2 from the ambient air to the intercellular spaces in the leaf, there are other resistances associated with the transport of dissolved CO_2 from the cell walls to the reaction sites in the chloroplasts. Since these resistances cannot be measured experimentally many attempts have been made to assess their importance in limiting higher plant photosynthesis by use of mathematical models (e.g., Chartier, 1970). These models are based on the assumptions that this *intra*cellular transport can be treated as if it were by passive diffusion alone and therefore obeys Fick's law, while photosynthesis at the chloroplast level obeys classical reaction kinetics, obtained with single enzyme reactions *in vitro*. Unfortunately, experimental support for either assumption is lacking. Available data rather infer that intracellular CO_2 transport involves enzyme-catalyzed reactions and also that photosynthesis at the chloroplast level may be subject to several kinds of regulation (Bassham, 1971), suggesting that it may not obey simple Michaelis-Menten enzyme kinetics. Several aspects of intracellular transport and utilization of CO_2 by plants have recently been treated in a review by Raven (1970).

4. Photosynthetic Adaptation

Photosynthetic adaptation may be defined as an adjustment of photosynthetic characteristics in such a manner that it enables the plant to photosynthesize more efficiently under the conditions prevailing in particular environment. The presence of differences in photosynthetic characteristics per se therefore does not necessarily mean that they represent adaptations. As with other adaptations, one can distinguish between two kinds on photosynthetic adaptations: (1) induced directly by the environment in which the plant is grown, within the limits set by the genotype, and (2) genetically determined as a result of a genetic adjustment to a particular habitat through natural selection. It should be realized that it is the ability to adapt which is inherited: the final adaptation itself is not necessarily inherited. Thus, the photosynthetic characteristics of two different species, or genotypes, may well be identical when the plants are grown in *one* common environment, but they may be markedly different when the plants are grown in *another* common environment. Where different species or genotypes exhibit profound differences in photosynthetic characteristics that would confer a selective advantage in their particular environments regardless of the conditions under which the plants were grown, these adaptations are commonly said to be genetically fixed. It seems probable that whenever such adaptations are encountered they involve a qualitative differentiation such as that which exists between C_3 and C_4 plants.

Photosynthetic efficiency is a broad term which may mean different things depending on the environment under consideration. In a habitat where light intensity is very low, it is obviously the efficiency with which the plant is capable of absorbing and utilizing light of low intensity which is important. How much water it spends in the process is at most of only secondary importance. One must also take into account the biosynthetic cost of producing and maintaining the photosynthetic apparatus itself. Thus, if two plants have the same light-absorbing efficiency and maximum quantum yields but the total investment in proteins and other constituents of the photosynthetic apparatus differ, then it is obviously the plant which has invested the least that is the more efficient of the two.

The criteria for photosynthetic efficiency in the other extreme of habitats, xerophytic sites in hot deserts, are obviously quite different. Here radiant energy is abundant but water supply is limited and the high thermal load on the plant together with low humidity of the atmosphere result in a very high water vapor pressure gradient between the leaf and the atmosphere. The water loss for a given resistance to gaseous diffusion thus becomes extremely high. In this environment the efficiency with which the plant is capable of fixing CO_2 in terms of water loss through transpiration is likely to be of overriding importance. The maximum quantum yield at low light intensities, and perhaps even the capacity for light-saturated photosynthesis under optimum conditions of temperature and water supply, can be expected to have little influence on the success of the plant in this environment. High *actual* rate of light-saturated photosynthesis and maximum water use efficiency are mutually exclusive since the former requires that the stomatal resistance to gaseous diffusion be low whereas the latter requires that it be high. However, high efficiency of utilization of low intercellular CO_2 concentrations is required both for high water use efficiency and for high light-saturated rate at a given stomatal diffusion resistance. Hence these two characteristics are often correlated.

In mesophytic environments with ample water supply and moderate irradiances and temperatures photosynthetic efficiency can probably be considered solely in terms of the capacity for both light-limited and light-saturated photosynthesis in relation to the metabolic energy and inorganic nutrients required to produce and maintain the photosynthetic apparatus. High photosynthetic efficiency in this type of habitat requires that the minimum resistance to gas diffusion be low.

4.1 Adaptations Involving Quantitative Differences

4.1.1 ADAPTATION TO CONTRASTING LIGHT CLIMATES

Light, the driving force of photosynthesis, shows a great variation in intensity among different habitats occupied by higher plants. Recent studies in a Queensland rain forest have shown that higher plants are capable of net photosynthesis and sustained growth in habitats where the average total daily quantum flux incident on the plants is only 22 μeinsteins cm^{-2} day^{-1}, including that contributed by sunflecks (Björkman and Ludlow, 1972). This compares with a daily flux of over 5000 μeinsteins cm^{-2} day^{-1} on top of the rain forest canopy, or approximately 7000 μeinsteins cm^{-2} day^{-1} in Death Valley, California, during the summer months, when several species carry out most of their photosynthesis and growth (Björkman *et al.*, 1972c). Thus, the daily quantum flux may differ by a factor of at least 200 between habitats occupied by higher plants.

In spite of the extremely low quantum flux on the rain forest floor, leaves of the two species investigated, *Alocasia macrorrhiza* (Araceae) and *Cordyline rubra* (Liliaceae) were able to fix CO_2 at a low but positive rate for 90% of the daylight hours, both on clear and on heavily overcast days. The quantum flux required to balance respiration was extremely low, 0.05–0.20 nanoeinstein cm^{-2} sec^{-1}, which is only a few percent of that required by plants growing in open, sunny habitats. This is probably primarily attributable to the very low rates of respiratory CO_2 release in the rain forest plants. The mean daily efficiency of utilization of *incident* quanta for *net* CO_2 uptake by *Alocasia* and *Cordyline* was 0.07 mole CO_2 per einstein which corresponds to a net conversion of incident light energy (400–700 nm) to chemical energy in carbon products of about 16%, a high value for any photosynthetic organism. Other measurements showed that photosynthesis in this habitat is not measurably limited by CO_2 concentration and temperature but that light is the sole limiting factor.

In contrast to the high efficiencies of utilization of light of low intensities for photosynthesis the light-saturated rates were low. This is a well-known characteristic of shade plants, and the first reports on differences in light-saturated rates of photosynthesis between shade and sun plants were published at least fifty years ago. The mechanisms underlying these differences have remained largely unknown, however. Stomatal diffusion resistance has been invoked as an important factor since shade plants do often have higher minimum resistances than sun plants (Holmgren *et al.*, 1965; Holmgren, 1968). While this is also true of the rain forest plants investigated by Björkman and Ludlow (1972), it is clear that the stomatal diffusion resistance was not an important factor in limiting the light-saturated rate in normal air.

This is illustrated in Fig. 9, which shows the light-saturated rate of photosynthesis as a function of the reciprocal of stomatal diffusion resistance for the rain forest species *Alocasia macrorrhiza* and the mesophytic sun species *Atriplex patula* ssp. *hastata*, grown at three different light regimes under controlled conditions in the phytotron. The high light regime is approximately equal to the quantum flux that the plant receives in its native habitat on the central California coast. The stomatal conductance values (reciprocal of resistance) that the plants typically assumed under conditions of light saturation and normal air are indicated with arrows. Clearly, a further increase in stomatal conductance for each plant would result in only a small increase in the light-saturated photosynthetic rate. Evidently, the rain forest plants, as well as the sun plant *Atriplex*, grown under different light levels, have so adjusted their stomatal conductances that these do not impose a major restriction on photosynthesis at the cellular level. It follows that the low capacity for light-saturated photosynthesis of the shade species, or of *Atriplex* grown under low light levels, cannot be attributed to restrictions on CO_2 dif-

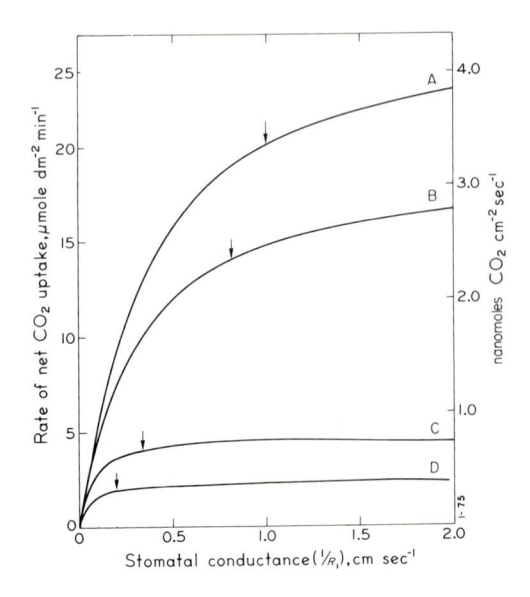

FIG. 9. Rate of light-saturated CO_2 uptake as a function of stomatal conductance in leaves of the rain forest species *Alocasia macrorrhiza*, grown in its native habitat (curve D), and in the sun species, *Atriplex patula* ssp. *hastata*, grown at three different levels of light intensity (curve A, high light; B, intermediate light; C, low light). Rates are calculated for a constant CO_2 concentration of 0.032% external to the leaves. Arrows indicate the actual stomatal conductances assumed by the leaves at light saturation and 0.032% CO_2. (Redrawn from Björkman et al., 1972a,b.)

fusion between the external air and the intercellular spaces in the leaves. Estimates of the relative *intra*cellular diffusion paths, based on phase-contrast microscopy of living and fixed leaf tissues of the different plants, indicated that the diffusion paths between the cell walls and the chloroplasts in the shade plants would be rather shorter than in the sun plants (Goodchild et al., 1972). Thus, on this basis it would appear unlikely that the differences in the capacity for light-saturated photosynthesis can be explained by a much higher intracellular resistance to the diffusive transport of dissolved CO_2 in the shade plants.

The chloroplasts of extreme shade plants, such as those used in the above studies, are characterized by well-developed grana stacks, which may reach prodigious proportions (Fig. 10). In contrast to the situation in most sun species, these grana stacks are oriented in more than one plane. Moreover, the proportion of grana lamellae to stroma lamellae is much higher than in sun plants. This striking granal development in the shade plants is reflected in the ratio of soluble protein to chlorophyll. Surveying a considerable number of shade and sun species grown in their native habitats Goodchild et al. (1972) showed that the protein:chlorophyll ratio was, on the average, seven times greater in the sun plants (cf. Table I). Thus, both the protein:chlorophyll ratio and the ultrastructural characteristics of the chloroplasts indicate that shade plants invest more in the production of an effective light-harvesting system in relation to protein which is involved in the conversion of photoproducts to chemical energy bound in the final products of photosynthesis. The absolute level of chlorophyll per unit chloroplast of cell volume is substantially greater in the shade plants whereas the opposite is true of protein. A number of studies have shown that in shade plants and in sun plants grown at low light levels, the ratio of the activity of RuDP carboxylase to chlorophyll is much lower than in sun plants grown at high light levels (Section 3.3.1). The activity of this enzyme in C_3 plants is closely correlated with the capacity for light-saturated photosynthesis *in vivo*. The content of "fraction I protein" (which is very probably the same entity as RuDP carboxylase) has also been found to be closely related to RuDP carboxylase activity, indicating that differences in amount rather than in turnover number are responsible for the observed differences in activity. Thus, there seems to be little doubt that the capacity of the carboxylation reaction is an important component in photosynthetic adaptation to contrasting light climates.

The studies by Boardman et al. (1972) and Björkman et al. (1972b) demonstrate, that differences between sun and shade plants is not limited to the carbon fixation system, but also include the capacity of the photosynthetic electron transport chain (cf. Section 3.2). Interestingly, the light-saturated rates of electron transport on a chlorophyll basis was about 10-fold greater in the sun species than in the extreme shade species. Growth of the sun species under a low light intensity resulted in a 3- to 4-fold decrease in this ratio. These values are quite similar to those obtained with RuDP carboxylase activity in the same plant materials (cf. Section 3.3.1), both on a relative and on an absolute basis. In addition, they show a close correlation with light-saturated capacity of photosynthesis *in vivo* (Fig. 11). Moreover, the differences in the capacity for electron transport which include that driven by photosystem I as well as that driven by photosystem II, were related to changes in the amount of several carriers of the electron transport chain (Section 3). It is evident from these studies that photosynthetic adaptation to different light climates cannot be ascribed to one single limiting factor but that it involves changes in a number of component processes of photosynthesis, all of which can be expected to affect the ability of the plants to photosynthesize efficiently in their respective contrasting native environments. One can consider this as evidence of a good adjustment of the process

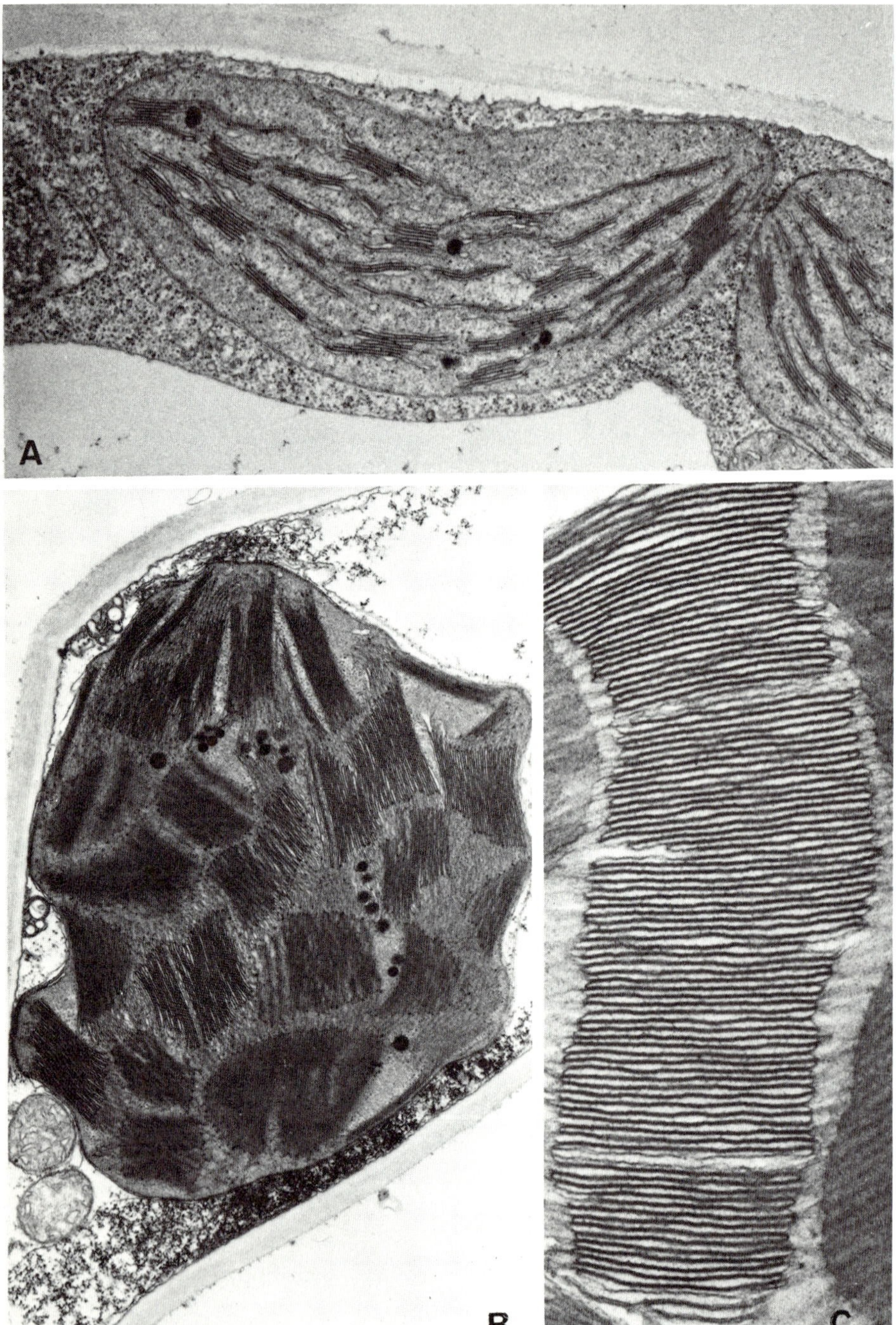
A
B
C

as a whole since it infers that no single leaf factor exclusively limits the overall rate under the conditions prevailing in the native habitat.

Growth of a sun species such as *Atriplex patula* under low light intensities results in changes in the characteristics of the photosynthetic apparatus so that they tend toward those of species whose distribution is limited to shaded habitats (Fig. 11). However, in none of the characteristics analyzed did the low-light-grown *Atriplex* come near those of the rain forest species. This could possibly be explained by the fact that the amount of light received by these *Atriplex* plants was considerably higher than that received by the rain forest plants although it was the lowest that could sustain growth. Sun plants such as *Atriplex* are incapable of growth at light levels approaching those tolerated by the

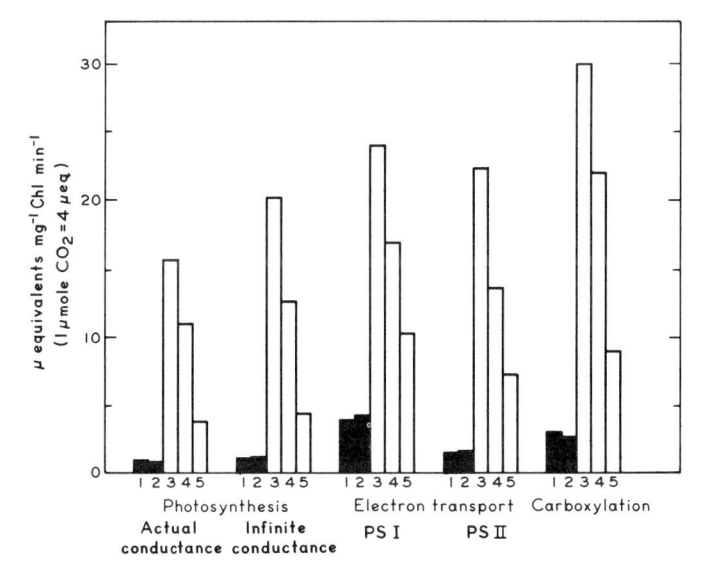

FIG. 11. Capacity for light-saturated CO_2 uptake *in vivo* and chloroplast reactions *in vitro* for the "shade" plants *Alocasia macrorrhiza* and *Cordyline rubra*, grown in their native rain forest habitats, and for the "sun" plant *Atriplex patula* ssp. *hastata*, grown at three different levels of light intensity. All rates are expressed as microequivalents of electrons transferred per milligram of chlorophyll per minute. It has been assumed that the fixation of 1 μmole of CO_2 corresponds to 4 μeq. Bars: 1, *Alocasia*; 2, *Cordyline*; 3, 4, and 5, *Atriplex* at high, intermediate, and low light levels, respectively. (Drawn from data of Boardman *et al.*, 1972, and Björkman *et al.*, 1972b.)

FIG. 10. Electron micrographs showing grana arrangements in chloroplasts of the "sun" plant *Atriplex patula* ssp. *hastata*, grown at a high light intensity (A), the "shade" plant *Alocasia macrorrhiza*, grown in its native rain forest habitat (B), and a large granum of the latter species (C). A, ×18,600; B, ×15,500; C, ×46,000. (From Björkman *et al.*, 1972b, and Goodchild *et al.*, 1972.)

shade species and this may in itself suggest that the former plants lack the genetic ability to produce a photosynthetic apparatus which is as efficient at low light as those of the rain forest plants. Conversely, other evidence indicates that species limited in nature to densely shaded habitats lack the genetic ability to produce a photosynthetic apparatus that is as efficient at high light intensities as that of sun plants.

Several studies indicate that such differences in adaptability are not restricted to unrelated species but exist also between different ecotypes (ecologically adapted races) within single biological species that occupy both densely shaded and open sunny habitats. Björkman and Holmgren (1963, 1966) investigated the potential range of photosynthetic adjustment among such ecotypes. The rate of photosynthetic CO_2 uptake as a function of light intensity during measurement was found to be strongly influenced by the light intensity under which the plants were grown but the response was strikingly different between ecotypes from shaded and those from sunny habitats. In the "sun" ecotypes the capacity for photosynthesis at high and moderate light intensities was considerably higher when the plants were grown in strong light than when they were grown in weak light, a response which these sun *ecotypes* have in common with *species* restricted to sunny habitats. In sun ecotypes of *Solidago virgaurea* the higher light-saturated rate obtained when the plant was grown in strong light was also in this case paralleled by a higher activity of RuDP carboxylase, a greater protein content (Björkman, 1968b), and a decreased stomatal resistance to CO_2 diffusion (Holmgren, 1968). In contrast to ecotypes from sunny habitats, those from shaded habitats failed to yield higher light-saturated photosynthetic rates when they were grown at high light intensities. The activity of RuDP carboxylase remained about the same as when the clones were grown in weak light and so did the resistance of stomata to CO_2 diffusion. Moreover, growing these shade ecotypes at high light intensities resulted in a reduction of the quantum yield at low light intensities, a response which was not found with ecotypes from sunny habitats.

Intraspecific differentiation similar to that in *Solidago* was also found between sun and shade ecotypes of *Solanum dulcamara* by Gauhl (1968, 1969). The existence of differences in the response of light-saturated photosynthesis and RuDP carboxylase to light intensity during growth between populations of *Dactylis glomerata* is also indicated by the studies of Eagles and Treharne (1969). Gauhl (1968, 1969) was able to show that even fully expanded leaves of sun clones of *Solanum* that had developed under a low light intensity retained the ability to increase their capacity for light-saturated photosynthesis upon transfer to high light intensities. The increase in photosynthesis was paralleled by an increase in protein synthesis. This synthesis was accompanied by increases in "fraction I protein" and RuDP carboxylase activity. Leaves of *Solanum* clones from a shaded habitat lacked the ability to increase their capacity for light-saturated photosynthesis when transferred to high light intensities and the contents of total protein, fraction I protein, and RuDP carboxylase activity did not increase. Instead, inhibition of the quantum yield occurred.

Although chlorophyll bleaching does take place at advanced stages of inhibition by strong light, the reduction in quantum yield is not caused by destruction of light-harvesting pigments per se. Strong light causes an inactivation of the photochemical reaction centers as a result of an excess of excitation energy being transferred to them (Jones and Kok, 1966a,b; cf. Section 3.4). A smaller number of light-harvesting pigment molecules per reaction center in sun plants would provide an attractive explanation for their greater ability to resist photoinhibition of the reaction centers since at any given incident light intensity each reaction center would then receive less excitation energy. However, this hypothesis is not supported by the experimental results obtained by Boardman *et al.* (1972) and Björkman *et al.* (1972b). There was no evidence that the ratio of light harvesting pigments to reaction centers differed appreciably between the extreme shade and sun plants investigated by these workers. Moreover, the degree of inactivation of photochemical reactions by strong light was no greater in chloroplast isolated from shade plants than from sun plants when no exogenous electron acceptors were added during the exposure to photoinhibitory light. These results indicate that the greater resistance to photoinhibition found with intact photosynthesizing leaves of sun plants is a consequence of their greater ability to increase their capacity for electron transport, CO_2 fixation, and CO_2 diffusion when they are grown in strong light. A high photosynthetic capacity at high light intensities would divert a greater fraction of the excitation energy to be used in photosynthesis and less excess energy would thus be left at the reaction centers to cause their inactivation.

1.1.2 ADAPTATION TO DIFFERENT THERMAL REGIMES

Temperature affects virtually all growth processes, including photosynthesis, and it differs widely between different habitats occupied by higher plants. Arctic tundra and subtropical deserts probably represent the extremes. However, it also varies during the growing season in a given habitat. The thermal load imposed on a winter-active plant in a desert habitat need not be greater than that imposed on a summer-active plant in the Arctic. When considering photosynthetic adaptation to temperature it should also be realized that it is the *leaf* temperature, prevailing during that period of the day when light intensity is relatively high, which is important. At low light intensities photosynthesis is little, if at all, affected by temperature.

At high irradiance levels, leaf temperature is a dependent variable of at least seven independent variables, all acting simultaneously, and it is only indirectly related to air temperature (see Gates, 1968). Among the major variables affecting leaf but not air temperature is the conductance of the stomata to gas diffusion, and this is controlled by the plant itself. The rate of transpiration and the temperature of a given leaf as functions of stomatal conductance is illustrated in Fig. 12. Clearly, even under conditions of constant heat load, constant wind speed, and constant air temperature, a given leaf may be considerably hotter or considerably cooler than the ambient air, or it may have the same temperature. In evaluating temperature relationships in different environments it is therefore important to take into account the total energy load imposed on the plant as well as the water supply and other factors that influence the loss of latent heat through transpiration. For example, Gates *et al.* (1964) were able to show that leaf temperatures of *Mimulus cardinalis*, growing at a low elevation in the Sierra Nevada of California, were up to 10°C cooler than air temperature. In contrast, leaf temperatures of *Mimulus lewisi*, growing at 3500 meter elevation were up to 10°C warmer than the air temperature. Thus, in spite of 20°C difference in air temperature be-

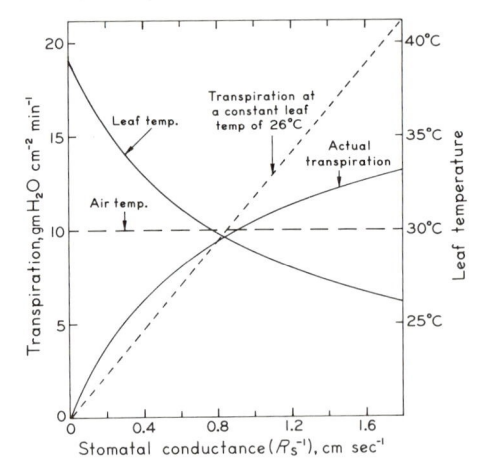

Fig. 12. The dependence of transpiration rate and leaf temperature on stomatal conductances for a leaf of given dimensions (5 cm long, 4 cm wide) and constant conditions of irradiance (1.2 cal cm^{-2} min^{-1}), air temperature (30°C), relative humidity (20%), and wind speed (2 m sec^{-1}). (Calculated from equations given by Gates, 1968.)

tween the two sites, leaf temperatures were about equal. Also, leaves of *Phragmites communis*, growing near a spring in Death Valley, California, had leaf temperatures up to 9°C cooler than those of other species growing in adjacent, more xerophytic sites, when air temperature was 46°C at both sites (Pearcy *et al.*, 1972).

Several investigations have shown that the temperature dependence curve for light-saturated photosynthesis is markedly influenced by the temperature regime under which a given plant is grown (Mooney and West, 1964; Billings *et al.*, 1971; Björkman and Gauhl, 1968; Mooney and Harrison, 1970). This is also true for C_4 plants (Björkman and Pearcy, 1971). Growth under low as compared with moderate temperatures generally results in a higher light-saturated rate in the lower temperature range and shifts the photosynthesis optimum toward a lower temperature. Growth under high temperatures has the opposite effect (Fig. 13). This plasticity may permit the photosynthetic apparatus to adjust to seasonal changes in thermal regime. The studies of Mooney and Shropshire (1967) and Mooney and Harrison (1970) also show that appreciable shifts in the temperature response curve can occur in less than several days, indicating that even mature leaves have a considerable capacity for photosynthetic acclimation to changes in the temperature regime.

It is also evident that the potential range for such environmentally induced temperature acclimation differs among species occupying habitats with contrasting thermal regimes. This is particularly striking between C_3 species from cool, moist environments and C_4 species from hot, semi-arid habitats (see Section 4.2, Fig. 16). However, it is also evident but less striking among different C_3 species and among ecotypes of the same

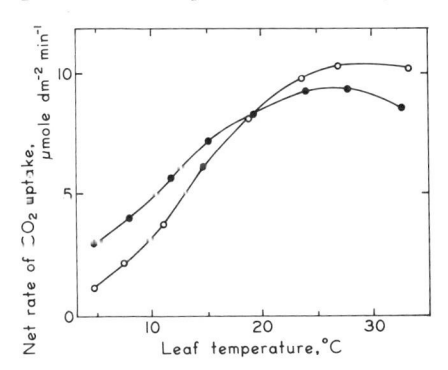

Fig. 13. Effect of growth temperature on the temperature dependence of light-saturated net photosynthesis for a clone of *Mimulus cardinalis* in air of normal CO_2 and O_2 concentration. ●—●, Grown at 10°C; ○—○, grown at 30°C. (Redrawn from Björkman *et al.*, 1969a.)

species (Billings *et al.*, 1971; Björkman *et al.*, 1969a; Mooney and Billings, 1961).

Very little is known about the mechanism underlying photosynthetic adaptation to temperature in plants having the same pathway of CO_2 fixation. The kinetics of the temperature-dependence curve of photosynthesis in normal air are highly complex which make interpretations difficult. At low temperatures, photosynthesis increases exponentially with temperature and the Arrhenius activation energy is constant, but as temperature is increased further the increase in photosynthesis becomes smaller than predicted by the Arrhenius equation (Fig. 14). Then it reaches a plateau or a broad peak, and finally declines, slowly at first, then precipitously. This steep fall is indicative of progressive thermal inactivation of the photosynthetic apparatus.

The early deviation of photosynthesis from the course predicted by the Arrhenius equation in air of normal CO_2 and oxygen content has been interpreted as indicating that CO_2 diffusion rather than biochemical steps are primarily limiting the rate of CO_2 uptake. There is no doubt that diffusion is often an important limiting factor, particularly when water supply is restricted and, as a result, stomatal resistance is high. However, in addition to diffusion becoming increasingly limiting as temperature is increased there are at least three interdependent factors that may also be responsible for the observed early deviation from the Arrhenius equation in C_3 plants. One important factor is the inhibitory effect of O_2, which is small or absent at low temperatures and which increases with temperature. As shown in Fig. 14, the temperature dependence is considerably greater when the O_2 inhibition is suppressed by a low O_2 concentration. Second, the solubility of CO_2 in aqueous media decreases with increasing temperature, and this may result in a decreased *intra*cellular CO_2 concentration. Third, if the affinity of the carboxylation enzyme for CO_2 decreases with increasing temperature *in vivo*, as it apparently does *in vitro* (Björkman and Pearcy, 1971), this would also influence the temperature dependence under normal atmospheric CO_2 concentration. At saturating CO_2 concentrations none of these factors would influence the rate of photosynthesis. It is interesting to note that under these conditions the temperature range in which photosynthesis shows a pronounced increase with temperature and obeys the Arrhenius equation is, indeed, considerably extended. As can also be expected, the effect of an increased CO_2 concentration on the temperature dependence of photosynthesis is much less pronounced in C_4 plants (cf. Section 4.2).

A decreased activation energy of enzyme-catalyzed reactions that are rate-limiting at low temperature would provide one possible mechanism

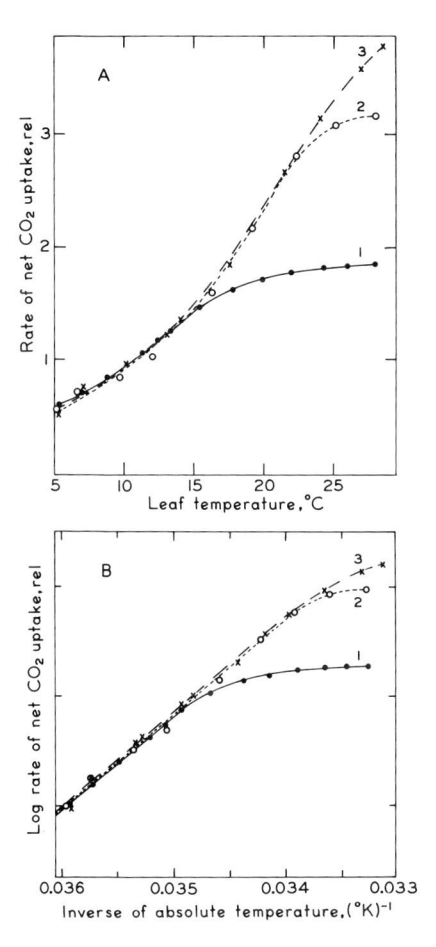

FIG. 14. Influence of O_2 and CO_2 concentration on the temperature dependence of net photosynthesis at light saturation for a given leaf of *Atriplex patula* ssp. *spicata*. (A) Conventional plot; (B) Arrhenius plot. Curve 1: 21% O_2, 0.03% CO_2; curve 2: 1.5% O_2, 0.03% CO_2; curve 3: 1.5% O_2, 0.07% CO_2. (Drawn from data of Björkman and Pearcy, 1971.)

for photosynthetic adaptation to cold environments. There is, however, as yet no experimental evidence that such differences do exist. Björkman and Pearcy (1971) compared the Arrhenius activation energy for *in vivo* photosynthesis at low temperatures and O_2 concentration in a wide diversity of species and ecotypes from habitats with greatly contrasting thermal loads. These plants were also grown under a wide range of temperature regimes. No appreciable differences in activation energies were found even though the absolute rate of photosynthesis and also the temperature dependence at higher temperatures varied greatly among the different species and growing conditions. The activation energy for RuDP carboxylase-catalyzed CO_2 fixation *in vitro*, determined with enzyme preparations from these plants, also did not vary appreciably. It is noteworthy that the activation energies for RuDP carboxylase *in vitro* and photosynthesis *in vivo* are similar (16–20 kcal mole^{-1} °K^{-1}). The activation energy for photosynthesis is unusually high for biological reactions, and its constancy among the species tested suggests that the same enzyme may be rate-limiting at *low* temperatures in all higher plants, including C_4 species. This enzyme could well be RuDP carboxylase, but in the absence of information of other potentially rate-limiting reactions at low temperature, it cannot be stated that this is necessarily the case.

There are examples in the literature of heterotrophic organisms in which the temperature dependence of the affinity of certain enzymes for their substrates differs so that maximum affinity occurs at a lower temperature in organisms adapted to cold than in those adapted to warm temperatures. This would certainly be a possible mechanism also in photosynthetic adaptation to contrasting thermal habitats but to the author's knowledge no comparative study of the temperature dependence of the Michaelis constants of photosynthetic enzymes has been reported to date.

There is some evidence that the amount of chloroplast protein and RuDP carboxylase activity tends to fall with increasing temperature for growth (Björkman and Gauhl, 1969; Björkman *et al.*, 1969a). This effect is particularly marked if unit leaf area is used as the basis for comparison. Thus, the extent by which photosynthesis is limited by the capacity of enzyme-catalyzed reactions in relation to CO_2 diffusion at a given temperature is likely to increase with increased growth temperature. This could at least in part explain why growth of plants at high temperatures results in a lower absolute photosynthetic rate per unit leaf area at *low* temperatures and why it extends the range in which photosynthesis increases with increasing temperature. However, it does not explain the observation that high growth temperature results in an increased absolute photosynthetic rate per unit leaf area at *high* temperatures. Neither does it explain why plants, genotypically adapted to hot environments, are capable of a more efficient photosynthesis at high temperatures than plants adapted to cold environments under conditions of equal resistances to gaseous diffusion.

Information is also lacking on the mechanisms which permit the photosynthetic apparatus of plants adapted to hot environments to function effectively at leaf temperatures which result in a permanent damage to it in plants adapted to cold or temperature habitats.

4.2 Adaptive and Evolutionary Aspects of C_4 Photosynthesis and Crassulacean Acid Metabolism

4.2.1 ADAPTATION

Plants possessing C_4 photosynthesis and succulent plants exhibiting CAM are found in all continents of the world with the exception of Antarctica and Arctic regions of Eurasia and North America. In general they are found in habitats with high thermal loads, and they are often dominant in areas where this is combined with drought. Succulents possessing capacity for CAM differ from C_4 plants in that they tend to be restricted to habitats where there is a pronounced diurnal variation in temperature with cold nights. When water is abundant, or the diurnal temperature variation small, these succulents may cease to carry out CAM. C_4 plants are common also in tropical and subtropical areas with warm nights, and in all probability they continue to fix CO_2 via the C_4 pathway in the light regardless of environmental conditions.

It is evident that the basic adaptive advantage of CAM is that it permits the plant to carry out CO_2 fixation during the night when the heat load is lowest and therefore the potential evapotranspiration is at its minimum. There are, however, some apparent disadvantages with this mechanism. In order for CO_2 fixation to occur at a reasonably high rate, malic acid must accumulate in massive amounts during the night, and the capacity for storage of this product may well be a limiting factor. This is indicated by the observation that the rate of CO_2 fixation tends to decline as the malic acid pool builds up in the dark (for references, see Ting et al., 1972). Carbohydrate storage may impose another problem. In order to supply the CO_2 acceptor, PEP, and the $NADPH_2$, required for the reduction of oxaloactic acid to malic acid, a carbohydrate pool must be retained in the leaves. These and related leaf anatomical factors may explain the very low maximum rates of CO_2 uptake, integrated over a 24-hour period, which is characteristic of many CAM plants. Thus, in the case of CAM plants, the ability to operate with a high water use efficiency when water supply is restricted may preclude a high photosynthetic capacity when water is abundant.

In C_4 photosynthesis there is a rapid turnover of malic and aspartic acid so there is no need to store these intermediates. Photosynthesis can therefore proceed at a sustained high rate throughout the day provided stomatal resistance is low. This is an obvious advantage when water supply is sufficient to permit a high rate of transpiration. However, since CO_2 fixation can occur only in the light, the plants cannot take advantage of the lower evaporative load during the night. Yet, as will be discussed in the following paragraphs, C_4 plants are generally capable of a high water use efficiency when water is limiting and are also capable of high photosynthetic rates when water supply is ample.

Already in 1960 to 1965, several studies in crop plants indicated that certain plants such as maize, sorghum, and sugar cane, which have an unusually high productivity, also possess unusual photosynthetic gas exchange characteristics. These include the ability of illuminated leaves to reduce the CO_2 concentration in an enclosed space to almost zero (Meidner, 1962; Moss, 1962); unusually high light intensities are required to saturate photosynthesis (Moss et al., 1961; Hesketh and Musgrave, 1962; Hesketh, 1963; Murata and Iyama, 1963; El-Sharkawy and Hesketh, 1965); and the optimum temperature for photosynthesis is unusually high. These and many other plants with similar characteristics are now known to possess the C_4 pathway of photosynthesis (for references, see Hatch et al., 1971).

Most comparative studies of photosynthetic gas exchange on C_3 and C_4 plants have been made on taxonomically unrelated species, grown in uncontrolled and varying conditions. As discussed in the preceding sections, considerable differences in the light intensity and temperature dependence of photosynthesis exist even among species having the same CO_2 fixation pathway, and these characteristics are also markedly influenced by growing conditions. Moreover, the point has often been made that the fact that C_4 plants occur in hot, arid environments does not in itself prove that it is the C_4 pathway that is responsible for their success in these habitats. Other adaptations might also have evolved, and these could be decisive in the success of these plants.

The ideal plant materials for studies whose purpose it is to assess the functional and adaptive significance of the C_4 pathway itself would be plants that have identical genetic constitutions with the exception of those specific genes that determine the operation of the C_4 pathway itself. Such ideal plant materials probably do not exist in nature. However, there do exist C_3 and C_4 species whose genetic constitutions are sufficiently similar to permit hybridization between them and which coexist in the same habitats even though their main distributions differ (Björkman et al., 1969b). These plants are therefore uniquely suited for comparative studies of the adaptive significance of the C_4 pathway. Two such species, *Atriplex rosea* L. (C_4) and *A. patula* ssp. *hastata* Hall and Clem. (C_3) have been extensively studied in the author's laboratory and some examples from this work are used here to illustrate the functional differences between C_3 and C_4 plants. In spite of its close relationship to a C_3 species, the C_4 species *A. rosea* possesses all of the diagnostic traits of a C_4 plant. These include high activities of the enzymes of the C_4 pathway, fixation of CO_2 into malic and aspartic acid, presence of a well-developed "Kranz" anatomy with separate mesophyll and bundle-sheath cell layers, low discrimination against carbon isotope ^{13}C to ^{12}C during photosynthesis, low CO_2 compensation point, absence of CO_2 release into CO_2-free air in the light, and absence of an inhibitory effect of atmospheric O_2 on net CO_2 fixation (Björkman and Gauhl, 1969; Björkman et al., 1969b, 1970, 1971; Boynton et al., 1970; Nobs et al., 1970; Pearcy and Björkman, 1970).

The comparative photosynthetic responses to light intensity, temperature, and CO_2 concentration of the two *Atriplex* species, when grown under identical conditions, are shown in Fig. 15, A, B, and C, respectively. Photosynthesis is in all cases expressed on the basis of unit leaf area. The amount of chlorophyll and protein per leaf area, however, are similar in the two species, so these bases may also be used more or less interchangeably with leaf area. As shown in Fig. 15A the photosynthetic rates are similar in the two species at low light intensities, but as light intensity is increased the rate of *A. patula* (C_3) saturates at relatively low intensities whereas the rate of *A. rosea* (C_4) continues to increase up to intensities equal to full sunlight. Thus, the capacity for photosynthesis at high light intensities is considerably greater in the C_4 species. The temperature dependence of photosynthesis at a high light intensity of 250×10^3 ergs cm^{-2} sec^{-1} (400–700 nm) is shown in Fig. 15B. At low temperatures the photosynthetic rate differs little between the two species,

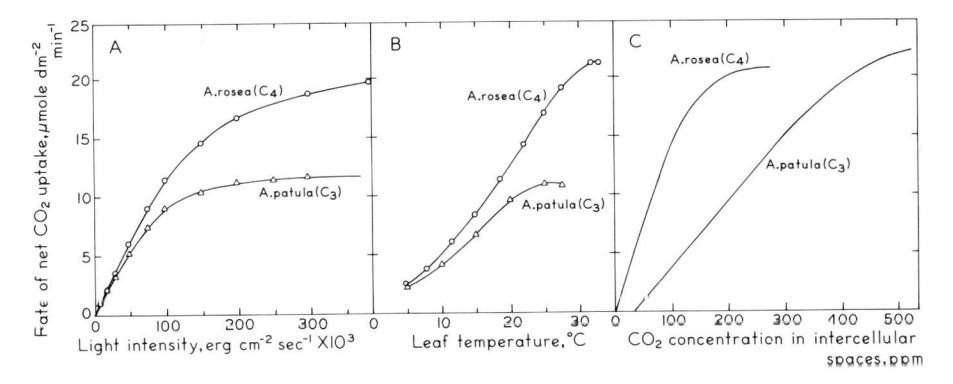

Fig. 15. Comparative photosynthetic characteristics of the C_3 species *Atriplex patula* ssp. *hastata* and the related C_4 species *A. rosea* in air with 21% O_2. The plants were grown under identical controlled conditions of 25°C day/20°C night, 110×10^3 ergs cm^{-2} sec^{-1} (400–700 nm) for 16 hr/day and ample water and nutrient supply. Photosynthetic rates of single attached leaves are given as functions of (A) light intensity at 27°C leaf temperature, 21% and 0.03% CO_2, (B) leaf temperature at 250×10^3 ergs cm^{-2} sec^{-1} and 0.03% CO_2, and (C) intercellular space CO_2 concentration at 250×10^3 ergs cm^{-2} sec^{-1} and 27°C leaf temperature. (Redrawn from Björkman et al., 1970, and O. Björkman, unpublished data.)

but as temperature is increased the C_4 species becomes increasingly superior, and at a leaf temperature of 30°C its rate is twice that of the C_3 species. Figure 15C shows the rate of photosynthesis at 27°C and 250×10^3 ergs cm^{-2} sec^{-1} as a function of the intercellular space CO_2 concentration. As indicated by the steeper initial slope of the curve in *A. rosea*, the photosynthetic apparatus of this C_4 species is considerably more efficient in utilizing low CO_2 concentrations than the C_3 species. Slatyer (1970) found similar differences in the CO_2 dependence between the C_4 species *A. spongiosa* and the C_3 species *A. hastata*. The CO_2 concentration required for saturation of photosynthesis to occur is also much lower in *A. rosea* (C_4), but the CO_2 saturated rate is not higher than in *A. patula* (C_3). That C_3 plants may be capable of at least as high photosynthetic rates at CO_2 saturation as C_4 plants is also indicated by the data of El-Sharkawy and Hesketh (1965).

Under conditions of saturating CO_2 concentrations, the light and temperature response curves for photosynthesis in *A. rosea* and *A. patula* are much more similar than they are under normal CO_2 concentrations. Thus the photosynthetic responses to these factors of *A. rosea* (C_4) in normal air are similar to those of *A. patula* (C_3) at elevated CO_2 concentrations. This is consistent with the view that the C_4 pathway acts as a CO_2-concentrating mechanism (cf. Section 3.3.2). In a similar manner the differences between the two species in their responses to light, temperature, and CO_2 concentration, present in air of normal oxygen concentrations, become much smaller if the oxygen concentration is reduced to 1% or 2%. This is also consistent with the concept of the C_4 pathway serving as a CO_2-concentrating mechanism, since a high CO_2 concentration suppresses the inhibitory effect of oxygen in *A. patula* and other C_3 plants (cf. Section 3.4). Regardless of the way by which the C_4 pathway achieves these effects, it is clear that it enables the plant to minimize the restrictions imposed on photosynthesis by the high concentration of oxygen and low concentration of CO_2 in the present atmosphere.

It is therefore obvious that the selective advantage of the C_4 pathway is greatest where the limitations imposed by these two factors is maximal, i.e., under conditions of high irradiance, high leaf temperature, and low intercellular space CO_2 concentrations. Since under a high irradiance a decrease in stomatal conductance inevitably results in both increased leaf temperature (Fig. 12) and a decreased intercellular space CO_2 concentration, it follows that the adaptive advantage of the C_4 pathway is particularly great in environments that are both hot and arid. Water loss through transpiration is proportional to stomatal conductance and is the same regardless of species. However, because the C_4 pathway enables the plant to photosynthesize at a higher rate at any given intercellular space

CO_2 concentration that occurs under natural conditions, the C_4 plant will be capable of a more efficient water use. This does not mean, of course, that the actual water use efficiency is always higher in C_4 plants. Since, as a rule, transpiration decreases more than photosynthesis when stomatal conductance is decreased in C_3 as well as in C_4 plants, the water use efficiency tends to increase with decreased stomatal conductance in both types of plants. Consequently, a C_3 plant with a low stomatal conductance may have a higher water use efficiency than a C_4 plant with a high stomatal conductance. However, for any given leaf temperature and rate of water loss, the rate of photosynthesis will invariably be substantially greater in the C_4 plant.

Slatyer (1970) demonstrated that the differences in photosynthetic water use efficiency between *Atriplex spongiosa* (C_4) and *A. hastata* (C_3) was reflected in similar differences in the water use efficiency for dry matter accumulation during growth. Downes (1969) showed that C_4 grasses have higher water use efficiencies for growth than C_3 grasses. As can be expected, this difference increased with increasing growth temperature. Field data collected by Shantz and Piemeisel (1927) for a great number of plants, both grasses and dicots, reveal that the C_4 species had a substantially higher water use efficiency for growth.

The *Atriplex rosea* and *A. patula* plants used in the above comparisons were grown at moderately high irradiance and temperature and adequate water supply which permit both species to grow well. If the plants are grown at higher temperature the differences between the two species become even more pronounced (Björkman and Pearcy, 1971). A recent field study of *Tidestromia oblongifolia*, an herbaceous perennial which carries out almost all of its photosynthesis and growth during the extremely hot summer in its native habitat on the floor of Death Valley, California, illustrates how remarkably well adapted to such extreme conditions a C_4 plant can be (Björkman et al., 1972c). This plant was able to photosynthesize at a very high rate throughout the day, and the rate reached its maximum at noon when irradiance was at a maximum and leaf temperatures were 46°C to 50°C. Controlled experiments *in situ* showed that in full noon sunlight the optimum temperature for net photosynthesis occurred at 47°C (Fig. 16), a temperature at which C_3 plants from temperate environments have ceased to photosynthesize and which causes permanent damage to their photosynthetic apparatus. Optimum temperatures exceeding 47°C have been previously reported only for thermophilic blue-green algae, occurring in hot springs. At 47°C, photosynthesis in *T. oblongifolia* was essentially directly proportional to light intensity up to full noon sunlight. Thus, even at these very high photosynthetic rates and resulting low intercellular space CO_2 concentrations,

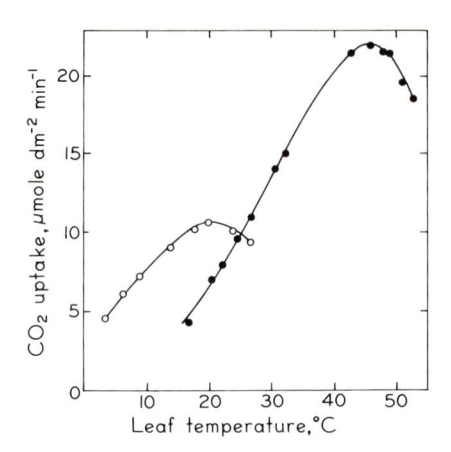

FIG. 16. Dependence of net CO_2 uptake on leaf temperature for the C_4 species *Tidestromia oblongifolia* (●—●) at an irradiance of 1.33 cal cm^{-2} min^{-1} (350–2400 nm, natural sunlight). Measurements were made in the field in Death Valley, California. (Redrawn from Björkman et al., 1972c.) For comparison, the temperature dependence of light-saturated photosynthesis for a subalpine ecotype of the C_3 species *Deschampsia caespitosa* (○—○) is also shown. (From R. W. Pearcy, unpublished data.)

CO_2 was evidently not severely limiting photosynthesis. There was no evidence that the plant was under severe water stress even though the plant water potential fell to −25 bars at noon.

These results indicate the existence of a much greater potential for adaptive differentiation in photosynthetic characteristics among higher plants than could be expected from previous information. While it is likely that the extraordinary photosynthetic performance of *T. oblongifolia* is partly attributable to the presence of the C_4 pathway, the results strongly suggest that other adaptive mechanisms have also evolved, resulting in a greater thermal stability of the photosynthetic apparatus than is present in plants occupying cooler environments.

4.2.2 EVOLUTION

Present evidence strongly indicates that both C_4 photosynthesis and capacity for CAM are relatively recent events in evolution. No algae, bryophytes, ferns, gymnosperms, or even the more primitive angiosperms, have been found to possess these pathways. They are, however, rather widespread among the more highly developed orders of angiosperms. In addition, these pathways can be considered to be addenda to Calvin-Benson cycle. There is no evidence that they would have conferred any advantage under primitive atmospheres with high CO_2 or low O_2 concentrations, nor that they would do so even in present cool, moist, or densely shaded habitats. Incidentally, neither is there any evidence that the C_4 pathway would be disadvantageous in temperate environments, and C_4 plants also do occur in relatively cool habitats, such as coastal strands and salt marshes.

In the period immediately following the discovery of the C_4 pathway it was thought that the differentiation in C_3 and C_4 plants represented a deep evolutionary split in the plant kingdom. This implied that the differentiation had to occur only once during evolution. Since that time the C_4 pathway has been found in at least ten major plant families of both monocots and dicots (Table II), nearly one hundred genera, and several hundred species. All the families that contain C_4 plants also contain C_3 plants, and a few contain CAM plants. Moreover, at least eleven genera contain both C_3 and C_4 plants (Table II); at least one of these genera (*Euphorbia*) has CAM species as well. This taxonomical distribution demonstrates that both C_4 photosynthesis and CAM have arisen polyphyletically from C_3 plants. The genus *Atriplex* is of particular interest in this context. Hall and Clements (1923), who were working with primarily North American atriplexes, on the basis of embryonic characteristics divided the genus into two subgenera: *Euatriplex* and *Obione*. It is now known that each one of these subgenera contain both C_3 and C_4 species. In each subgenus the more primitive species are C_3 plants whereas the majority, but not all, of the more highly evolved species are C_4 plants. This suggests that the C_4 pathway has evolved independently, at least twice, and possibly several times, even within this single genus.

The successful hybridization between C_4 and C_3 species of *Atriplex* (Björkman et al., 1969b, 1971; Nobs et al., 1970) proved that the genetic

TABLE II
FAMILIES AND GENERA HAVING BOTH C_3 AND C_4 SPECIES

Monocotyledons		Dicotyledons	
Families	Genera	Families	Genera
Cyperaceae	*Cyperus*	Azioaceae	
	Scirpus	Amaranthaceae	*Althernera*
Gramineae	*Panicum*	Chenopodiaceae	*Atriplex*
			Bassia
			Kochia
			Sueda
		Compositae	
		Euphorbiaceae	*Euphorbia*
		Nyctaginaceae	*Boerhavia*
		Portulacaceae	
		Zygophyllaceae	*Zygophyllum*

diversity between C_3 and C_4 species need not be great. Analyses of segregating populations also indicated that the number of genes determining each one of the major components of the C_4 pathway, such as high activity of PEP-carboxylase and Krantz-type anatomy, is small (Boynton et al., 1970; Pearcy and Björkman, 1970). However, for effective and integrated operation of C_4 photosynthesis it is not sufficient that the components of this pathway are present, but they must also be properly coordinated and spatially compartmented (Björkman et al., 1970, 1971). It is likely, therefore, that deficiency of a particular enzyme or aberration in its intra- or intercellular location will result in a photosynthetic performance which would have a selective disadvantage rather than an advantage over the C_3 progenitor.

One key question in evaluating the complexity of the evolution of the C_4 pathway is the extent to which the evolution of unique enzyme species is required. Of the enzymes known to operate in C_4 photosynthesis only pyruvate, P_i, dikinase appears to be totally lacking in C_3 plants. All the other enzymes, including PEP carboxylase, have functional counterparts in C_3 plants, but the activity of these enzymes is much higher in C_4 plants. Clearly, the evolution of the C_4 pathway would be much simpler if the increased enzyme activities were caused by an enhanced expression of genes already present in the C_3 progenitor. However, recent comparative studies of enzymes in a number of C_3 and C_4 species of *Atriplex* and in hybrids between them indicate that this might not be so. Hatch et al. (1972) showed that PEP carboxylase and several other enzymes of the C_4 pathway have distinctly different electrophoretic and chromatographic profiles than those isolated from C_3 plants. In the case of PEP carboxylase it is clear that the kinetic properties also differ (Hatch et al., 1972; Ting and Osmond, 1973a). For example, the affinity of PEP-carboxylase for PEP was 5-fold higher in the C_4 atriplexes and several other C_4 species than in the C_3 species investigated by Ting and Osmond. Moreover, the various enzyme forms, characteristic of the C_3 and the C_4 species of *Atriplex*, were transferred to the F_1 hybrids between these two types of species and then segregated in the F_2 and F_3 individuals. This gives further support to the view that at least some of the enzymes necessary for the operation of the C_4 pathway are unique alloenzyme entities, evolved specifically for C_4 photosynthesis. Further studies are required, however, to determine the extent to which the high activity of these alloenzymes in the C_4 plants stem from genetic control of protein structure.

The similarities between the biochemical pathways of CO_2 fixation in C_4 photosynthesis and CAM suggest the possibility of a common origin. Since CAM plants lack the leaf anatomical elaboration of C_4

plants, CAM has been implicated as an evolutionary link between C_3 and C_4 species. However, even though some families contain both C_4 and CAM plants, the latter are absent from several major families, including all grasses. Similarly, no CAM plants have been found within the vast majority of genera known to contain C_4 plants. It is also noteworthy that although CAM plants possess high PEP carboxylase activities the kinetic properties of this enzyme in these plants are similar to those of the C_3, not of the C_4, enzyme (Ting and Osmond, 1973b). On the basis of present information it seems likely that CAM and C_4 photosynthesis have both arisen independently a number of times in the relatively recent evolutionary history of higher plants (cf. Karpilov, 1970).

Comparative studies of higher plant photosynthesis started to gain momentum only a few years ago. The photosynthetic characteristics of a number of plants from ecologically diverse habitats are now subject to intensive investigations in many laboratories, and serious attempts are being made to integrate studies on the whole-plant level with those on the cellular and molecular levels. This combination of approaches undoubtedly will prove to be very fruitful, and information on functional, adaptive, and evolutionary aspects of photosynthesis is likely to increase progressively during the next few years.

REFERENCES

Aagaard, J., and Sistrom, W. (1972). *Photochem. Photobiol.* **15**, 209–225.
Anderson, J., Boardman, N., and Spencer, D. (1971a). *Biochim. Biophys. Acta* **245**, 253–258.
Anderson, J., Woo, K., and Boardman, N. (1971b). *Biochim. Biophys. Acta* **245**, 398–408.
Baldry, C. W., Bucke, C., and Coombs, J. (1971). *Planta* **97**, 310–319.
Bassham, J. A. (1971). *Science* **172**, 526–534.
Bazzaz, M., and Govindjee (1971). *Plant Physiol.* **47**, Suppl., 189 (abstr.).
Ben-Amotz, A., and Avron, M. (1972). *Plant Physiol.* **49**, 240–243.
Berry, J. (1970). *Carnegie Inst. Wash., Yearb.* **69**, 649–655.
Berry, J. (1971). *Carnegie Inst. Wash., Yearb.* **70**, 526–530.
Berry, J., Downton, J., and Tregunna, E. B. (1970). *Can. J. Bot.* **48**, 777–786.
Billings, W. D., Godfrey, P. J., Chabot, D. F., and Bourque, D. P. (1971). *Arctic Alpine Res.* **3**, 277–289.
Björkman, O. (1966a). *Physiol. Plant.* **19**, 618–633.
Björkman, O. (1966b). *Carnegie Inst. Wash., Yearb.* **65**, 454–459.
Björkman, O. (1967). *Carnegie Inst. Wash., Yearb.* **66**, 220–228.
Björkman, O. (1968a). *Physiol. Plant.* **21**, 1–10.
Björkman, O. (1968b). *Physiol. Plant.* **21**, 84–99.
Björkman, O. (1971). *Carnegie Inst. Wash., Yearb.* **70**, 520–526.
Björkman, O., and Gauhl, E. (1968). *Carnegie Inst. Wash., Yearb.* **67**, 479–482.
Björkman, O., and Gauhl, E. (1969). *Planta* **88**, 197–203.

Björkman, O., and Holmgren, P. (1963). *Physiol. Plant.* **16**, 889–914.
Björkman, O., and Holmgren, P. (1966). *Physiol. Plant.* **19**, 854–859.
Björkman, O., and Ludlow, M. (1972). *Carnegie Inst. Wash., Yearb.* **71**, 85–94.
Björkman, O., and Pearcy, R. W. (1971). *Carnegie Inst. Wash., Yearb.* **70**, 511–520.
Björkman, O., Nobs, M. A., and Hiesey, W. H. (1969a). *Carnegie Inst. Wash., Yearb.* **68**, 614–620.
Björkman, O., Gauhl, E., and Nobs, M. A. (1969b). *Carnegie Inst. Wash., Yearb.* **68**, 620–633.
Björkman, O., Pearcy, R. W., and Nobs, M. A. (1970). *Carnegie Inst. Wash., Yearb.* **69**, 640–648.
Björkman, O., Nobs, M. A., Pearcy, R. W., Boynton, J., and Berry, J. (1971). *In* "Photosynthesis and Photorespiration" (M. D. Hatch, C. B. Osmond, and R. O. Slatyer, eds.), pp. 105–119. Wiley (Interscience), New York.
Björkman, O., Ludlow, M., and Morrow, P. (1972a). *Carnegie Inst. Wash., Yearb.* **71**, 94–102.
Björkman, O., Boardman, N., Anderson, J., Thorne, S., Goodchild, D., and Pyliotis, N. (1972b). *Carnegie Inst. Wash., Yearb.* **71**, 115–135.
Björkman, O., Pearcy, R. W., Harrison, A. T., and Mooney, H. A. (1972c). *Science* **175**, 786–789.
Black, C. C., Jr., and Mayne, B. (1970). *Plant Physiol.* **45**, 738–741.
Boardman, N., Anderson, J., Thorne, S., and Björkman, O. (1972). *Carnegie Inst. Wash., Yearb.* **71**, 107–114.
Bowes, G., and Ogren, W. (1972). *J. Biol. Chem.* **247**, 2171–2176.
Bowes, G., Ogren, W., and Hageman, R. H. (1971). *Biochem. Biophys. Res. Commun.* **45**, 716–722.
Bowes, G., Ogren, W., and Hageman, R. H. (1972). *Crop Sci.* **12**, 77–79.
Boynton, J., Nobs, M. A., Björkman, O., and Pearcy, R. (1970). *Carnegie Inst. Wash., Yearb.* **69**, 629–632.
Bucke, C., and Long, S. (1971). *Planta* **99**, 199–210.
Bulley, N., Nelson, C. D., and Tregunna, E. (1969). *Plant Physiol.* **44**, 678–684.
Chartier, P. (1970). *In* "Prediction and Measurement of Photosynthetic Productivity," pp. 307–315. Pudoc, Wageningen.
Chollet, R., and Ogren, W. (1972). *Biochem. Biophys. Res. Commun.* **46**, 2062–2066.
Clayton, R. (1965). "Molecular Physics in Photosynthesis." Ginn (Blaisdell), Boston, Massachusetts.
Coombs, J. (1971). *Proc. Roy. Soc., Ser. B* **179**, 221–235.
Downes, R. W. (1969). *Planta* **88**, 261–273.
Downes, R. W., and Hesketh, J. D. (1968). *Planta* **78**, 79–84.
Downton, J. (1971). *In* "Photosynthesis and Photorespiration" (M. Hatch, C. Osmond, and R. O. Slatyer, eds.), pp. 3–17. Wiley (Interscience), New York.
Downton, J., Berry, J., and Tregunna, E. (1970). *Z. Pflanzenphysiol.* **63**, 194–198.
Eagles, C. F., and Treharne, K. J. (1969). *Photosynthetica* **3**, 29–38.
Edwards, G., Lee, S., Chen, T., and Black, C. C., Jr. (1970). *Biochem. Biophys. Res. Commun.* **39**, 389–395.
Edwards, G., Kanai, R., and Black, C. C., Jr. (1971). *Biochem. Biophys. Res. Commun.* **45**, 278–285.
El-Sharkawy, M., and Hesketh, J. D. (1965). *Crop Sci.* **5**, 517–521.
Emerson, R., and Lewis, C. M. (1943). *Amer. J. Bot.* **30**, 165–178.
Fock, H., and Egle, K. (1966). *Beitr. Biol. Pflanz.* **42**, 213–239.
Fock, H., Schaub, H., and Hilgenberg, W. (1969). *Planta* **86**, 77–83.

Fock, H., Hilgenberg, W., and Egle, K. (1972). *Planta* **106**, 355–361.
Forrester, M. L., Krotkov, G., and Nelson, C. D. (1966a). *Plant Physiol.* **41**, 422–427.
Forrester, M. L., Krotkov, G., and Nelson, C. D. (1966b). *Plant Physiol.* **41**, 428–531.
French, C. S. (1971). *Proc. Nat. Acad. Sci. U. S.* **68**, 2893–2897.
French, C. S., and Berry, J. (1971). *Carnegie Inst. Wash., Yearb.* **70**, 495–498.
Gates, D. (1968). *Aust. J. Sci.* **31**, 67–74.
Gates, D., Hiesey, W., Milner, H., and Nobs, M. (1964). *Carnegie Inst. Wash., Yearb.* **63**, 418–430.
Gauhl, E. (1968). *Carnegie Inst. Wash., Yearb.* **67**, 482–487.
Gauhl, E. (1969). *Carnegie Inst. Wash., Yearb.* **68**, 633–636.
Gauhl, E., and Björkman, O. (1969). *Planta* **88**, 187–191.
Gibbs, M. (1969). *Ann. N. Y. Acad. Sci.* **168**, 356–368.
Goldsworthy, A. (1969). *Nature (London)* **224**, 501–502.
Goldsworthy, A. (1970). *Bot. Rev.* **36**, 321–340.
Goodchild, D., Björkman, O., and Pyliotis, N. (1972). *Carnegie Inst. Wash., Yearb.* **71**, 102–107.
Hall, A. E. (1970). Ph.D. Thesis, University of California, Davis.
Hall, H. M., and Clements, F. E. (1923). *Carnegie Inst. Wash., Publ.* **326**.
Halldal, P. (1968). *Biol. Bull.* **134**, 411–424.
Hatch, M. D. (1971). *Biochem. J.* **125**, 425–432.
Hatch, M. D., and Slack, C. R. (1966). *Biochem. J.* **101**, 103–111.
Hatch, M. D., Slack, C. R., and Johnson, H. S. (1967). *Biochem. J.* **102**, 417–422.
Hatch, M. D., Osmond, C. B., and Slatyer, R. O., eds. (1971). "Photosynthesis and Photorespiration." Wiley (Interscience), New York.
Hatch, M. D., Osmond, C. B., Troughton, J. H., and Björkman, O. (1972). *Carnegie Inst. Wash., Yearb.* **71**, 135–141.
Hesketh, J. D. (1963). *Crop Sci.* **3**, 493–496.
Hesketh, J. D. (1967). *Planta* **76**, 371–374.
Hesketh, J. D., and Musgrave, R. B. (1962). *Crop Sci.* **2**, 311–315.
Highkin, H., Boardman, N., and Goodchild, D. (1969). *Plant Physiol.* **44**, 1310–1320.
Holmgren, P. (1968). *Physiol. Plant.* **21**, 676–698.
Holmgren, P., Jarvis, P. G., and Jarvis, M. S. (1965). *Physiol. Plant.* **18**, 557–573.
Huang, A., and Beevers, H. (1972). *Plant Physiol.* **50**, 242–248.
Jackson, W., and Volk, R. J. (1970). *Annu. Rev. Plant Physiol.* **21**, 385–432.
Jones, L. W., and Kok, B. (1969a). *Plant Physiol.* **41**, 1037–1043.
Jones, L. W., and Kok, B. (1966b). *Plant Physiol.* **41**, 1044–1049.
Kanai, R., and Black, C. C., Jr. (1972). In "Net Carbon Assimilation in Higher Plants" (C. C. Black, Jr., ed.), pp. 75–93. Cotton, Inc., Raleigh, North Carolina.
Karpilov, Y. S. (1970). In "The Photosynthesis of Xerophytes," pp. 3–19. Acad. Sci. Moldavian S. S. R., Kishnev (in Russian).
Keck, R. W., and Boyer, J. S. (1972). *Plant Physiol.* **49**, Suppl., 152 (abstr.).
Keck, R. W., Dilley, R., Allen, C., and Biggs, S. (1970). *Plant Physiol.* **46**, 692–698.
Kortschak, H. P., Hartt, C. E., and Burr, G. O. (1965). *Plant Physiol.* **40**, 209–213.
Laetsch, W. M. (1971). In "Photosynthesis and Photorespiration" (M. D. Hatch, C. B. Osmond, and R. O. Slatyer, eds.), pp. 323–349. Wiley (Interscience), New York.
Laetsch, W. M., and Kortschak, H. P. (1972). *Plant Physiol.* **49**, 1021–1023.
Lichtenthaler, H. (1971). *Z. Naturforsch. B* **26**, 832–842.
Lorimer, G., Andrews, T. J., and Tolbert, N. E. (1972). *Fed. Proc., Fed. Amer. Soc. Exp. Biol.* **31**, 461 (abstr.).

Ludwig, L. J., and Canvin, D. T. (1971). *Plant Physiol.* **48**, 712–719.
McCree, K. (1971–1972). *Agr. Meteorol.* **9**, 191–216.
McNaughton, S. J. (1967). *Science* **156**, 1363.
Mayne, B., Edwards, G., and Black, C. C., Jr. (1971). In "Photosynthesis and Photorespiration" (W. Hatch, C. Osmond, and R. Slatyer, eds.), pp. 361–371. Wiley (Interscience), New York.
Medina, E. (1970). *Carnegie Inst. Wash., Yearb.* **69**, 655–662.
Medina, E. (1971). *Carnegie Inst. Wash., Yearb.* **70**, 551–559.
Meidner, H. (1962). *J. Exp. Bot.* **13**, 284–293.
Meidner, H., and Mansfield, T. A. (1968). "Physiology of Stomata." McGraw-Hill, New York.
Mooney, H. A., and Billings, W. D. (1961). *Ecol. Monogr.* **31**, 1–29.
Mooney, H. A., and Harrison, A. T. (1970). In "Prediction and Measurement of Photosynthetic Productivity," pp. 411–417. Pudoc, Wageningen.
Mooney, H. A., and Shropshire, F. (1967). *Oecol. Plant.* **2**, 1–13.
Mooney, H. A., and West, M. (1964). *Amer. J. Bot.* **51**, 825–827.
Moss, D. N. (1962). *Nature (London)* **193**, 587.
Moss, D. N., Musgrave, R. B., and Lemon, E. R. (1961). *Crop Sci.* **1**, 83–87.
Mousseau, M., Coste, F., and de Kouchkovsky, Y. (1967). *C. R. Acad. Sci., Ser. D* **264**, 1158–1161.
Murata, Y., and Iyama, J. (1963). *Proc. Crop Sci. Soc. Jap.* **31**, 315–322.
Myers, J., and Graham, J. (1971). *Plant Physiol.* **48**, 282–286.
Neales, T. F., Treharne, K. J., and Wareing, P. F. (1971). In "Photosynthesis and Photorespiration" (M. D. Hatch, C. B. Osmond, and R. O. Slatyer, eds.), pp. 89–96. Wiley (Interscience), New York.
Nobs, M. A., Björkman, O., and Pearcy, R. W. (1970). *Carnegie Inst. Wash., Yearb.* **69**, 625–629.
Ogren, W. L., and Bowes, G. (1971). *Nature (London) New Biol.* **230**, 159–160.
Osmond, C. B., and Björkman, O. (1972). *Carnegie Inst. Wash., Yearb.* **71**, 141–148.
Osmond, C. B., and Harris, B. (1971). *Biochim. Biophys. Acta* **234**, 270–282.
Parlange, J. Y., and Waggoner, P. (1970). *Plant Physiol.* **46**, 337–342.
Pearcy, R. W., and Björkman, O. (1970). *Carnegie Inst. Wash., Yearb.* **69**, 632–640.
Pearcy, R. W., Berry, J., and Bartholomew, B. (1972). *Carnegie Inst. Wash., Yearb.* **71**, 161–164.
Poincelot, R. (1972). *Plant Physiol.* **50**, 336–340.
Raven, J. A. (1970). *Biol. Rev. Cambridge Phil. Soc.* **45**, 167–221.
Schmid, G., and Gaffron, H. (1971). *Photochem. Photobiol* **14**, 451–464.
Shantz, H. L., and Piemeisel, L. M. (1927). *J. Agr. Res.* **34**, 1093–1190.
Sheridan, R. (1972). *Plant Physiol.* **50**, 336–340.
Slack, C. R. (1969). *Phytochemistry* **8**, 1387–1391.
Slatyer, R. O. (1970). *Planta* **93**, 175–189.
Smillie, R., and Entsch, B. (1971). In "Methods in Enzymology" (A. San Pietro, ed.), Vol. 23, Part A, p. 504. Academic Press, New York.
Sutton, B., and Osmond, C. B. (1972). *Plant Physiol.* **50**, 360–365.
Tiezen, L. L., and Helgager, J. A. (1968). *Nature (London)* **219**, 1066–1067.
Ting, I. P. (1971). In "Photosynthesis and Photorespiration" (M. D. Hatch, C. B. Osmond, and R. O. Slatyer, eds.), pp. 169–185. Wiley (Interscience), New York.
Ting, I. P., and Osmond, C. B. (1973a). *Plant Physiol.* **51**, 439–447.
Ting, I. P., and Osmond, C. B. (1973b). *Plant Physiol.* **51**, 448–453.
Ting, I. P., Johnson, H. B., and Szarek, S. R. (1972). In "Net Carbon Dioxide

Assimilation if Higher Plants" (C. C. Black, Jr., ed.), pp. 26–53. Cotton, Inc., Raleigh, North Carolina.
Tolbert, N. E. (1971). *Annu. Rev. Plant Physiol.* **22**, 45–74.
Tolbert, N. E., and Yamazaki, R. K. (1970). *Ann. N. Y. Acad. Sci.* **168**, 325–341.
Tregunna, E. B., Krotkov, E., and Nelson, C. D. (1966). *Physiol. Plant.* **19**, 723–733.
Troughton, J. H. (1969). *Aust. J. Biol. Sci.* **22**, 815–827.
Wareing, P. F., Khalifa, M. M., and Treharne, K. J. (1968). *Nature (London)* **220**, 453–454.
Wild, A., Zickler, H. O., and Grahl, H. (1971). *Planta* **97**, 208–223.
Williams, G. J. (1971). *Photosynthetica* **5**, 139–145.
Wojcieska, U. B., Ogren, W. L., and Hageman, R. H. (1972). *Plant Physiol.* **49**, Suppl., 40 (abstr.).
Wolf, F. T. (1970). *Advan. Front. Plant Sci.* **26**, 161–231.
Woo, K., Anderson, J., Boardman, N., Downton, J., Osmond, C. B., and Thorne, S. (1970). *Proc. Nat. Acad. Sci. U. S.* **67**, 18–25.
Woolhouse, H. W. (1967). *Hilger J.* **9**, 7–12.
Zelitch, I. (1971). "Photosynthesis, Photorespiration, and Plant Productivity." Academic Press, New York.

Chapter 2

ANALYSIS OF PHOTOSYNTHESIS IN GREEN ALGAE THROUGH MUTATION STUDIES

Norman I. Bishop

Department of Botany, Oregon State University, Corvallis, Oregon

1. Introduction 65
2. Methodology 66
3. Types of Mutations Directly Influencing the Photosynthetic Mechanism . 67
   3.1 Mutations Affecting Enzymes of the Calvin Cycle 68
   3.2 Mutations Influencing the Photosynthetic Electron Transport System . 70
   3.3 Mutations Directly Affecting Photophosphorylation 84
   3.4 Mutations Affecting Photosystem II 85
References 93

1. Introduction

Photosynthesis involves a multitude of reactions ranging from extremely rapid photochemical processes to those concerned with rather general aspects of the cell's physiology. Within these broad boundaries would be included events such as the mechanisms for the synthesis of chlorophyll, carotenoids, components of the electron transport system of photosynthesis, essential enzymes, and so on. To achieve a thorough understanding of such a complex process as photosynthesis by attempting to fix all the various parameters affecting the overall reaction, except for a restricted few, is by-and-large impossible. However, much progress has been made in studies of complicated biological phenomena by the use of mutant organisms; the list of biochemical specialties to which this technique has been applied is too large to detail here. It is sufficient only to recall the successful application of mutant analysis to elucidate sundry biochemical pathways in bacteria and fungi and for our current knowledge of the molecular biology of genetics in order to appreciate the usefulness of this approach. Prior to the 1960's, mutation studies on the photosynthetic process were largely sporadic and based largely upon naturally occurring mutations. When the methodology of microbiology and the techniques of mutant induction were utilized for studies of the biosynthesis of the chlorophylls and carotenoids by Granick (1948, 1949) and Sistrom and Stanier (Sistrom et al., 1956; Sistrom and Clayton, 1964), the way was opened for a mutational analysis of photosynthesis.

The various ways in which selected mutants of the green algae *Scenedesmus obliquus* and *Chlamydomonas reinhardi* have been employed to study certain fundamental aspects of photosynthesis will be detailed herein. The types of mutants obtained with these algae (*Chlamydomonas*) are, for the most part, nuclear mutants which show classic patterns of Mendelian inheritance and lend themselves more adequately to studies on photosynthesis. Both nuclear and extranuclear genes are affected in a number of mutants of higher plants; the latter type of mutation, the so-called plastome mutations, give maternal patterns of inheritance, which generally lead to strongly altered patterns of pigmentation and chloroplast structure. Relatively few studies have been made on this type of mutation as to their direct effects on the photosynthetic process, and only limited mention will be made of them. It is not the intention of this article to give a comprehensive review of the genetics of the numerous and diversified mutations influencing the photosynthetic process; for a more complete coverage of this aspect of the subject, the reader is referred to the excellent monograph of Kirk and Tilney-Basset (1967) and to the recent review articles on the genetics of the mutations of *Chlamydomonas* by Levine (1969) and Levine and Goodenough (1971).

2. Methodology

For induction and isolation of pigment and photosynthetic mutants of algae the techniques employed have been identical, for most purposes, to those utilized in classic genetic studies in microbiology. Since *Scenedesmus* and *Chlamydomonas* grow either autotrophically or heterotrophically (when supplied with the appropriate energy and carbon source), it is possible to maintain, with relative ease, a wide variety of mutants completely devoid of photosynthetic activity. The mutagenic agents commonly utilized, such as X ray, ultraviolet irradiation, and various chemical mutagens, are equally effective in producing altered photosynthetic phenotypes. The various methods for induction, isolation, and identifica- of mutants of the photosynthetic process have been published in detail for *Scenedesmus* (Bishop, 1971), *Chlamydomonas* (Levine, 1971), *Chlorella* (Granick, 1971), and *Euglena* (Schiff et al., 1971) and need not be considered in further detail.

3. Types of Mutations Directly Influencing the Photosynthetic Mechanism

Because of the involved interrelationship between the translational and transcriptional events of the nuclear and chloroplast DNA, RNA, and ribosomes, which are involved in the regulation of the many factors and events necessary for photosynthesis, it is now apparent that an extreme range of mutations in which photosynthesis is affected can be obtained. In general, two large groups are apparent: in the first the photosynthetic apparatus, including both structural and biochemical entities, remains intact except for an individual component. Ideally, the deletion of this single component is caused by the mutation of one gene. Although many photosynthetic mutants of algae now exist which appear to suffer from the loss of a specific enzyme or electron transport component, the evidence which specifically demonstrates that this is the single cause for the many abnormalities that a given mutant displays is incomplete. Attention will be given in later sections to the few mutant forms that may represent single gene mutations. In Table I the general

TABLE I
GENERAL CHARACTERISTICS OF CERTAIN PHOTOSYNTHETIC MUTANTS OF
Scenedesmus AND *Chlamydomonas*

Mutants	Altered component	Reference
Scenedesmus obliquus		
a[1], 4, 5, 10, 11, 15, 40	Plastoquinone	Bishop and Wong (1971)
42, 67-70, 74, 75, 87	Q(C-550?)	Erixon and Butler (1971b)
27, 34, 48, 50, 56, 64	Cytochrome 553	Powls et al. (1969)
8	P700	Butler and Bishop (1963), Weaver and Bishop (1963)
26	Chloroplast membrane protein?	Powls et al. (1969)
Chlamydomonas reinhardi		
ac-115, ac-141, F-34	Cytochrome 559	Smillie and Levine (1963)
hFd-91	Q(C-550?)	Epel and Butler (1972)
	Plastoquinone	Smillie and Levine (1963)
ac-20	Chloroplast ribosomes	Goodenough and Levine (1970)
ac-20	Unknown component "M"	Levine and Gorman (1966)
ac-80a, F-1	P700	Givan and Levine (1969)
F-60	Phosphoribulokinase	Levine (1969)
F-54	Terminal steps in photophosphorylation	Sato et al. (1971)
ac-206	Cytochrome 553	Gorman and Levine (1966)
ac-208	Plastocyanin	Gorman and Levine (1966)

features of several photosynthetic mutants of *Scenedesmus* and *Chlamydomonas* which have had fairly thorough examination are presented. Definitive characteristics of each general grouping will be discussed subsequently and in greater detail.

The second major type of photosynthetic mutant is typified by a gross alteration of chloroplast structure and often a decreased chlorophyll content. This type of mutation is most often observed in the plastome mutants, e.g., *N. tabacum* (Homann and Schmid, 1967), *Oenothera* (Dolzmann, 1968), and in many pigment mutants of higher plants (von Wettstein, 1959). Similar mutants also occur in several mutants of *Chlamydomonas* and *Scenedesmus*, where the normal capacity to form chlorophyll during heterotrophic growth, but not during myxotrophic growth, has been lost (Hudock and Levine, 1964; Ohad *et al.*, 1967a,b; Bishop and Senger, 1972; Senger and Bishop, 1972a). However, the altered structure of the chloroplast in these mutants is restored during growth in the light, and the fundamental nature of this mutant type is most likely not related to those seen in the typical plastome mutant of higher plants. Nevertheless, an important feature brought out through studies on algal mutants is that the production of chlorophyll and the formation of structure are closely interrelated phenomena. The failure of mutants to synthesize appropriate chloroplast membranes is manifested not only by a grossly altered chloroplast ultrastructure, but also by diminished amounts of chlorophyll and other components of the chloroplast which are normally considered to be chloroplast-membrane bound. A number of derepression–repression or feedback control mechanisms undoubtedly function in the overall control of the synthesis of normal chloroplast membrane.

3.1 Mutations Affecting Enzymes of the Calvin Cycle

It is currently held that two major pathways for carbon dioxide fixation can function in photosynthesis, i.e., the Calvin cycle and/or the C_4 dicarboxylic acid pathway. Although both of these mechanisms function in some higher plants, evidence for the participation of the C_4 pathway in the photosynthesis of unicellular algae is lacking or incomplete. Photosynthetic mutant strains of *Chlamydomonas* have been isolated by Levine and co-workers which show greatly decreased activities of two of the principal enzymes of the Calvin cycle, i.e., phosphoribulokinase and ribulose diphosphate carboxylase. Studies on the mutant, F-70, by Moll and Levine (1970) have shown that isolated chloroplast particles perform a variety of reactions, such as the Hill reaction with (2,6-dichlorophenolindophenol) DCIP or $NADP^+$ as the electron acceptor and water as the electron donor, the photoreduction of $NADP^+$ with ascorbate–DCIP as the electron donor system, and the light-induced absorbancy changes of cytochromes 553 and 559 and P700. Thus, the mutant, although incapable of photosynthesis, still has the capacity for the photoreduction of NADP⁺, and with rates equivalent to that of the wild-type cells. The inability of the mutant to perform photosynthesis apparently is due to either the inability of the mutant to synthesize phosphoribulokinase or that the enzyme is synthesized in an inactive form. Of thirteen chloroplast-type enzymes studied, only the activity of this single enzyme appeared altered. It seems clear from Moll and Levine's studies that the control of the synthesis of this enzyme is through at least one nuclear gene, but the decision as to whether it is a structural gene that is affected must await actual isolation and characterization of the protein fraction representative of the phosphoribulokinase. Since it is well established that this enzyme is a chloroplast enzyme (Latzko and Gibbs, 1968), future experiments with the F-70 strain concerning the role played by the nucleus in determining structure and regulating synthesis of components of the chloroplast will be of special interest.

Earlier studies by Levine and Togasaki (1965) on a photosynthetic mutant of *Chlamydomonas*, ac-20, demonstrated that the loss of photosynthetic competence was associated only with the loss of ribulose disphosphate (RuDP) carboxylase activity since chloroplast particles of this mutant catalyzed seemingly normal rates of NADP⁺ photoreduction. However, the conclusion that the ac-20 mutant was incapable of complete photosynthesis only because of a deletion of the ribulose carboxylase now seems to have been premature. Subsequent findings have shown that this is not the principal cause of the mutation syndrome of ac-20. Specifically, Togasaki and Levine (1970), Levine and Paszewski (1970), and Goodenough and Levine (1970), through much more comprehensive studies, have shown that this mutant, under appropriate conditions of growth, fails to develop normal chloroplast ribosomes. As a consequence, this mutant forms only about 50% of the normal amount of chlorophyll, shows reduced levels of plastoquinone, cytochrome b-559, and the fluorescence quencher Q; furthermore, the ultrastructure of its chloroplast is also noticeably altered. These are changes in addition to the lowered level of the RuDP carboxylase. The seemingly simple assessment of the cause of the mutation of ac-20, as originally made, serves to illustrate one of the major problems of mutational analysis, i.e., that while mutant induction and isolation present relatively few difficulties, the final assessment of the specific cause of a given mutation is often complicated and time-consuming. The newer knowledge about the fundamental cause of the mutation in ac-20 does provide additional information about the function of chloroplast ribosome.

No other information concerning specific mutations of the enzymes involved in either the Calvin cycle or the C_4 pathway has yet appeared. A variety of mutants of higher plants, both genome and plastome in nature, exist for *Oenothera, Zea mays, Lycopersicon, Antirrhinum, Nicotiana,* and *Arabidopsis,* but except for the mutants of *Oenothera,* no efforts have been made to establish whether any of the other mutants concern photosynthetic carbohydrate metabolism.

3.2 Mutations Influencing the Photosynthetic Electron Transport System

The most common method for representing a general reaction mechanism for photosynthesis is through the so-called Z formulation within which there are two photoreactions (PS I and PS II) sensitized by two different pigment systems (Chl I and Chl II). It is believed that the two photosystems operate in series through an electron transport system connecting them. Such a scheme is presented only for the purpose of discussion of the several mutants known to have altered or missing components of the photosynthetic electron transport system (Fig. 1).

While this conception of the mechanism of photosynthesis fits most of the experimental data from many laboratories, a few recent experiments performed with chloroplasts which had been extensively modified by isolation procedures, detergent treatment, sucrose density centrifugation, etc., provided data which did not fit the conceptualized design of the two-light reaction mechanism. Based upon their recent data, Arnon and colleagues (Arnon et al., 1968, 1970, 1971; Knaff and Arnon, 1969) proposed that three light reactions are required for photosynthesis: two light reactions are required within photosystem II (2a and 2b) which operate in a conventional series formulation, and a separate light reaction which forms only ATP through a cyclic photophosphorylation and operates in parallel to the other photosystems. The principal experimental evidence for proposing this new formulation is based upon the absence of enhancement of $NADP^+$ photoreduction by PS I and PS II wavelengths of light, the observation of $NADP^+$ photoreduction by chloroplast particles in the absence of added ferredoxin, an efficient oxidation of cytochrome b-559 by PS II light and the observed enhancement of carbon dioxide reduction by chloroplasts wherein the enhancement supposedly occurs because of the additional ATP generated by the separate PS I reaction. Although these findings are of primary interest, none of the information reported to date from studies on the photosynthetic mutants of *Scenedesmus* and *Chlamydomonas* supports this newer concept.

As depicted in part in Fig. 1, the photosynthetic electron transport system (PETS) includes the many reactions involved between water as the electron donor, and $NADP^+$ as the terminal electron acceptor. A

2. PHOTOSYNTHETIC MUTATIONS IN GREEN ALGAE

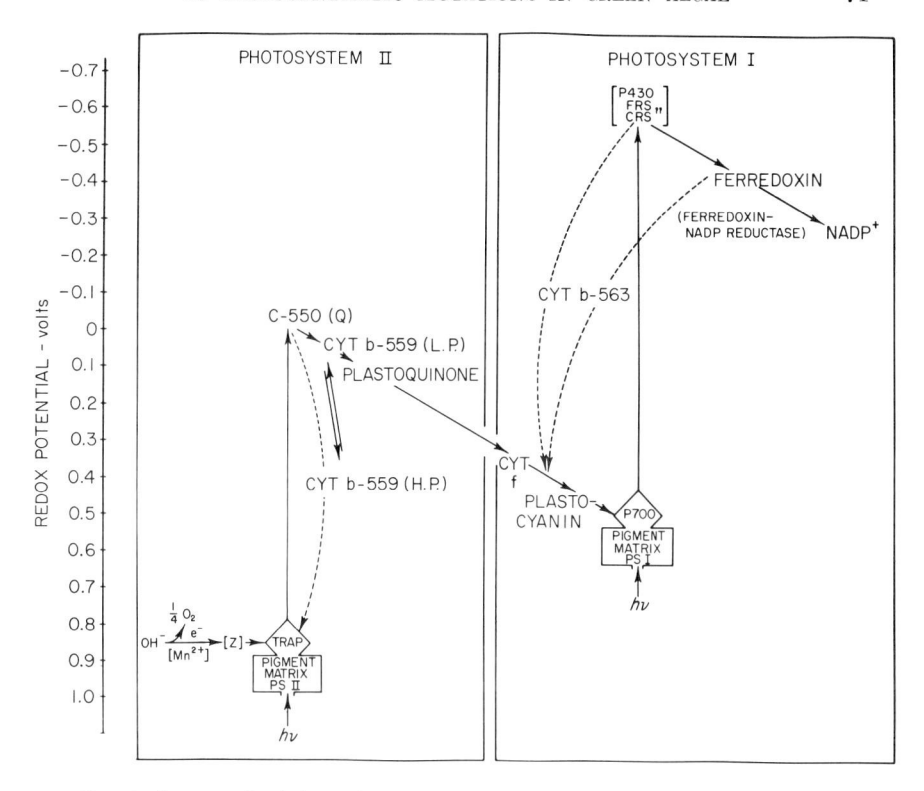

Fig. 1. A generalized formulation for the light-induced electron flow in photosynthetic tissue. Direction of electron flow is indicated by the arrows, and a scale of redox potential is shown on the left. Further consideration on this scheme, as it would be applicable to various algal mutants, is provided in the text. L.P., H.P. = low potential, high potential. Q = fluorescence quencher.

number of mutant strains have been described for *C. reinhardi* (Levine, 1971), *Scenedesmus obliquus* (Bishop, 1964, 1971) and *Vicia faba* (Heber and Gottschalk, 1963) which are blocked at specific locations in this pathway. Most of these mutants fit the prescribed conditions for being true photosynthetic mutants in that they possess nearly normal amounts of the photosynthetic pigments and, insofar as studied, normal chloroplasts. Because they have selective deletions in the electron transport chain, special use has been made of individual mutants in assessing the role of a particular component in the electron transport scheme. Employing the scheme depicted in Fig. 1 as a convenient guide, but hopefully not a restrictive one, the characteristics of the various mutants showing deletions in the electron transport system will be described. Where pertinent, findings with the various mutants will be evaluated in terms of the conventional Z scheme (series formulation) and the more recent formulation of Arnon.

3.2.1 Ferredoxin-NADP+ Oxidoreductase

No specific mutation of either algae or higher plants has yet been described which is specifically deficient in this important component of the PETS. Current information suggests that the synthesis of this component is under nuclear control which increases the feasibility of mutant induction at this site. However, the problem remains as to whether or not a mutation within the genome controlling flavoprotein synthesis would produce lethal phenotypes. An earlier report by Heber and Gottschalk (1963) showed that chloroplasts of a mutant of *Vicia faba* could not photoreduce NADP+ but possessed normal PS II activity and the capacity for cyclic photophosphorylation. Since the mutant contained the water-soluble components normally required for NADP+ photoreduction, i.e., ferredoxin and the NADP+ oxidoreductase, it was concluded by these authors that an insoluble chloroplast factor required for NADP+ reduction was absent in the mutant. Previous review articles have interpreted these findings as indicating that the *Vicia faba* mutant was deficient in either ferredoxin or the ferredoxin–NADP+ oxidoreductase. Unfortunately, further studies on this now interesting mutant have not been reported. From current information it would appear that this mutant might be deficient in the ferredoxin-reducing substance (FRS) rather than ferredoxin or the oxidoreductase.

3.2.2 Ferredoxin

No mutant of green plants is known which lacks ferredoxin. Attempts in the author's laboratory to induce and isolate such a mutant in *Scenedesmus obliquus* have been routinely unsuccessful. Ferredoxin, like the NADP+ reductase, is one of several of the easily solubilized components of the chloroplast whose synthesis is believed to be under the control of the nucleus. Thus, it would be anticipated that a single-gene mutation within the genome controlling the synthesis of ferredoxin should give rise to mutants deficient in or lacking activity of this component. However, since ferredoxin is involved in several other key reactions of the cell's physiology, such as nitrate and sulfate reduction, such mutants would most likely be lethal or at least not detectable with the techniques currently employed.

3.2.3 Ferredoxin- and Cytochrome-Reducing Substance

It had been previously believed that the first recognizable chemical reductant formed in the energy conversion act inherent to PS I was ferredoxin. Recent studies by several laboratories now suggest the participation of an additional cofactor between ferredoxin and the photochemistry of PS I. The most extensively studied substances, which perhaps react at this position, are the cytochrome-reducing-substance (CRS) and the ferredoxin-reducing-substance (FRS) which have been extracted from a variety of plant sources. Some of the biochemical properties of these compounds and their possible relation to each other have been reviewed by Bishop (1971), Trebst (1972), and Yokum and San Pietro (1972). It is still questionable whether FRS or CRS function as true electron carriers or whether they represent important structural entities of either PS I or PS II. No specific mutant has yet been identified that is deficient in either of these factors. The mutant of *Vicia faba*, which is apparently lacking an insoluble cofactor of PS I (Heber and Gottschalk, 1963), may lack FRS (or CRS); if this mutant is still in existence, additional studies might help to resolve the nature of the mutation and to further our understanding of the possible involvement of other factors in PS I.

Many photosynthetic mutants of both *Chlamydomonas* and *Scenedesmus* exist which are deficient in PS I activity but do not show losses of known components associated with this system. As the methodology for isolation, purification, and activity analysis become better defined, it should be feasible to examine these mutant strains for possible deletion of FRS or CRS. Obviously, the availability of mutant strains deficient in these compounds, which are currently of questionable importance, would help resolve their role in the photosynthetic apparatus.

3.2.4 THE REACTION CENTER OF PS I: P700

In photosynthesis the energy conversion process takes place through the cooperation of a large number of light-absorbing molecules which transfer their excess energy to a smaller number of photochemically active molecules; these, in turn, interact with the photochemically inactive molecules of such a unit. This unit, as comprised of the photochemically active pigments and the associated reactants, represents in part the so-called reaction center of PS I. From studies of light-induced absorbancy changes which occur in the region of 700 nm (Kok, 1956, 1963; Kok and Hoch, 1961), it has been determined that a form of chlorophyll, P700, is the photochemically active component of the reaction center of PS I. Excellent support for its earlier proposed role in photosynthesis has come from studies on mutants of *S. obliquus* (mutant 8) and *C. reinhardi* (ac-80a and F-1). From the series formulation for photosynthetic electron transport (Fig. 1), certain logical predictions can be made as to the types of partial reactions of photosynthesis mutants lacking P700 should possess. Reactions which are exclusive PS I reactions, such as the reduction of NADP⁺ with either water or DCIP-ascorbate as the electron donor systems, should be reactions requiring the cooperation of both photosystems. Similarly, cyclic photophosphorylation should require the participation of P700. On the contrary, those partial reactions of photosynthesis requiring only PS II, such as the reduction of potassium ferricyanide or p-benzoquinone, should continue unimpaired.

The recent interpretation by Arnon and colleagues (1970, 1971) and Rurainski et al. (1970, Rurainski and Hoch, 1972), that P700 functions in a separate light reaction which is viewed as a mechanism principally for cyclic photophosphorylation, i.e., that P700 is not involved in the pathway of electron transport from water to NADP, requires a more critical assessment of the characteristics of whole cell and chloroplast reactions of the mutant types known to lack P700. It is pertinent to state that the work from the two laboratories mentioned above has not taken into consideration the characteristics of the P700-deficient mutant strains in the formulation of their newer concept for the functioning of P700 in the mechanism of photosynthesis.

According to the interpretation of Arnon the photoreduction of ferredoxin and NADP is accomplished only by PS II according to the following reaction scheme:

$$H_2O \rightarrow PS_{IIb} \rightarrow C\text{-}550 \rightarrow \text{plastocyanin} \rightarrow PS_{IIa} \rightarrow \text{ferredoxin} \rightarrow NADP^+$$

This formulation states that two photoreactions (PS_{IIb} and PS_{IIa}) occur within PS II and that neither cytochrome f nor P700 is required for the photoreduction of NADP⁺. The electron transport chain of PS I is accordingly viewed as proceeding in a cyclic fashion as follows:

$$PS\ I \rightarrow \text{ferredoxin} \rightarrow \text{cytochrome } b_6 \rightarrow \text{cytochrome } f \rightarrow PS\ I$$

(Photophosphorylation would accompany both of the light reactions.) Some of the features of this concept of a three-light-reaction mechanism have been reviewed recently by Park and Sane (1971).

From studies on the relaxation kinetics of the light-induced absorbancy changes attributable to P700 and cytochrome f, Rurainski et al. (1970; Rurainski and Hoch, 1972) have proposed that P700 and the sites of NADP⁺ reduction are located in different light reactions. They also propose that magnesium ion controls the distribution of excitation energy to the separate systems which normally compete for incoming quanta of short wavelengths.

The general features of whole cell reactions of mutant 8 of *Scenedesmus* and mutant ac-80a of *Chlamydomonas* are strictly comparable; neither is capable of reducing carbon dioxide either through normal photosynthesis or through photoreduction (where hydrogen gas functions as the electron donor, through the mediation of an hydrogenase, in a pure PS I type reaction). Whole cells of chloroplast fragments of the two mutants possess Hill reaction activity with a variety of electron acceptors (Kok and Datko, 1965; Givan and Levine, 1969) although the capacity of this reaction appears to be decreased by about 50% at saturating light intensities. Kok and Datko (1965) suggested that this lowered saturation rate in the *Scenedesmus* mutant might result because of the loss of an interreaction between P700 and PS II which would normally have produced high quantum efficiency. Givan and Levine's findings on the ac-80 mutant demonstrate, however, that although the light saturated rate of the Hill reaction is lowered in this mutant (with ferricyanide as the Hill oxidant), the efficiency at low light intensities is comparable to that of the wild type. As an explanation for the Hill reaction characteristics of ac-80, they have suggested that two sites for $Fe(CN)_6^{3-}$ exist in the normal cell wherein one of them requires P700 and the other is strictly PS II dependent. The mutant thus would show decreased rates at saturating intensities, but not at low, i.e., identical quantum efficiency to the wild-type strain. This interpretation implies that the loss of P700 would in no way alter PS II itself.

For further assessment of the possible secondary implications resulting from the loss of P700 to chloroplast reactions, it would be necessary to know more about the biochemical and biophysical properties of P700. Although our knowledge in this vein is extremely limited, it can be proposed that P700 represents a specific complex between chlorophyll a and a special protein. Mutations of algae which lack P700 have not been shown to be deficient in chlorophyll and, thus, are most likely altered in some protein essential for the formation of the complex, P700. The recently published findings of Gregory *et al.* (1971) on the polyacrylamide gel electrophoresis patterns of detergent extracts of the *Scenedesmus* mutant demonstrated that not only is a green complex missing (complex I) but so also is the corresponding protein. This chlorophyll–protein complex was easily detectable in the normal strain and in a PS II type mutant (mutant 11). A plausible cause of the mutation syndrome seen in mutant 8 might be either (1) the inability of the mutant to form the protein which normally associates with a small portion of the total chlorophyll to give the PS I reaction center, or (2) that the protein is synthesized but is so altered that the complex cannot be formed. In either case, a mutation which is altered in an essential chloroplast matrix protein might also cause secondary manifestation in the ultrastructure of the chloroplast membranes which in turn would affect PS II activity such that the efficiency of PS II reactions would be decreased. The purification of complex I by Bailey and Kreutz (1969) and Thornber (1968, 1971), and the demonstration that it specifically contains P700, should allow for a more intensive investigation of the nature of the protein(s) essential for the formation of an active P700–protein complex.

More detailed studies on the capacity of isolated chloroplast particles of the P700 mutants of *Scenedesmus* and *Chlamydomonas* to perform a series of partial reactions of photosynthesis have shown that little or no photoreduction of NADP$^+$ takes place when water serves as the electron donor. Pratt and Bishop (1968) demonstrated further that under conditions where chloroplast particles from *Scenedesmus* mutant 8 reduced DCIP and cytochrome c, no reduction of NADP$^+$ occurred with either water or DCIP–ascorbate as the electron donor. These data were interpreted at that time as providing further evidence for the essential role of P700 in the reduction of NADP$^+$. The conventional light-induced oxidation and reduction of cytochromes 553 and 559 by PS I and PS II wavelengths of light were studied in the ac-80 *Chlamydomonas* mutant by Givan and Levine (1969), and they observed that, although PS II wavelength light caused a normal reduction of both cytochromes, no apparent PS I catalyzed photooxidation of them could be induced. Considering the observations on chloroplast particles from the two P700 mutants together, it is apparent that the loss of P700 prevents the normal flow of electrons through the photosynthetic electron transport chain as would be predicted by the standard Z formulation (Fig. 1).

Somewhat contradictory findings on *Scenedesmus* mutant 8 chloroplast particles have been reported by Gee et al. (1969); these authors reported that such particles, in the presence of ferredoxin, DCIP, and ascorbate, photoreduced NADP$^+$ at a rate comparable to chloroplast particles from the normal *Scenedesmus*. They suggested that their findings indicated a direct photochemical interaction of PS II in the reduction of NADP$^+$. Their failure to report whether or not the photoreduction of NADP$^+$ by mutant 8 chloroplast particles was inhibited by dichlorophenyl dimethyl urea (DCMU) and their use of light intensities (white light?) far in excess of that employed by Pratt and Bishop prevent a direct comparison of the findings from the two laboratories. It is known that the mutation characteristic of mutant 8 can be temporarily reversed (Gee et al., 1969) if the mutant is grown under myxotrophic conditions. It is assumed, however, that the findings reported by these authors were obtained with chloroplasts derived from dark-grown cultures even though it was not so stated. Obviously, the discrepancies noted for the studies of chloroplast reactions of mutant 8 are important ones and must be resolved in order to evaluate critically the role of P700 in NADP$^+$ photoreduction.

If, as predicted by the current formulation of Arnon and colleagues, the photoreduction of NADP⁺ proceeds exclusively through PS II type reactions and independently of P700, then it would be expected that mutants lacking P700 should still perform some photosynthesis. One of the essential criteria employed in the isolation of the mutants that are now known to be deficient in P700 was that they would be photosynthetically incompetent. Detailed studies on the *Scenedesmus* and *Chlamydomonas* mutants have shown them to be almost completely devoid of photosynthesis; $^{14}CO_2$ fixation studies by Gee *et al.* (1969) on *Scenedesmus* mutant 8 showed that most of the compounds labeled were predominantly those derived from dark CO_2 fixation. In illuminated samples, sufficient levels of ATP should be available since it is known that chloroplast particles of mutant 8 and of ac-80 perform an acyclic photophosphorylation with ferricyanide as cofactor. However, cyclic photophosphorylation, as measured by either the phenazine methosulfate (PMS)-catalyzed process or as the anaerobic light-induced assimilation of glucose is absent in both mutant 8 and mutant ac-80 (Pratt and Bishop, 1968; Givan and Levine, 1969; Gee *et al.*, 1969). The phosphorylation capacities of these mutants has previously been reviewed by Levine (1969), and the subject need not be elaborated on further at this time. The summary remark must be made, however, that the loss of P700 in the algae *Scenedesmus* and *Chlamydomonas* results in the complete loss of the capacity for cyclic photophosphorylation in addition to the loss of the capacity for NADP⁺ photoreduction.

It was previously mentioned that mutant forms lacking P700 do not show the characteristic antagonistic effects of PS I and PS II wavelengths of light on the redox status of cytochrome 553 and cytochrome b-559. From the series formulation concept, such a finding indicates a loss of coupling between PS I and PS II. If, according to Arnon's formulation, the only function of the photosystem containing P700 is to make additional ATP available through a cyclic photophosphorylation mechanism, then it is not clear why the redox status of cytochrome b-559, for example, should be influenced by PS I wavelengths of light in normal cells but not in the P700-deficient mutants.

Related observations on the variable yield fluorescence of *Scenedesmus* mutant 8 have been made in the author's laboratory, and some of the data are summarized in Fig. 2. This procedure is a useful and valuable tool for evaluating the degree of coupling between the two photosystems; its application and interpretation has been extensively documented previously (Butler, 1966) and need not be considered in detail here. Simply stated, the quantum yield of fluorescence emanating from PS II of whole cells can be modified according to the redox status of the fluores-

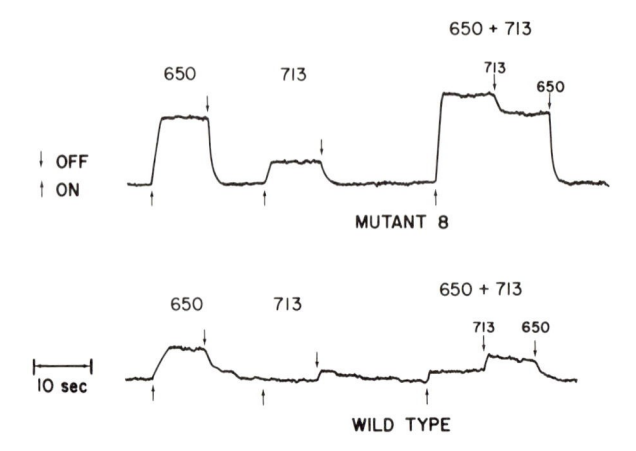

FIG. 2. Comparison of the influence of PS I and PS II wavelengths of light on the variable-yield fluorescence of wild-type and mutant 8 cells of *Scenedesmus obliquus*. Five microliters of cells were resuspended in 3 ml of phosphate buffer, pH 6.5, just prior to fluorescence measurements. Fluorescence was induced by an exciting wavelength of 436 nm (100 ergs sec^{-1} cm^{-2}). Intensities of the 650 and 712 nm light beams were 10^3 ergs sec^{-1} cm^{-2}. Fluorescence was monitored at 686 nm. For additional experimental details, see Senger and Bishop, 1972b).

cence quencher, Q; when in the oxidized form, light absorbed by PS II causes its reduction to QH (nonquenching form), and in the dark or in the presence of wavelengths of light absorbed by PS I, QH is reoxidized. The fluorescence patterns of the normal and mutant 8 strains of *Scenedesmus*, as shown in Fig. 2, clearly show that the loss of P700 does not prevent the increase in the fluoresence yield caused by 650 nm wavelength light; i.e., PS II is present and functional as previously noted, but the quenching effect of 712 nm wavelength light on the fluorescence induced by 436 nm excitation is lost. These findings only offer additional support for the role of P700 in the classic series formulation of photosynthesis.

The results of the recent and elegant experiments of Arntzen et al. (1972) on the types of photoreactions performed by stroma and grana membranes of spinach chloroplasts are also inconsistent with the interpretations of Arnon et al. (1970) and Rurainski et al. (1970; Rurainski and Hoch, 1972).

From the evidence currently available from mutant studies on unicellular algae, only some of which has been discussed here, no overwhelming reason is seen for abandoning the concept of the role of P700 in photosynthesis as predicted by the series formulation. It is appropriate to remark, however, that much of the research on the chloroplast particle derived from the P700-deficient mutants is lacking in the depth and thoroughness which would allow for a direct comparison to those made on higher plant chloroplast subfractions. Conversely, it may be inappropriate to attempt to make direct comparisons between studies on mutant forms of algae and on subchloroplast particles of higher plants derived by all the means and accoutrements of the biochemist.

3.2.5 CYTOCHROME b-563 (CYTOCHROME b_6)

The presence of two or more b-type cytochromes in chloroplasts of higher plants or algae has been extensively documented by a variety of researchers. The most predominant one is the low potential cytochrome b-563, and it is generally accepted that it functions in a cyclic electron transport system indigenous to PS I. Space does not permit an elaborate documentation of the various lines of evidence that support this deduction; the reader may be referred to many recent reviews concerning plant cytochromes (e.g., Hind and Olson, 1968; Bendall and Hill, 1968; Bishop, 1971). Of the existing photosynthetic mutants of algae that have been studied, none is known which suffers a deletion of cytochrome b-563; it is certainly questionable, however, whether serious attention has been directed toward the isolation of such mutant phenotypes. From studies with antibiotics and with the mutant of *Chlamydomonas* deficient in chloroplast ribosomes, ac-20 (Surzycki et al., 1969), evidence is available which suggests that the synthesis of cytochrome b-563 is under nuclear control and, hence, suggests that it should be feasible to induce mutants deficient in content or activity of the b-type cytochromes. Whether or not it would be possible to obtain mutants which would be specifically deficient in cytochrome b-563, but possess the other b-type cytochrome (b-559), seems doubtful, although cytochrome b-559 synthesis does seem to depend upon chloroplast ribosomes whereas that of cytochrome b-563 does not. The existence of b-type cytochrome mutants would certainly facilitate further evaluation of the potential roles of these components in the electron transport system of photosynthesis.

3.2.6 PLASTOCYANIN

The copper-containing acidic protein plastocyanin is seemingly ubiquitously distributed in the chloroplast of all higher plants and apparently also in most algae. That this component plays an essential role in the electron transport system of photosynthesis seems unquestionable; the problem of localizing its precise site of action, however, still presents a dilemma for students of photosynthesis. Some of the difficulties in interpreting the variety of responses found with plastocyanin in reactions of chloroplast or chloroplast subparticles have, in part, been reviewed recently by Bishop (1971). The position of plastocyanin in the photosynthetic electron transport system, as depicted in Fig. 1, is formulated principally upon the evidence offered from studies on a plastocyanin-deficient mutant of *Chlamydomonas* (strain ac-208) and on PS I particles isolated from chloroplast by detergent treatment (Trebst and Elstner, 1965; Wessels, 1968; Hind, 1968).

The data obtained by Gorman and Levine (1966; Levine and Gorman, 1966) with the plastocyanin-deficient strain (ac-208) have been extensively and repeatedly quoted in support of the localization of plastocyanin in the electron transport chain as depicted in Fig. 1. Because of the extremely varied opinions as to the precise site of function of the copper-protein in photosynthesis, it would seem that more extensive studies would have been made with ac-208 by Levine or by other investigators of photosynthesis. However, no additional findings or reevaluations of the characteristics of this mutant strain have been published. It should be recalled that in addition to the ac-208 strain of *Chlamydomonas*, Gorman and Levine reported on a suppressed strain of ac-208, which although lacking plastocyanin, still performed photosynthesis at approximately 50% of the rate of the wild-type strain. Gorman and Levine suggested that this finding might be interpreted to mean that the presence of plastocyanin was not obligatory for photosynthesis. To escape the dilemma that this statement suggests in relation to the conclusions drawn from studies on the unsuppressed strain, ac-208, these authors proposed that perhaps the plastocyanin of the suppressed form may be modified structurally and, therefore, less stable. Consequently, the techniques normally employed for the extraction and purification of plastocyanin would have resulted in its complete denaturation and, as a result, the mutant form would falsely appear to be lacking plastocyanin. Proof of this hypothesis would require the demonstration that the whole cells of the suppressed mutant actually contain plastocyanin; no such efforts seem to have been made.

It is pertinent to indicate an additional apparent discrepancy in the data of Gorman and Levine. Specifically, isolated chloroplast particles of both mutant forms not only are unable to support NADP+ photoreduction with either water or DCIP-ascorbate as the electron donor system, but also show only about 10% of the wild-type's rate of the Hill reaction with DCIP as the Hill oxidant. From studies on mutant forms of *Scenedesmus* lacking cytochrome 553 (see Section 3.2.7), it has been shown that deletion of this component does not significantly affect PS II activity, at least with p-benzoquinone as the Hill oxidant. From fluorescent yield studies it is also known that the *Scenedesmus* mutants possess an intact PS II. The absorbancy changes attributed to the reduction of cytochrome 559 and cytochrome 553, which are considered to be an index of PS II activity, are still extant in ac-208 and indicate that PS II activity is unaffected in spite of the findings on the Hill reaction. Obviously, there exist a number of internal discrepancies within the studies on the *Chlamydomonas* mutants which must be settled before these mutant forms can be so indiscriminately and authoritatively employed to localize the site of action of plastocyanin. Arnon's proposal that plastocyanin functions only in the PS II reactions, and not in PS I, is, seemingly, in direct contradiction to the findings on mutants ac-208 and ac-208(sup).

Based in part upon their findings that plastocyanin was not essential for $NADP^+$ photoreduction with artificial electron donors of PS I, Arnon et al. (1968) and Tsujimoto et al. (1969) proposed that plastocyanin functions only in PS II type reactions, not in PS I. Obviously, this interpretation is not supported by the findings on mutant ac-208 and ac-208(sup). Recent studies by Baszynski et al. (1971), which demonstrated an absolute requirement of PS I reactions for plastocyanin, are all in direct contradiction to those of Arnon and colleagues. Also Arntzen et al. (1972) demonstrated rather convincingly that plastocyanin is present both in the grana and the stroma membranes of spinach chloroplasts.

3.2.7 Cytochrome-553 (Cytochrome f)

In 1951 Hill and Scarisbrick demonstrated the existence of two new cytochromes, c-553 and b-563, in the chloroplasts of higher plants. During the intervening time much information has accumulated from chloroplast studies which show that cytochrome 553 functions exclusively in PS I and serves also to couple the electron transport systems of PS I and PS II (Schuldiner and Ohad, 1969; Hiller and Boardman, 1971; Anderson and Boardman, 1964; Boardman, 1968).

Several photosynthetic mutants of *Chlamydomonas* and *Scenedesmus* have been shown to be deficient in cytochrome 553 and studies on them have provided additional evidence for the specific function of this cytochrome in the electron transport and photophosphorylation systems of photosynthesis. Intact cells of the ac-206 mutant of *Chlamydomonas* have insignificant rate of carbon dioxide fixation (either by photosynthesis or by photoreduction with hydrogen gas as the electron donor), and isolated chloroplast particles show only limited rates of photoreduction of $NADP^+$. With DCIP–ascorbate as the electron donor system, nearly normal rates of $NADP^+$ photoreduction proceeded; also Hill reaction rates of chloroplast particles with DCIP and ferricyanide as oxidants compare favorably to those of the normal strain (Gorman and Levine, 1966; Levine, 1971).

Six photosynthetic mutants of *Scenedesmus* have been described by

Bishop (1972) which show varying deficiencies of cytochrome f. The features outlined above for whole cell and chloroplast particle reactions of the ac-206 mutant of *Chlamydomonas* are essentially identical for the *Scenedesmus* mutants. Thus a summary statement can be made for the mutants lacking cytochrome f: loss of this important constituent of the electron transport system causes a specific deletion of at least the capacity for the PS I-catalyzed photoreduction of NADP$^+$. Studies on these mutant phenotypes provided additional evidence that does not support Arnon's (Arnon *et al.*, 1970) contention that NADP$^+$ reduction can occur through an exclusive PS II sensitized reaction in which cytochrome-553 is not required. At the time of the analysis on the activities of chloroplast particles of the various cytochrome-deficient mutants, sufficient attention may not have been given to their limited capacity to photoreduce NADP$^+$ with water as the electron donor system in comparison with their limited capacity for *in vivo* carbon dioxide reduction. Perhaps a more systematic analysis of the photochemical activities of isolated chloroplast particles of these mutant forms would be necessary. However, the presence of an active PS II in these mutant forms with all of the Hill reaction oxidants tested, except for NADP$^+$, seems to provide adequate evidence against the existence of an exclusive PS II-driven photoreduction of NADP$^+$ independent of cytochrome-553.

Similar conclusions can be deduced from evaluation of the characteristics of the variable yield fluorescence of the *Scenedesmus* mutants. In Section 3.2.4 the general implication of such studies was provided and information was given for the mutant lacking P700 (mutant 8). Essentially identical patterns have also been noted with whole cells of mutant 50, i.e., the loss of cytochrome-553 results in a loss of coupling between the two photosystems (Bishop, 1972) as would be normally visible in variable-yield fluorescence analysis. Those studies also provided evidence for a functional PS II. Also, Levine has reported (1971) that the light-induced reduction of cytochrome b-559 could be initiated in the ac-206 mutant of *Chlamydomonas*.

Studies on the phosphorylation capacity of chloroplast particles of cytochrome-deficient mutants of both *Scenedesmus* and *Chlamydomonas* have shown that their ability to form ATP through the mechanism of the PMS-catalyzed cyclic photophosphorylation is not impaired (Bishop, 1972; Levine, 1971). The true significance and relation to findings on PMS-catalyzed photophosphorylation to the *in vivo* mechanism of photophosphorylation has often been questioned. An alternate method of determining the *in vivo* capacity of the PS I-catalyzed cyclic photophosphorylation has been suggested by Tanner and Kandler (1969). This procedure involves the determination of the anaerobic photoassimilation of glucose by algal cells; Tanner and Kandler have shown that this reaction is exclusively catalyzed by PS I. When this assay procedure was applied to the cytochrome-deficient mutant forms of *Scenedesmus*, it became evident that the loss of cytochrome-553 was also correlated with the inability of the cells to accumulate glucose anaerobically. In Table II, values are compared for the rates of PMS-catalyzed photophosphorylation by isolated chloroplast particles and of anaerobic glucose assimilation by whole cells of the cytochrome-553 deficient strains of *Scenedesmus*. Since it is believed that the anaerobic photoassimilation of glucose is dependent upon the ATP generated through cyclic photophosphorylation, it appears that the loss of cytochrome-553 is correlated with the decreased capacity for photophosphorylation. On the contrary, the PMS-catalyzed photophosphorylation by isolated chloroplast particles showed comparable rates for the mutants and the normal strain. It appears, consequently, that the mechanism for cyclic photophosphorylation (PMS-catalyzed) does not require cytochrome-553; this finding is in agreement with those of Gorman and Levine (1966) for the ac-208 strain of *Chlamydomonas*. It would seem, moreover, that the PMS-catalyzed cyclic photophosphorylation is not representative of a true *in vivo* mechanism. It appears more likely that the normal flow of electrons associated with the *in vivo* mechanism is through cytochrome-553 and is further mediated through the cytochrome reductase activity of the NADP⁺ oxidoreductase as was originally suggested by Forti and Zanetti (1969).

An additional point of interest can be extracted from the studies

TABLE II

COMPARISON OF THE RATES OF GLUCOSE ASSIMILATION IN NORMAL AND MUTANT STRAINS OF *Scenedesmus*

Algal strain	Glucose uptake mg/hr	(% Wild type)	PMS-catalyzed photophosphorylation[a]	
Wild type	1.91	(100)	100	(100)
Mutant 27	0.97	(51.8)	92	(88.3)
Mutant 34	0.34	(20.4)	90	(85.3)
Mutant 48	0.24	(12.5)	111	(104)
Mutant 50	0.21	(11.0)	131	(112)
Mutant 56	0	(0)	116	(100)
Mutant 8	0	(0)	0	(0)
Mutant 11	1.31	(68.5)	116	(112)

[a] Rates of phosphorylation as micromoles of ATP esterified per hour per milligram of chlorophyll. In parentheses, percent wild type. PMS = phenazine methosulfate.

on these mutant forms. Although PS I activity of these mutants, as measured either with whole cells or chloroplast particles, is inhibited, no comparable loss of PS II activity has been noted with a variety of Hill reaction oxidants. Although extensive evaluation of the kinetics of the Hill reaction of chloroplast particles of these mutants with the different Hill oxidants has not been made, it appears that DCIP, ferricyanide, and p-benzoquinone are reduced with comparable efficiencies. Consequently, the point(s) where electrons are diverted from PS II for the reduction of externally supplied oxidants precedes the normal site of electron acceptance by cytochrome-553; i.e., this component is not generally required for PS II activity.

3.3 Mutations Directly Affecting Photophosphorylation

In the preceding sections, the various known photosynthetic mutants of *Chlamydomonas* and *Scenedesmus* have been described. Most of them have specific defects within the electron transport system associated with PS-I. As a consequence of the loss of an electron carrier, many of these mutants also show altered photophosphorylation activities. However, only limited information is available on mutant strains which possess an intact electron transport system but lack the capacity for photophosphorylation. It is predictable that mutants which cannot reduce carbon dioxide either through photosynthesis or photoreduction might suffer from genetic lesions within the biochemical apparatus of phosphorylation. Such mutants, or chloroplasts from them, would be expected to have an intact electron transport system and to show normal fluorescence (variable yield) and light-induced absorbancy changes (at least for cytochrome-553 and cytochrome-559). A mutant of *Chlamydomonas* (F-54) has recently been described by Sato et al. (1971), which is apparently affected in one of the terminal steps of photophosphorylation. Although this mutant lacks NADP⁺ reducing capacity and photophosphorylation capacity, it was demonstrated that the former activity could be induced by the addition of the uncouplers, gramicidin D and nigericin. It was deduced that these uncouplers caused the breakdown of an accumulated "intermediate" of the phosphorylation sequence. Further studies demonstrated that the mutant strain contained an active, but not a latent, Ca^{2+}-dependent ATPase whereas the same enzyme of the normal strain is typical in that it requires heat treatment prior to its manifestation as an ATPase. Additional studies on the light-induced pH changes in mutant F-54 and on the lack of effect of arsenate on these changes caused these authors to conclude that the genetic lesion affecting the ATP-synthesizing apparatus has resulted in the alteration of a component involved in the terminal steps of phosphorylation.

While most of the information so far gathered with this mutant must be regarded as introductory, it is obvious that an alternate method is now available to study one of the more fascinating and difficult problems of biology, i.e., ATP synthesis.

3.4 Mutations Affecting Photosystem II

Over the past twenty-five years, intensive studies on the mechanism of photosynthesis have allowed for a better understanding of the general overall process. But as would be anticipated, certain aspects of the process are less understood than others. This is particularly true as concerns the sequence of reactions involved in the photolysis of water and the associated electron transport system hereafter referred to as photosystem II. Several mutants of *Scenedesmus* and *Chlamydomonas* have been of special use for studies of this system. As is indicated in Table I, 18 mutants have been isolated and described, in part, which cannot perform photosynthesis because of the loss of PS II. In this section the known qualities of these mutants will be described in an attempt to outline a "general" view of this photosystem. For this purpose it will be appropriate to divide PS II into two parts, i.e., the reducing and oxidizing portions.

As a working model the reducing section can be described as follows:

$$\text{Cytochrome-553} \leftarrow \text{plastoquinone} \leftarrow \begin{array}{c}\text{"Q"}\\ \text{C-550}\end{array} \leftarrow \begin{array}{c}\text{Chl-II}\\ |\\ \boxed{e^-}\end{array}$$

and the oxidizing side as:

$$\begin{array}{c}\text{"Q"}\\ \text{C-550}\end{array} \leftarrow \begin{array}{c}\text{Chl-II}\\ |\\ \boxed{\text{PS II pigment matrix}}\end{array} \leftarrow \begin{array}{c}\text{"Z"}\\ [\text{cytochrome-559}]\\ ??\end{array} \underset{Cl^-}{\overset{Mn^{2+}}{\leftarrow}} H_2O$$

The recent studies of Knaff and Arnon (1969a) revealed the previously undetected absorbancy change at 550 nm, which apparently is not caused by cytochromes and is involved in PS II reactions. This substance has been termed C-550 by Knaff and Arnon (1969a) and P-546 by Bendall and Sofrova (1971). Detailed studies on this absorbancy change in chloroplast or chloroplast subfractions as liquid nitrogen temperature by Erixon and Butler (1971a) and by Bendall and Sofrova on the apparently associated oxidation and reduction of cytochrome b-559 have allowed for further elaboration of PS II.

By redox titration of C-550 and of the variable-yield of fluorescence of isolated chloroplast, Erixon and Butler (1971a) have tentatively equated C-550 with substance Q, the previously mentioned quencher of

PS II fluorescence. Thus, in the essential core reactions of PS II, the b-559 component possibly serves as the electron donor for C-550; i.e., cytochrome b-559 and C-550 may be equatable with the previously hypothetical intermediates of PS II, "Z" and "Q," respectively. Evidence obtained by the laboratories of Arnon and of Butler have provided further evidence that the redox potential of cytochrome b-559 can be +0.37 V rather than the +0.09 to −0.055 V values as reported by Fan and Cramer (1970) and Hind and Nakatani (1970). The identification of a b-type cytochrome showing two different redox potentials has resulted in considerable speculation about whether there exist two separate cytochromes b-559's that function in PS II or whether the less-positive one represents an artifact of preparation wherein the high-potential form has been modified. Biochemical "common sense" would suggest the latter reason to be more likely; the recent purification of a chloroplast cytochrome b-559 by Garewal et al. (1971) may offer the opportunity to evaluate the possible existence of different b-type cytochromes-559. Wada and Arnon (1971) have demonstrated that apparently three redox forms of cytochrome b-559 (a high-potential, a medium-potential, and low-potential form) occur in chloroplasts; these authors suggested that it is only the high-potential form that is required for PS II activity, since increasing loss of Hill reaction activity is associated with an increased conversion of the high-potential to the low-potential form.

Bendall and Sofrova have recognized that even the redox potential of the b-559 (high potential) is insufficient to allow this component to participate singly in the oxidation of water. They speculated that even higher oxidizing potentials might be attained *in vivo* through the interaction of the cytochrome b-559 (high potential) with manganese ions. They also hypothesized that "super"-high potential form of cytochrome b-559, or rather discrete steps in its generation, may be analogous to Kok's carrier, S, whose generation and use appears to undergo a series of four one-equivalent oxidation steps before reaction with water.

3.4.1 MUTATIONS ON THE REDUCING SIDE OF PS II

With this background information of some of the recent findings concerning the electron transport system of PS II, it is now possible to examine some of the characteristic mutants of *Scenedesmus* and *Chlamydomonas* known to lack PS II activity. The general features of whole-cell reactions of these mutants are in general identical and include the following: (1) They cannot photosynthesize but perform a normal photo-reduction of carbon dioxide with hydrogen as the electron donor (Bishop, 1962, 1964; Levine, 1969). (2) They show a high unquenchable fluorescence with no evidence for the existence of a variable-yield component (Butler and Bishop, 1963; Bishop and Wong, 1971). (3) Of these mutant forms examined, the deletion of PS II results in the loss of the slow EPR signal and the disappearance of the major portion of the delayed light emission (Bishop, 1964; Weaver and Bishop, 1963; Levine and Piette, 1962; Bertsch et al., 1967). (4) The PS I-catalyzed oxidation of cytochrome 553 and cytochrome 564 is still present, but absorbancy changes due to cytochrome 559 are absent. P700 is present and functional (Levine, 1969).

Studies on chloroplast particles isolated from certain of these strains, notably *Scenedesmus* mutant 11 and ac-115 of *Chlamydomonas*, have shown a complete loss of Hill reaction activity with $NADP^+$, DPIP, ferricyanide, or p-benzoquinone as the Hill oxidants. The PS I catalyzed reduction of $NADP^+$, with ascorbate–DPIP as the electron donor system, and cyclic photophosphorylation are unaffected by the mutation. The PS II-catalyzed reduction of DPIP with diphenylcarbazide as the electron donor system is absent in mutant 11 of *Scenedesmus* (Bishop and Wong, 1971). In general, it seems clear that most of the PS II mutants so far studied have the reducing side of the mechanism affected.

Attempts to determine which of the many components of this photosystem are affected by the mutation have yielded somewhat contradictory results. Bishop and Wong (1971) showed that the PS II mutants of *Scenedesmus* generally showed a marked decrease in their content of plastoquinone A and a measurable deficiency in chlorophyll and carotenoids. As indicated in Table II, the major component affected in the PS II mutants of *Chlamydomonas* is the cytochrome b-559. However, earlier publication of Levine and co-workers (Smillie and Levine, 1963; Levine and Smillie, 1962) stressed the importance of their findings that mutant ac-115 and ac-141 were about 85% deficient of plastoquinone. Rather greater emphasis has been accorded to the additional finding that the *Chlamydomonas* mutants lack detectable cytochrome b-559, while the reported deficiency of plastoquinone has been completely neglected in more recent publications. On the other hand, Bishop and Wong did not observe any major differences in the cytochrome content or composition in three PS II mutants. However, the methodology employed by them for cytochrome analysis was inadequate to detect cytochrome b-559 reproducibly. Unpublished findings from the author's laboratory indicate that when more mild techniques of chloroplast isolation and extraction of chlorophyll are employed, *Scenedesmus* mutant 11 seems also to be deficient in at least the high potential form of cytochrome b-559.

An additional common feature of the *Chlamydomonas* and *Scene-*

desmus mutants is their high fluorescence yield, which lacks a variable-yield component. This characteristic has been interpreted as indicating that either the fluorescence quencher Q, or perhaps C-550 (see above), is absent or remains in the unquenching form, QH, or that the reaction center of PS II, which is responsible for the variable-yield fluorescence, is missing. Erixon and Butler (1971b) and Epel and Butler (1972) have demonstrated that the PS II mutants of *Scenedesmus* and *Chlamydomonas* lack C-550 and an ascorbate-reducible cytochrome b-559 and cytochrome b-564. Thus it would appear that an additional consequence of this type of mutation is the deletion of the fluorescence quencher, Q, or C-550.

One of the more intriguing aspects of the information so far obtained with the PS II mutants of *Scenedesmus*, as well as the similar *Chlamydomonas* mutants, is the nearly identical behavior and characteristics of the many similar phenotypes. The features of lowered plastoquinone content, lowered total chlorophyll, altered complementation of the carotenoids, decreased content of at least the high-potential form of cytochrome 559, and the loss of C-550 suggest that the changes produced by the mutation, i.e., the loss of PS II activity, is not necessarily the result of the inability of the mutants to synthesize a single factor, such as plastoquinone, but rather to the loss of a discrete part of the PS II unit within the chloroplast membranes. However, preliminary studies on the chloroplast structure of the *Scenedesmus* PS II mutants have revealed no obvious alteration of structure.

To evaluate these mutant types further, it is essential that mention be made of studies of the effects of various agents that are known to influence PS II activity of chloroplasts in a manner somewhat analogous to that of the mutations just described. These studies include the action of ultraviolet irradiation and the influence of various lipases and proteases.

The mechanism through which ultraviolet irradiation inhibits the Hill reaction and associated electron transport has been a subject of study in a number of laboratories (Bishop, 1961; Mantai and Bishop, 1967; Mantai *et al.*, 1970; Shavit and Avron, 1963; Jones and Kok, 1966; Yamashita and Butler, 1968; Malkin and Jones, 1968). In general, the results obtained by these many investigators were in agreement in that they demonstrated that ultraviolet irradiation primarily inhibited PS II. Jones and Kok (1966) concluded that the reaction centers of PS II were principally inactivated, and Yamashita and Butler suggested that a site between water and PS II was mainly affected by ultraviolet irradiation. Mantai and Bishop demonstrated that the effects of ultraviolet irradiation and lipase or protease treatment on PS II activity were quite similar and suggested that the primary action of ultraviolet irradiation was a disruption of the structural integrity of the lamellar membranes. The most recent findings of Erixon and Butler (1971a) have revealed that ultraviolet irradiation destroys C-550 and both oxidizes and denatures cytochrome b-559 so that it is no longer reducible by ascorbate. Since Butler and colleagues suggested that the absorbancy change which represents C-550 is perhaps only a shift of an absorption band from 547 to 543 nm without an appreciable change in the intensity of the band, it may be that C-550 is only an indicator of the state of a pigment–membrane protein intimately associated with PS II (Butler and Okayama, 1971). Disruption of this pigment–protein complex may be the *principal* effect of UV irradiation on PS II; secondary effects are undoubtedly produced by longer irradiation times which produce changes also in PS I-type reactions (Mantai et al., 1970; Erixon and Butler, 1971a). Associated with the alteration of C-550 and cytochrome b-559 is a continuous decline of the variable-yield component of the chloroplast fluorescence. This phenomenon of decreased variable-yield fluorescence associated with UV irradiation was observed by other investigators, but no clear interpretation could be made of this behavior of isolated chloroplasts. However, as Mantai et al. (1970) indicated, irradiated algal cells show a less complicated fluorescence pattern that may be more easily interpreted.

In a series of elegant experiments on the delayed light emission by *Chlorella*, in which it was shown that this emission consists of two separate components, Bonaventura and Kindergan (1971) showed that ultraviolet irradiation preferentially inactivated the fast component.

Treatment of isolated chloroplasts with a variety of lipases have shown that PS II is preferentially inhibited. Okayama (1964) found that evolution of oxygen, but not the DCIP–ascorbate supported photoreduction of NADP$^+$, was inhibited by pancreatic lipase digestion of chloroplast; this treatment also caused a decrease in the fluorescence yield of the chloroplast. Mantai et al. (1970) extended and confirmed these findings and suggested that the primary cause of the lipase inactivation of PS II and the decrease of the variable-yield fluorescence was due to disruption of chloroplast membrane structure. Somewhat earlier, Greenblatt et al. (1960) observed that lipase caused the separation of lamellae in *Euglena* chloroplasts; subsequent studies by Bamberger and Park (1966) showed that the inner surface of thylakoids appeared to be removed preferentially by lipase. Recent studies by Butler and Okayama (1971) and Okayama et al. (1971) on the light-induced fluorescence yield changes and absorbancy changes in chloroplasts of spinach and *Chlamydomonas* (at $-196°C$) have further extended the concept of structural alteration as being the fundamental cause of lipase inactivation of chloroplast reactions. Similar to the findings of Butler and colleagues on UV inactivation, it was noted that pancreatic lipase treatment of spinach chloroplasts destroyed the absorption band due to C-550 and eliminated the variable-yield component of fluorescence. However, with *Chlamydomonas* chloroplasts, although the C-550 absorbancy change and the variable-yield fluorescence component were eliminated, the invariant fluorescence was stimulated in contrast to the findings with spinach chloroplasts; this difference is important in evaluation of the PS II mutants and will be discussed later. Lipase treatment also was found to alter the cytochrome b-559 of spinach chloroplasts to the extent that it was reducible by dithionite but not by ascorbate, i.e., apparently the high-potential form had been converted to the low-potential form. With *Chlamydomonas* chloroplasts, the cytochrome b-559 was similarly altered, but in addition a portion of it appeared to have been denatured or destroyed. Also, the lipase treatment caused the loss of the slow-decaying EPR signal normally associated with PS II and, as shown in a later publication, also the delayed fluorescence (or delayed light emission) (Butler, 1971).

The hydrolytic activity of trypsin on spinach chloroplast also appears to attack PS II activity preferentially (Mantai, 1969, 1970; Selman and Bannister, 1971). Similar to lipase treatment of spinach chloroplast, trypsin digestion apparently destroys the variable-yield component of the fluorescence.

When these findings on the effects of the various inhibitor treatments of PS II are considered in relation to what is known about the PS II mutants of *Chlamydomonas* and *Scenedesmus*, a somewhat clearer pattern for the potential cause of this type of mutant becomes evident. As a provisional explanation for the cause of this type of mutation, it is suggested that the principal genetic alteration may be within essential proteins of the pigment–lipoprotein matrix of PS II which leads to secondary changes in several chloroplast components. These would include alteration of the redox status of cytochrome b-559 from a high-potential to a low-potential form and a decrease in the total amount of plastoquinone and chloroplast pigments synthesized by the mutants. The suggested change in the redox status of the cytochrome b-559 has not yet been thoroughly evaluated in either *Scenedesmus* or *Chlamydomonas* because of the difficulty of resolving the low redox form from the cytochrome b-563 in most extracts of algal chloroplasts.

The increased fluorescence yield and the associated loss of the variable-yield component of the fluorescence of the mutants is similar to that seen with lipase-treated chloroplasts of *Chlamydomonas* (Okoyama et al., 1971).

Increased attention is currently being directed toward the role of membranes in energy conservation in mitochondrial reactions. Chance (1972), for example, has suggested that the observed changes in midpoint potentials of certain mitochondrial cytochromes may be influenced by changes in membrane structure. Perhaps subtle changes in the constitution of the lipoprotein matrix of chloroplast membranes, might sufficiently alter the liganding environment of the heme moiety of cytochrome b-559 to cause the apparent loss of the high potential form of cytochrome b-559. But such interpretations, or flights of fancy, must only be considered premature; until more definitive information is available concerning the role of cytochrome b-559 in PS II type reaction it is impossible to consider what might be the function of high and low redox states of this component of the chloroplast.

3.4.2 Mutations on the Oxidizing Side of PS II

It is envisaged that reactions involved in the photolysis of water and the transfer of electrons through the reaction center of this photosystem, Chl II, to the terminal electron acceptor, Q or C-550, may comprise the steps involved on the oxidizing side of PS II as was indicated earlier. Until recently this aspect of the mechanism of photosynthesis was the least understood portion. Progress toward its elucidation has come principally from the laboratories of B. Kok and G. Cheniae, S. Malkin, P. Joliot and H. Witt.

From the earlier flash experiments of Allen and Franck (1955) on the kinetics of oxygen production by dark-adapted algal cells, a number of investigators have established that the kinetics of the production of oxygen (initial rather than steady state) in short saturating flashes of light can be accounted for by a model that contains a number of reaction centers which undergo activation in a linear sequence prior to the evolution of a molecule of oxygen. On the oxidizing side of PS II, each trapping center is assumed to operate independently where four quanta are required to raise a center at the ground state to the final excited state in four successive steps:

$$S_0 \xrightarrow{h\nu} S_1 \xrightarrow{h\nu} S_2 \xrightarrow{h\nu} S_3 \xrightarrow{h\nu} S_0 + O_2$$

The intermediate "Z" in the earlier formulation is representative of the component which perhaps interacts with Chl II to provide the kinetics observed in the early stages of oxygen production. Joliot's interpretation (Joliot and Joliot, 1972) suggests that there are twice as many electron donors (Z) as there are electron acceptors (Q), and the formation of molecular oxygen requires that Z donate two electron through Chl II to Q, thus becoming Z^{2+}. The generation of this state, and its stability, is most likely the determining factor for the observed kinetics of oxygen production. Z^{2+} may perhaps be the "strong" oxidant produced by PS II which would be responsible for the removal of electron from water (of OH⁻).

Comprehensive studies by Homann (1967) and Cheniae and Martin (1968) on the role of manganese in PS II have confirmed and extended the hypothesis previously held that this substance functions in oxygen evolution and on the oxidizing side of PS II. Repeated analyses have shown that there exists 3 Mn/200 PS II chlorophyll molecules and that the removal of two-thirds of this pool results in the loss of oxygen evolution. However, a DCMU-sensitive photoreduction of NADP⁺ with ascorbate or phenylenediamine (and a number of other artificial electron donors) still occurs, but at a reduced rate. The findings indicate at least two separate pools of manganese which function as separate units in the transfer of electrons of Chl II. The existence of separate functional pools of manganese might result in kinetics of oxygen evolution suggesting two or more different intermediates between water and Chl II. The question whether manganese (or a portion of the total pool), cytochrome b-559 (high potential), or other unknown constituents of chloroplasts can be equated with Z remains unanswered at the present time.

Attempts to obtain mutants blocked on the oxidizing side of PS II have been made in the author's laboratory and that of Levine. We have not yet obtained a mutant of *Scenedesmus* or *Chlorella* that has the characteristics expected of this type of mutant. Butler, Epel, and Levine (1972) and Epel and Butler (1972) have recently reported on the characteristics of six mutant strains of *Chlamydomonas* which appear to be mutated on the oxidizing side of PS II. These mutants are principally characterized by their low fluorescence, lack of a light-inducible variable-yield fluorescence, and chloroplasts of these mutants show both a variable-yield fluorescence and a DCMU-sensitive photoreduction of NADP⁺ with electron donors which provide electrons directly to PS II. Since these mutants show a normal photoreduction of C-550 and the photooxidation of cytochrome b-559 at low temperatures, it is assumed that they have a functional PS II reaction center. Chloroplast particles from the mutant strains contain equivalent amounts of the ascorbate-reducible cytochrome b-559 and C-550, and all the cytochrome b-559 participates in the low temperature photoreaction. In contrast to the wild-type chloroplast fragments, mutant chloroplast particles contain twice the amount of high potential cytochrome b-559 as C-550 and with only half of the b-559 participating in the low-temperature photoreaction. No difference was noted in the low-fluorescence mutants in their contents of cytochrome f-553, low potential cytochrome b-559 and cytochrome b-564 in comparison to the wild-type *Chlamydomonas*. These authors propose that there are two functionally distinct pools of ascorbate-reducible cytochrome b-559 both of which function in water photolysis; the mutants lack this capacity because one of the pools has been lost. Their conclusions are supported by the previously mentioned action of lipase on PS II and on the cytochrome b-559 pool. They offer the interesting speculation that perhaps the mutation has resulted in modification of certain chloroplast membranes, which in turn has produced an apparent alteration of the redox potential of the cytochrome b-559. The latter hypothesis is much more favored by our own work and predelection. The type of mutants described by Epel, Butler, and Levine may provide an excellent experimental tool for further analysis of the most difficult aspect of the mechanism of photosynthesis.

Most of the evidence linking the cytochrome b-559 to reactions concerned with PS II has been obtained with either isolated chloroplasts or various subparticles of the chloroplast and at low temperatures. The findings from such studies have been interpreted as showing that this cytochrome plays a direct role in the transfer of electrons from water to at least the terminal electron acceptor of PS II. The results with the two types of mutants of PS II have been interpreted in this fashion as indicated in the preceding pages. However, the function of cytochrome b-559 is still in a questionable state. Based on studies made at room temperature and other more general physiological conditions Hiller *et al.* (1971) [as well as at liquid nitrogen temperatures (Boardman *et al.*, 1971)], concluded that cytochrome b-559 functions on a side pathway connected to the reaction center of PS II, not in the main coupled pathway between PS I and PS II. Essentially the same conclusion has been reached by Ke *et al.* (1972), Boehme and Cramer (1971, 1972), and Cramer and Boehme (1972). Thus the precise role of cytochrome b-559, either in the high- or low potential form, in the mechanism of PS II cannot be succinctly stated at this time. To engage in oversimplification of its role, as in the PS II mutants blocked on the oxidizing side of PS II, may be extremely premature.

REFERENCES

Allen, F. L., and Franck, J. (1955). *Arch. Biochem. Biophys.* **58**, 124.
Anderson, J. M., and Boardman, N. K. (1964). *Aust. J. Biol. Sci.* **17**, 93.
Arnon, D. I., Tsujimoto, H. Y., McSwain, B. D., and Chain, R. K. (1968). *In* "Comparative Biochemistry and Biophysics of Photosynthesis" (K. Shibata *et al.*, eds.), p. 113. Univ. Park Press, State College, Pennsylvania.

Arnon, D. I., Chain, R. K., McSwain, B. D., Tsujimoto, H. Y., and Knaff, D. B. (1970). *Proc. Nat. Acad. Sci. U. S.* **67**, 1404.
Arnon, D. I., Knaff, D. B., McSwain, B. D., Chain, R. K., and Tsujimoto, H. Y. (1971). *Photochem. Photobiol.* **14**, 397.
Arntzen, C. J., Dilley, R. A., Peters, G. A., and Shaw, E. R. (1972). *Biochim. Biophys. Acta* **256**, 85.
Bailey, J. L., and Kreutz, W. (1969). *Progr. Photosyn. Res.* **1**, 149.
Bamberger, E. S., and Park, R. B. (1966). *Plant Physiol.* **41**, 1591.
Baszynski, T., Brand, J., Krogmann, D. W., and Crane, F. L. (1971). *Biochim. Biophys. Acta* **234**, 537.
Bendall, D. S., and Hill, R. (1968). *Annu. Rev. Plant Physiol.* **19**, 167.
Bendall, D. S., and Sofrova, D. (1971). *Biochim. Biophys. Acta* **234**, 371.
Bertsch, W., Azzi, J. R., and Davidson, J. B. (1967). *Biochim. Biophys. Acta* **143**, 129.
Bishop, N. I. (1961). *Quinones Electron Transp., Ciba Found. Symp., 1960* p. 385.
Bishop, N. I. (1962). *Nature (London)* **195**, 55.
Bishop, N. I. (1964). *Rec. Chem. Progr.* **25**, 181.
Bishop, N. I. (1971). In "Methods in Enzymology" (A. San Pietro, ed.), Vol. 23, Part A, pp. 130–143. Academic Press, New York.
Bishop, N. I. (1972). *Proc. Int. Congr. Photosyn. Res. 2nd, 1971* Vol. 1, p. 549.
Bishop, N. I., and Senger, H. (1972). *J. Plant Cell Physiol.* **13**, 633.
Bishop, N. I., and Wong, J. (1971). *Biochim. Biophys. Acta* **234**, 433.
Boardman, N. K. (1968). *Advan. Enzymol.* **30**, 1.
Boardman, N. K., Anderson, J. M., and Hiller, R. G. (1971). *Biochim. Biophys. Acta* **234**, 126.
Boehme, H., and Cramer, W. A. (1971). *FEBS Lett.* **15**, 349.
Boehme, H., and Cramer, W. A. (1972). *Biochemistry* **11**, 1155.
Bonaventura, C., and Kindergan, M. (1971). *Biochim. Biophys. Acta* **234**, 249.
Butler, W. L. (1966). *Curr. Top. Bioenerg.* **1**, 49.
Butler, W. L. (1971). *FEBS Lett.* **19**, 125.
Butler, W. L., and Bishop, N. I. (1963). *Nat. Acad. Sci.—Nat. Res. Counc., Publ.* **1145**, 91.
Butler, W. L., and Okayama, S. (1971). *Biochim. Biophys. Acta* **245**, 237.
Butler, W. L., Epel, B., and Levine, R. P. (1972). *Biochim. Biophys. Acta* **275**, 395.
Chance, B. (1972). In "Horizons of Bioenergetics" (A. San Pietro and H. Gest, eds.), p. 75. Academic Press, New York.
Cheniae, G. M., and Martin, I. F. (1968). *Biochim. Biophys. Acta* **153**, 819.
Cramer, W. A., and Boehme, H. (1972). *Biochim. Biophys. Acta* **246**, 358.
Dolzmann, P. (1968). *Z. Pflanzenphysiol.* **58**, 300.
Epel, B. L., and Butler, W. L. (1972). *Biophys. J.* **12**, 922.
Erixon, K., and Butler, W. L. (1971a). *Biochim. Biophys. Acta* **253**, 483.
Erixon, K., and Butler, W. L. (1971b). *Photochem. Photobiol.* **14**, 427.
Fan, H. N., and Cramer, W. A. (1970). *Biochim. Biophys. Acta* **216**, 200.
Forti, G., and Zanetti, G. (1969). *Progr. Photosyn. Res.* **3**, 1213.
Garewal, H. S., Singh, J., and Wasserman, A. R. (1971). *Biochem. Biophys. Res. Commun.* **44**, 1300.
Gee, R., Saltman, P., and Weaver, E. C. (1969). *Biochim. Biophys. Acta* **189**, 106.
Givan, A. L., and Levine, R. P. (1969). *Biochim. Biophys. Acta* **189**, 404.
Goodenough, U. W., and Levine, R. P. (1970). *Sci. Amer.* **233**, 22.
Gorman, D. S., and Levine, R. P. (1966). *Plant Physiol.* **41**, 1648.

Granick, S. (1948). *J. Biol. Chem.* **172**, 717.
Granick, S. (1949). *Harvey Lect.* **44**, 220.
Granick, S. (1971). In "Methods in Enzymology" (A. San Pietro, ed.), Vol. 23, Part A, pp. 168–171. Academic Press, New York.
Greenblatt, C. L., Olson, R. A., and Engel, E. K. (1960). *J. Biophys. Biochem. Cytol.* **7**, 235.
Gregory, R. P. F., Raps, S., and Bertsch, W. (1971). *Biochim. Biophys. Acta* **234**, 330.
Heber, U., and Gottschalk, W. (1963). *Z. Naturforsch. B* **18**, 36.
Hill, R., and Scarisbrick, R. (1951). *New Phytol.* **50**, 98.
Hiller, R. G., and Boardman, N. K. (1971). *Biochim. Biophys. Acta* **253**, 449.
Hiller, R. G., Anderson, J. M., and Boardman, N. K. (1971). *Biochim. Biophys. Acta* **245**, 439.
Hind, G. (1968). *Biochim. Biophys. Acta* **153**, 235.
Hind, G., and Nakatani, H. Y. (1970). *Biochim. Biophys. Acta* **215**, 223.
Hind, G., and Olson, J. M. (1968). *Annu. Rev. Plant Physiol.* **19**, 249.
Homann, P. H. (1967). *Plant Physiol.* **42**, 997.
Homann, P. H., and Schmid, G. H. (1967). *Plant Physiol.* **42**, 138.
Hudock, G. A., and Levine, R. P. (1964). *Plant Physiol.* **39**, 889.
Joliot, P., and Joliot, A. (1972). *Proc. 2nd Int. Congr. Photosyn. Res.* **1**, 26.
Jones, L. W., and Kok, B. (1966). *Plant. Physiol.* **41**, 1044.
Ke, B., Vernon, L. P., and Chaney, T. H. (1972). *Biochim. Biophys. Acta* **256**, 345.
Kirk, J. T. O., and Tilney-Basset, R. A. E. (1967). "The Plastids." Freeman, San Francisco, California.
Knaff, D. B., and Arnon, D. I (1969a). *Proc. Nat. Acad. Sci. U. S.* **63**, 963.
Knaff, D. B., and Arnon, D. I. (1969b). *Proc. Nat. Acad. Sci. U. S.* **64**, 715.
Kok, B. (1956). *Biochim. Biophys. Acta* **22**, 399.
Kok, B. (1963). *Nat. Acad. Sci.—Nat. Res. Counc., Publ.* **1145**, 45.
Kok, B., and Datko, E. A. (1965). *Plant Physiol.* **40**, 1171.
Kok, B., and Hoch, G. E. (1961). In "Light and Life" (W. D. McElroy and B. Glass, eds.), p. 379. Johns Hopkins Press, Baltimore, Maryland.
Latzko, E., and Gibbs, M. (1968). *Z. Pflanzenphysiol.* **59**, 184.
Levine, R. P. (1969). *Annu. Rev. Plant Physiol.* **20**, 523.
Levine, R. P. (1971). In "Methods in Enzymology" (A. San Pietro, ed.), Vol. 23, Part A, pp. 119–129. Academic Press, New York.
Levine, R. P., and Goodenough, U. W. (1971). *Annu. Rev. Genet.* 397.
Levine, R. P., and Gorman, D. S. (1966). *Plant. Physiol.* **41**, 1293.
Levine, R. P., and Paszewski, A. (1970). *J. Cell Biol.* **44**, 540.
Levine, R. P., and Piette, L. H. (1962). *Biophys. J.* **2**, 369.
Levine, R. P., and Smillie, R. M. (1962). *Proc. Nat. Acad. Sci. U. S.* **48**, 417.
Levine, R. P., and Togasaki, R. K. (1965). *Proc. Nat. Acad. Sci. U. S.* **53**, 987.
Malkin, S., and Jones, L. W. (1968). *Biochim. Biophys. Acta* **162**, 297.
Mantai, K. E. (1969). *Biochim. Biophys. Acta* **189**, 449.
Mantai, K. E. (1970). *Plant Physiol.* **45**, 563.
Mantai, K. E., and Bishop, N. I. (1967). *Biochim. Biophys. Acta* **131**, 350.
Mantai, K. E., Wong, J., and Bishop, N. I. (1970). *Biochim. Biophys. Acta* **197**, 257.
Moll, B., and Levine, R. P. (1970). *Plant Physiol.* **46**, 576.
Ohad, I., Siekevitz, P., and Palade, G. E. (1967a). *J. Cell Biol.* **35**, 521.
Ohad, I., Siekevitz, P., and Palade, G. E. (1967b). *J. Cell Biol.* **35**, 553.

Okayama, S. (1964). *Plant Cell Physiol.* **5**, 145.
Okayama, S., Epel, B. L., Erixon, K., Lozier, R., and Butler, W. L. (1971). *Biochim. Biophys. Acta* **253**, 476.
Park, R. B., and Sane, P. V. (1971). *Annu. Rev. Plant Physiol.* **22**, 395.
Powls, R., Wong, J., and Bishop, N. I. (1969). *Biochim. Biophys. Acta* **180**, 490.
Pratt, L. H., and Bishop, N. I. (1968). *Biochim. Biophys. Acta* **153**, 664.
Rurainski, H. J., and Hoch, G. E. (1972). *Proc. Int. Congr. Photosyn. Res, 2nd, 1971* Vol. 1, p. 133.
Rurainski, H. J., Randles, H., and Hoch, G. E. (1970). *Biochim. Biophys. Acta* **205**, 254.
Sato, V. L., Levine, R. P., and Neumann, J. (1971). *Biochim. Biophys. Acta* **253**, 437.
Schiff, J. A., Lyman, H., and Russell, G. K. (1971). *In* "Methods in Enzymology" (A. San Pietro, ed.), Vol. 23, Part A, pp. 143–162. Academic Press, New York.
Schuldiner, S., and Ohad, I. (1969). *Biochim. Biophys. Acta* **180**, 165.
Selman, B. R., and Bannister, T. T. (1971). *Biochim. Biophys. Acta* **254**, 428.
Senger, H., and Bishop, N. I. (1972a). *Plant Cell Physiol.* **13**, 633.
Senger, H., and Bishop, N. I. (1972b). *Proc. Int. Congr. Photosyn. Res., 2nd, 1971* Vol. 1, p. 677.
Shavit, N., and Avron, M. (1963). *Biochim. Biophys. Acta* **66**, 187.
Sistrom, W. R., and Clayton, R. K. (1964). *Biochim. Biophys. Acta* **88**, 61.
Sistrom, W. R., Griffith, M., and Stanier, R. Y. (1956). *J. Cell. Comp. Physiol.* **48**, 459.
Smillie, R. M., and Levine, R. P. (1963). *J. Biol. Chem.* **238**, 4058.
Surzycki, S. J., Goodenough, U. W., Levine, R. P., and Armstrong, J. J. (1969). *Symp. Soc. Exp. Biol.* **24**, 13.
Tanner, W., and Kandler, O. (1969). *Progr. Photosyn. Res.* **3**, 1217.
Thornber, J. P. (1968). *Biochim. Biophys. Acta* **172**, 230.
Thornber, J. P. (1971). *In* "Methods in Enzymology" (A. San Pietro, ed.), Vol. 23, Part A, pp. 682–687. Academic Press, New York.
Togasaki, R. K., and Levine, R. P. (1970). *J. Cell Biol.* **44**, 540.
Trebst, A. (1972). *Proc. Int. Congr. Photosyn. Res., 2nd, 1971* Vol. 1, p. 399.
Trebst, A., and Elstner, E. (1965). *Z. Naturforsch. B* **20**, 925.
Tsujimoto, H. Y., McSwain, B. D., Chain, R. K., and Arnon, D. I. (1969). *Progr. Photosyn. Res.* **3**, 1241.
von Wettstein, D. (1959). *Brookhaven Symp. Biol.* **11**, 138.
Wada, K., and Arnon, D. I. (1971). *Proc. Nat. Acad. Sci. U. S.* **68**, 3064.
Weaver, E. C., and Bishop, N. I. (1963). *Science* **140**, 1095.
Wessels, J. S. C. (1968). *Progr. Photosyn. Res.* **1**, 128.
Yamashita, T., and Butler, W. L. (1968). *Plant Physiol.* **43**, 2037.
Yocum, C. F., and San Pietro, A. (1972). *Proc. Int. Congr. Photosyn. Res., 2nd, 1971* Vol. 1, p. 477.

Chapter 3

SEPARATION OF PHOTOSYNTHETIC SYSTEMS I AND II*†

Jeanette S. Brown

Department of Plant Biology, Carnegie Institution of Washington, Stanford, California

1. Introduction 97
2. Differentiation of the Photosystems 99
   2.1 Action and Absorption Spectra 99
   2.2 Fluorescence and Luminescence 99
   2.3 Electron Transport Activity 100
   2.4 Immunology 101
   2.5 Electron Paramagnetic Resonance (EPR) Spectroscopy 101
3. Morphological Separation of the Photosystems in Plants 101
   3.1 Mutations 101
   3.2 Heterocysts of Blue-Green Algae 102
   3.3 Bundle-Sheath vs Mesophyll Chloroplasts 102
   3.4 Within Chloroplasts 102
4. Physical Separation of the Photosystems *in Vitro* 105
   4.1 Fractionation Procedures 106
   4.2 Recombination 109
   References 110

1. Introduction

Photosynthesis is the conversion of light into chemical energy by green plants. The more restricted definition used here is the photolysis of water and reduction of phosphopyridine nucleotide (NADP) by light absorbed by or transferred to chlorophyll a located in organized, membranous structures called chloroplasts. The enhancement studies

* CIW DPB Publication No. 490.
† The zigzag formulation of the mechanism of photosynthesis which was suggested by Hill and figured by Duysens is conceived as consisting of two photosynthetic systems (PS) connected by electron carriers. Some workers prefer to designate these PS 2 and PS 1, and other workers prefer to designate them PS II and PS I. The choice is arbitrary, nothing specific being implied by a choice between the two. (Ed.)

by Emerson (1957), which were extended by Myers and French (1960), and the chromatic transient effects of Blinks (1960) indicated that at least two photoreactions with light of different colors are necessary for photosynthesis. One photoreaction is closer to the O_2 evolving step and mainly utilizes visible light of wavelengths shorter than 680 nm absorbed by such active pigments as fucoxanthin, phycobilins, chlorophylls b and c, and certain forms of chlorophyll a. The other is mediated preferentially by the longer wavelength forms of chlorophyll a. Hill and Bendall (1960) proposed a model, now called the Z scheme, that in a general way describes two light reactions (see Fig. 1). Duysens et al. (1961) suggested the terms "Systems 1 and 2" to describe the reactions initiated by the longer and shorter wavelengths of light, respectively.

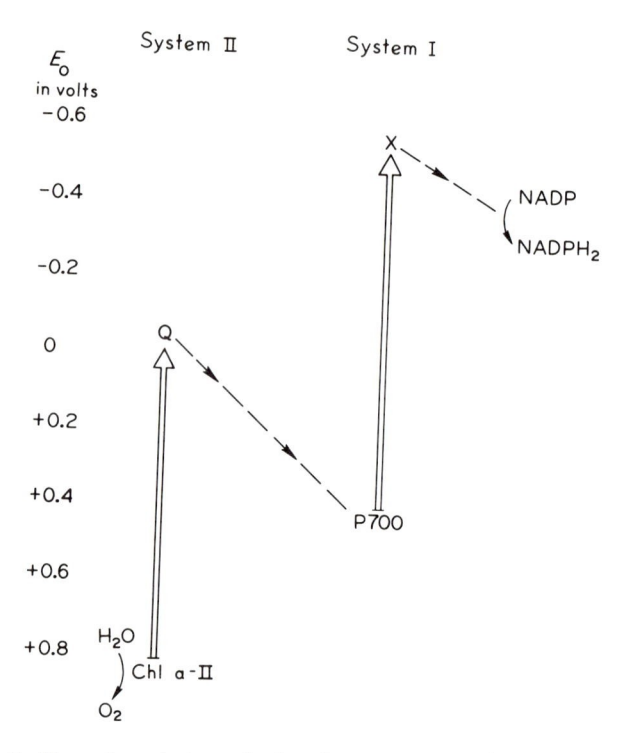

FIG. 1. Outline of a photosynthetic scheme encompassing two light reactions (Hill and Bendall, 1960; Duysens, 1963). Light energy absorbed by a special but as yet unidentified type of chlorophyll a, called Chl a-II, removes electrons from water and reduces an unknown substance, Q. The electrons then pass downward through a series of carriers, such as plastoquinone, cytochromes, and plastocyanin, to another special type of chlorophyll a called P700. Light reaction 1 oxidizes P700 and reduces the primary acceptor of photosystem I. X is able to reduce NADP in darkness.

This chapter is concerned with some of the ways in which various parts or components of the Z scheme can be distinguished and measured; the morphological separation of the photoreactions *in vivo* at the tissue, cell, and chloroplast levels; and the physical separation of the photosystems *in vitro*.

2. Differentiation of the Photosystems

2.1 Action and Absorption Spectra

Action spectra show the effectiveness of different colors of light for a particular reaction. Since only absorbed light can be effective, the action spectrum coincides with the absorption spectra of the photoreactive pigments. Action spectra for various steps in the Z scheme have shown that several different pigments and forms of chlorophyll are preferentially active in system I or II (French *et al.*, 1960; Sauer and Park, 1965; Fork and Amesz, 1969).

Because of this differential pigment action, it is possible to study reactions in either system I and II preferentially by the appropriate choice of wavelength of exciting light. It is generally thought that light energy absorbed by such pigments as phycoerythrin and chlorophyll b is transferred to one of the shorter wavelength-absorbing forms of chlorophyll a before finally passing to a reaction center for system II.

When the photosystems have been physically separated from each other, absorption spectra that show the relative amounts of the different pigments in each fraction may be used to indicate the degree of separation. This is illustrated in Fig. 2, which shows typical absorption spectra of spinach chloroplast particles preferentially active in either system II or I. Since chlorophyll b is more active in system II than in system I, the ratio of chlorophyll (Chl) a to b may be an indication of the relative amount of each photosystem.

2.2 Fluorescence and Luminescence

Most of both the fluorescence (Goedheer, 1972) and luminescence (Bertsch *et al.*, 1967) of plants comes from the chlorophyll a in system II. This is true also for the fluorescence from finely divided chloroplast particles (Brown, 1969). In live cells or active chloroplast particles, variable fluorescence can be observed from system II only (Park *et al.*, 1971). The variable fluorescence is a component of the total fluorescence that increases more slowly (several seconds) than the instantaneous or dark emission seen immediately after the light is turned on (Duysens and Sweers, 1963).

The fluoresence spectra of the chlorophyll in each of the two systems are also different (see Fig. 2). Generally speaking most of the

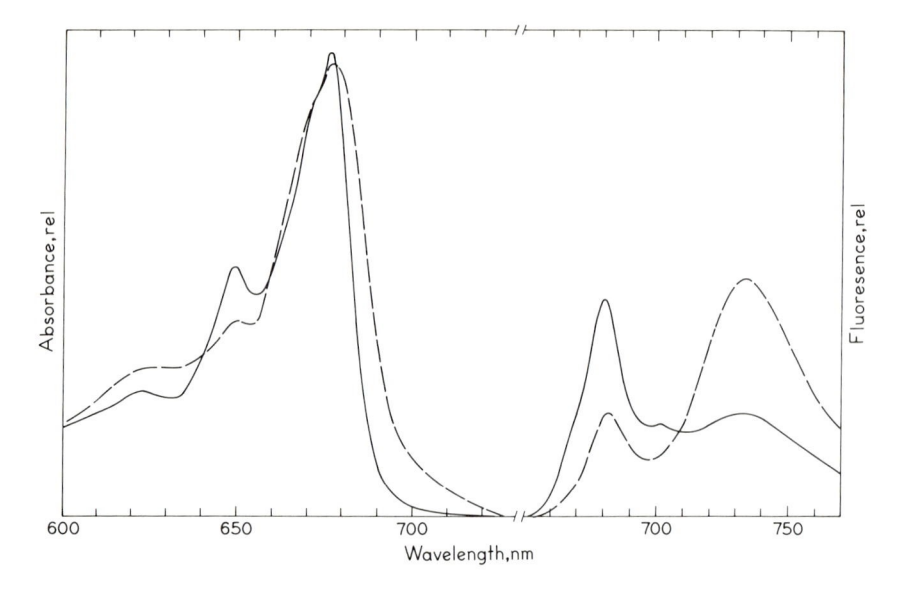

FIG. 2. Absorption and fluorescence spectra at −196°C of fractions (fraction 1, ———; fraction 2, —) prepared from spinach chloroplasts by extrusion through a French press and sucrose density gradient centrifugation (Brown, 1971).

emission from the chlorophyll in system II is near 685 nm whereas that from system I is at wavelengths longer than 710 nm. This long-wavelength emission from system I is greatly enhanced at low temperature (−190°C). An emission band near 695 nm seen only at −190°C in fresh preparations is probably from system II (Goedheer, 1972).

2.3 Electron Transport Activity

By the appropriate choice of electron donors, acceptors, and inhibitors, the activity of either system I or II can be measured separately. Many of these electron carriers, such as dichlorophenolindophenol (DCIP), enter the electron chain between Q and P700 as shown in Fig. 1. They may be reduced by system II or oxidized by system I. Inhibitors may block specific reactions in the chain. For instance, 3(3,4-dichlorophenyl)-1,1-dimethylurea (DCMU) specifically prevents the passage of electrons from Q to other members of the chain and thereby inhibits system II but not system I if an artificial electron donor is added.

The photoreduction of ferricyanide or DCIP measured spectrophotometrically is frequently used to assess the activity of system II, and the photooxidation of reduced mammalian cytochrome c, the photoreduction of NADP, or the oxidation of P700 may be used for system I

(San Pietro, 1971, 1972). Diphenylcarbohydrazide (DPC) is often used as an electron donor in system II when the O_2-evolving capacity has been lost (Vernon and Shaw, 1969).

2.4 Immunology

There have been several recent attempts to locate specific parts of the photosynthetic apparatus by immunological techniques. Radunz et al. (1971) made antibodies to pure chlorophyll a in rabbit blood serum. These antibodies were specifically adsorbed onto the membrane surfaces of chloroplast thylakoids, and system II activity was strongly inhibited whereas system I was not.

Briantais and Picaud (1972) had some success in their attempts to locate the two photochemical systems by preparing antibodies in rabbits against systems I and II fractions prepared with Triton (see Section 4.1). Their results indicate that system I may be located mainly in the external surface of the chloroplast lamellae and system II in the internal surface (see Section 3.4).

The immunological technique appears able to corroborate and extend conclusions obtained by very different methods. Currently the group with Professor A. Trebst at the Ruhr University is using antibodies in conjunction with its studies of different components of the photosynthetic processes.

2.5 Electron Paramagnetic Resonance (EPR) Spectroscopy

A light-induced rapid EPR signal with a narrow band at $g = 2.0025$ comes from P700 and is a specific indicator for photosystem I (Warden and Bolton, 1972). A slower, broader signal near $g = 2.0046$ is also light-induced but takes hours to decay in the dark and may be correlated with system II (Weaver and Weaver, 1972).

3. Morphological Separation of the Photosystems in Plants

3.1 Mutations

The study of various parts of the photosynthetic mechanism has been aided by the production of mutants that lack one or more components of the electron transport chain. Mutants of the algae *Chlamydomonas* or *Scenedesmus* perform some of the reactions of system II but not system I, or vice versa (Levine, 1969). *Scenedesmus* mutant No. 8 (Bishop, 1962) lacks system I activity and also lacks a specific chlorophyll–protein complex (Gregory et al., 1971).

The plastome mutants of *Oenothera* developed by Stubbe (1960) have been useful material for studies of partial reactions of photosynthesis. These are plastid mutants manifest in certain leaves of otherwise green plants. Some of these mutant plastids have only system I, and others only system II (Fork and Heber, 1968).

A more detailed description of the role of mutants in the study of the photosynthetic apparatus is given by Bishop in Chapter 2 of this volume.

3.2 Heterocysts of Blue-Green Algae

Another example of morphological differentiation of the photosystems is in the heterocysts of some blue-green algae. These algae form an enlarged cell at intervals in a chain of cells. Heterocysts occur at regular intervals in *Anabaena* and comprise about 5% of the normal cells. Donze et al. (1972) found that these cells contain photosystem I, but not photosystem II.

3.3 Bundle-Sheath vs Mesophyll Chloroplasts

Plants that have the C_4-dicarboxylic acid pathway of CO_2 fixation contain distinctly different chloroplast types in their bundle-sheath and mesophyll cells (Laetsch, 1971; see also Björkman, Chapter 1 of this volume). The bundle-sheath chloroplasts are greatly enlarged, and in some but not all cases the lamellar structure shows little if any stacking or grana (see Fig. 3 and discussion in Section 3.4). Anderson et al. (1971) found that these bundle-sheath chloroplasts possess very little photosystem II activity, but they can carry out the reactions of photosystem I. The chlorophyll a in these chloroplasts shows more of the long-wavelength absorption associated with system I (French and Berry, 1971; Black and Mayne, 1970). Downton et al. (1970) treated thin sections of C_4 plants with the dye tetranitro-blue tetrazolium (TNBT) and observed that the agranal bundle-sheath chloroplasts lacked the ability for noncyclic electron flow.

However, Andersen et al. (1972) maintain that photosystem II is present in bundle-sheath chloroplasts, but that a component essential to the coupling between the systems may be lacking. Although this controversy has not yet been resolved, it is apparent that plants are able to modify the expressed activity of the photosystems in certain of their chloroplasts. The relationship between grana formation and the location of the photosystems is discussed in Section 3.4.

3.4 Within Chloroplasts

The apparatus of each of the photosystems may be in part morphologically or structurally separated within chloroplasts that contain grana. Figure 3 is an electron micrograph of a chloroplast showing

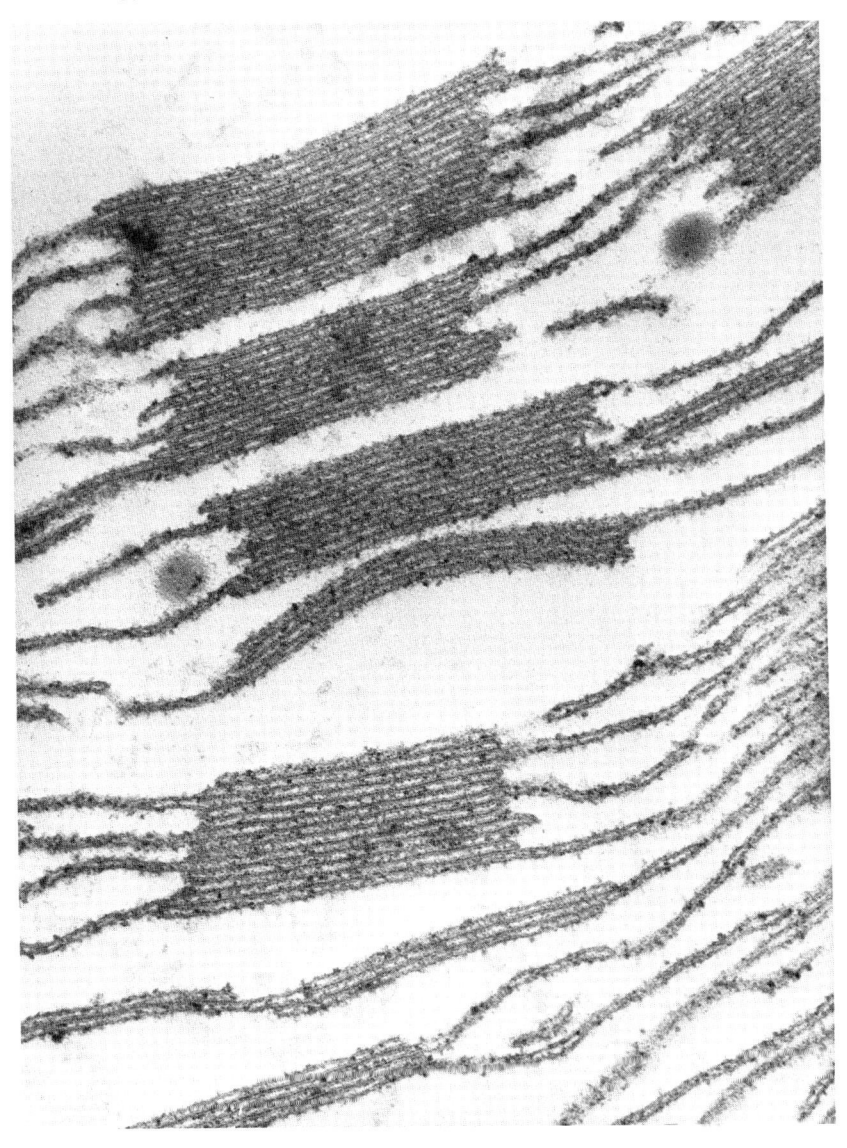

Fig. 3. Electron micrograph of a spinach chloroplast washed with 0.35 M KCl. Total magnification, 140,000. (Courtesy R. B. Park, University of California, Berkeley.)

stacked lamellae or grana and intergrana or stroma amellae. Jacobi and Lehmann (1968) were the first to note that chloroplast fractions enriched in photosystem II contained largely stacks of lamellae or grana, whereas the photosystem I fractions were composed of single lamellar strands. Subsequently, utilizing a different method of fractionation,

Sane et al. (1970) also observed that the grana fractions had both systems I and II activities whereas the stroma lamellae fractions had only system I activity.

Arntzen et al. (1971) compared the ability of various higher plant cells that differed in their proportions of grana and stroma lamellae to transport protons and phosphorylate. Their results showed that the greater the proportion of grana, the higher the amount of proton transport, whereas cells with larger amounts of stroma lamellae had high rates of ATP formation, but a reduced capacity for H^+ accumulation. A number of studies have correlated the appearance of photosystem II activity with the formation of grana (Boardman et al., 1971; Wiessner and Amelunxen, 1969). The lack of system II activity in the agranal bundle-sheath chloroplasts of C_4 plants was mentioned in Section 3.3.

A simplistic view of these results is that in chloroplasts that contain grana the complete photosynthetic apparatus is located in these grana, and that the activity in the intergrana or stroma lamellae may provide additional energy by a type of cyclic photophosphorylation. Section 4.1 contains additional discussion of this idea.

On the other hand, red and blue-green algae, Euglena, and certain mutants of green algae do not contain grana, but are competent in carrying out complete photosynthesis. When higher plant chloroplasts are prepared in low-salt media they lose their grana but not their system II activity (Izawa and Good, 1966). Weier et al. (1967) noted that the sites of TNBT reduction, a system II reaction, were distributed along the surface of both grana and stroma lamellae. Hall et al. (1971) did a cytochemical study with spinach chloroplasts whereby they localized the sites of ferricyanide photoreduction. They found discrete precipitates of Cu ferrocyanide on both stroma and grana lamellae indicating system II activity in both regions of the chloroplast. Hall et al. (1972) and Reeves and Hall (1972) have also demonstrated system II activity in both granal and agranal regions of chloroplasts in maize, sorghum, wheat, and rice. The conflict between the cytochemical results and the fractionation experiments could be explained if photosystem II in the grana is able to make an electron donor that can diffuse fairly rapidly along the membranes to the stroma region and there be used by system I to reduce ferricyanide.

Punnett (1971) was able to change the chloroplasts in Elodea leaves from granal to agranal by illumination in red light and to reverse the effect in blue light or darkness. The ability for photosynthetic enhancement was lost in the agranal state. The amount of enhancement appears to be one measure of the degree of coupling between the light reactions. Coupling can also be affected by various cations (Murata, 1969). Since both ions and preillumination influence grana formation and coupling, it may be reasoned that stacking of the lamellae is a structural modification which increases the efficiency of system II under certain environmental conditions.

Thus we have one set of experiments that indicates that the apparatus of system II is only in the grana; other results suggest that both photosystems are in all the lamellae, but that they are efficiently coupled for complete photosynthesis only in the stacked lamellae; and a cytochemical experiment showing system II activity in both grana and stroma lamellae. Part of the difficulty in understanding and correlating these different results is in the definition of a granum. As used here, a granum is a stack of several lamellae as shown in Fig. 3. This picture is typical for chloroplasts from most vascular plants except for bundle-sheath chloroplasts from some C_4 species. But some workers consider the oppression of 2 or 3 thylakoids or lamellae over a part or all of their surfaces to be sufficient for the granal condition. The lamellar structure of various kinds of chloroplasts is discussed by Kirk and Tilney-Bassett (1967). In normal *Euglena* the lamellae are paired, but in a system II-deficient mutant, the lamellae are separated from each other (Schwelitz et al., 1972).

The above discussion is about the possible location of the two photosystems within a chloroplast. Some work has been done to see in what arrangement the apparatus of each photosystem may be within a single lamella. Briantais and Picaud (1972) presented immunological evidence which suggested that, using the bilayer model for a lamella, the external part may be composed of system I units, and the internal of system II units.

Arntzen et al. (1969) compared electron micrographs of systems I or II enriched lamellar fragments using the freeze-etching technique. Their results suggested that the lamellar membranes have a binary structure with 110 Å particles embedded in one layer and 175 Å particles in the other. The intergrana lamellae having only system I activity also have only the smaller particles. The stacked lamellae in grana have both the large and small particles in opposing layers. The bilayer model of chloroplast structure was proposed earlier by Branton (1968) and has recently been extended by Mühlethaler (1972). An actual relationship between electron transport activity and the particles revealed in electron micrographs has yet to be proved. Experiments designed to isolate the particles will be discussed in Section 4.1.

4. Physical Separation of the Photosystems *in Vitro*

Experiments designed to fractionate chloroplasts have been going on for twenty-five years. When it was apparent that chlorophyll a exists as several natural forms absorbing at different wavelengths, many unsuccessful attempts were made to separate these forms without changing their absorption bands. Later when it was realized that the different chlorophylls are associated with the two photosystems in differing proportions, there was renewed activity to separate not only the chlorophylls, but also the photochemical systems.

4.1 Fractionation Procedures

Methods used to judge the extent of separation of system I from system II have been presented in Section 2. The rationale behind currently successful or partially successful fractionations of systems I and II may be found in the chloroplast ultrastructural research discussed in Section 3. Actually the studies of ultrastructure and fractionation attempts have been interwoven to a great extent—each complimenting the other.

Boardman (1970) has reviewed the various detergent and mechanical procedures used to fractionate chloroplasts. Much relevant information about isolation procedures may be found in *Methods in Enzymology*, Volume XXIII (San Pietro, 1971). Partial separations of the photosystems in a number of plants have been achieved with the aid of such detergents as digitonin, Triton X-100, deoxycholate, dodecyl sulfate and mechanical devices including the French pressure cell, sonicators, glass-bead shakers, simple grinders or blenders. We now realize that all these procedures separate the stroma lamellae containing only system I from the grana lamellae containing both systems I and II. For best results the chloroplasts should have grana and be suspended in a high-salt medium during disruption in order to keep the grana intact. The initial disruption should be relatively gentle in order not to break the grana into fragments similar in size to the stroma lamellae fragments; otherwise it will be difficult to separate them.

Since these partial fractionations discussed by Boardman (1970), several workers have been able to isolate smaller membrane fragments or particles with enriched system I or II activity only. These results are discussed below and summarized in Table I.

Vernon et al. (1971) have prepared photosystem I particles by first extracting the carotenoids from freeze-dried spinach chloroplasts and then treating with Triton. These particles have a chlorophyll a:P700 ratio of 30, indicating that they might be reaction center complexes for system I. However, their activity for NADP photoreduction is very low and chlorophyll a to b ratio only 4. Vernon suggests that these particles, called HP700, may be associated in the lamellar membranes with an accessory system I complex containing additional protein, light-

TABLE I
Properties of Photosystem I and II Fractions from Spinach Chloroplasts

Sample	Chl a:b	Absorption maximum	Chl:P700	Photosystem I[a] (μmoles mg⁻¹ Chl hr⁻¹)	Photosystem II[b] (μmoles mg⁻¹ Chl hr⁻¹)	Particle size (Å)
Photosystem I						
Stroma						
Wessels and Voorn (1972)	7.3	679	120	700	0	60 × 150
Arntzen et al. (1972)	5.6	—	100	754	3	50–500[c]
Grana						
Wessels and Voorn (1972)	3.4	678	250	500	0	60 × 150
Arntzen et al. (1972)	3.7	—	170	490	8	50–500[c]
HP700						
Vernon et al. (1971)	4	676	30	8	0	60 × 150
Photosystem II						
Accessory complex						
Wessels and Voorn (1972)	1.6	674–675	—	0	20	60
Vernon et al. (1972)	1.8	677	—	62	68	Membrane fragments
Reaction center complex						
Wessels et al. (1973)	7.2	675–676	—	0	220	<110
Vernon et al. (1972)	7.8	673	—	0	1100	110–120
Grana membranes						
Arntzen et al. (1972)	1.8	—	1730	43	122	Membranes[d]

[a] NADP reduction with dichlorophenolindophenol (DCIP) and ascorbate couple.
[b] DCIP reduction with DPC as electron donor.
[c] Tendency to aggregate.
[d] 175 Å particles seen in electron micrograph made with freeze-etch technique.

harvesting chlorophyll, β-carotene, cytochromes, plastoquinone, and lipids.

A system II particle has also been prepared by Triton treatment (Vernon et al., 1972). In addition to the Triton treatment (weight Triton to Chl = 25) the spinach membranes were extracted with 0.25 M sucrose and 0.02 M Tris at pH 8 and then passed through Bioglass 2500 to remove inactive chlorophyll complexes. The particles were very active (800–1000 μmoles mg^{-1} Chl hr^{-1}) in DCIP photoreduction with DPC and showed no system 1 activity. Since they had a high Chl a:b and relatively more β-carotene than xanthophylls, the authors suggested that these may be system II "reaction center complexes" which have been isolated from "accessory complexes" containing the Chl b and xanthophylls normally associated with system II. It is not evident why their absorption peak should be at 673 nm because other evidence would place the reaction center for system II nearer to 680 nm (Döring et al., 1969).

Recently, Wessels and Voorn (1972) have prepared rather pure system I and system II particles. They treated spinach chloroplasts with digitonin (weight ratio digitonin to chlorophyll = 15) and then isolated fractions by sucrose density-gradient centrifugation. These grana and stroma lamellar fractions were then further extracted with digitonin and recentrifuged. Photosystem I particles were obtained from both the original grana and stroma lamellae. Wessels favors the hypothesis that the complete electron transfer system with two photoreactions occurs in the grana, and that photosystem I in the stroma may function only in cyclic electron flow with phosphorylation. He also estimated that about 30% of the photosystem I chlorophyll is in the stroma region, and 70% in the grana.

Arntzen et al. (1972) have been doing experiments very similar to those of Wessels. Instead of two digitonin extractions, this group did their first disintegration mechanically with a French press, and then extracted the grana fraction with digitonin followed by sucrose density-gradient centrifugation to separate photosystems I and II from the grana.

The results of these three groups are compared in Table I. Both Wessels and Voorn (1972) and Arntzen et al. (1972) used digitonin and compared the system I particles from the stroma and grana lamellae. In both laboratories the system I particles from the stroma had a higher Chl a:b, lower Chl:P700 and higher activity for NADP reduction than did similar sized particles from the grana. Whether these differences are the result of the extraction procedures or inherent in the particles has not yet been determined. The HP700 system I particles were obtained by Triton treatment of carotenoid-extracted, freeze-dried chloroplasts. Although their P700 to chlorophyll ratio is very high, the Chl a:b and absorption maxima are low compared to those of the stroma particles obtained with digitonin. Vernon's suggestion that HP700 may be reaction center complexes needs further experimental evidence.

With respect to photosystem II, Wessels *et al.* (1973) using digitonin and Vernon *et al.* (1972) using Triton have come to the concept that reaction center complexes may be associated *in vivo* with accessory complexes and that these can be separated from each other. The results of the two groups are similar except that the absorption maximum of the reaction centers produced with Triton is at a shorter wavelength than is the maximum for particles made with digitonin. The reaction centers prepared with Triton have 5 times greater system II activity than do the digitonin-produced centers. The system II fraction prepared with digitonin by Arntzen *et al.* (1972) appears to be similar to the accessory complexes of Wessels *et al.* and Vernon *et al.*

Huzisige *et al.* (1969) have prepared system I and II fractions by a combination of digitonin and Triton treatments followed by density gradient and differential centrifugation. Their system I particles reduced NADP with the ascorbate–DCIP couple with a rate of 37 μmoles mg^{-1} Chl hr^{-1} which is very low compared to the results in Table I. Their system II fraction had no system I activity and reduced DCIP with diphenyl semicarbazide as electron donor with a rate of 42 μmoles mg^{-1} Chl hr^{-1} (Huzisige and Yamamoto, 1972). This compares with the accessory complexes for photosystem II isolated by Wessels *et al.* and Vernon *et al.* The Chl a:b was about 7 for fraction 1 and 2 for fraction 2. The absorption maxima, Chl:P700 and particle sizes were not reported.

Current work is being carried out on the protein composition of fractions representing the photosystems (Vernon *et al.*, 1972; Remy, 1971; Gregory *et al.*, 1971). Obvious differences can be seen in electrophoretic patterns, and further characterization of these different proteins is in progress. Allen *et al.* (1972) found that the total glycerolipid composition of the mechanically produced stroma and grana fractions was similar, but the grana fraction contained about 40% more chlorophyll on a protein basis than the stroma lamellae.

4.2 Recombination

Briantais (1969) separated the photosystems with Triton and then recombined them again with some small success as evidenced by the Chl a:b and enhancement of O_2 uptake. Huzisige *et al.* (1969) also attempted to recombine their separated fractions, but the activity was very low. More recently Arntzen *et al.* (1972) were able to make a preparation from grana system I particles and system II membranes that would reduce NADP using DPC as an electron donor at a rate of 65 μmoles mg⁻¹ Chl hr⁻¹. This reaction requires the activity of both photosystems and is inhibited by DCMU. The recombination was aided by using high concentrations of each photosystem fraction and lecithin. No recombination occurred when stroma system I particles were substituted for the grana system I fraction. This is further evidence that the system I particles from the grana and stroma are inherently different rather than merely apparently different because of damage during isolation.

A cautionary comment should be made about the effects of various fractionation procedures upon the original state of the photosystems *in vivo*. Fork and Murata (1971) have compared the light-induced changes in cytochrome f and P700 in fractions prepared with the French press before and after the addition of Triton and in the presence of various electron donors. They found that the absorbance changes of the mechanically prepared fractions retained the rapid kinetics characteristic of intact chloroplasts whereas the addition of detergents abolished the rapid reaction between cytochrome f and P700. It was only after the addition of detergent that their preparations would oxidize such large molecules as mammalian cytochrome c and plastocyanin. These results indicate that detergents induce fairly drastic changes in the structure of the reaction centers. The conditions of both the plant material and the nature of the French press itself seem to influence the integrity of the fractions. Therefore, care must be exercised in extrapolating from reactions observed in fractions or reconstituted systems to the original situation *in vivo*.

REFERENCES

Allen, C. F., Good, P., Trosper, T., and Park, R. B. (1972). *Biochem. Biophys. Res. Commun.* **48**, 907.
Andersen, K. S., Bain, J. M., Bishop, D. G., and Smillie, R. M. (1972). *Plant Physiol.* **49**, 461.
Anderson, J. M., Woo, K. C., and Boardman, N. K. (1971). *Biochim. Biophys. Acta* **245**, 398.
Arntzen, C. J., Dilley, R. A., and Crane, F. L. (1969). *J. Cell Biol.* **43**, 16.
Arntzen, C. J., Dilley, R. A., and Neumann, J. (1971). *Biochim. Biophys. Acta* **245**, 409.
Arntzen, C. J., Dilley, R. A., Peters, G. A., and Shaw, E. R. (1972). *Biochim. Biophys. Acta* **256**, 85.
Bertsch, W., Azzi, J. R., and Davidson, J. B. (1967). *Biochim. Biophys. Acta* **143**, 129.
Bishop, N. (1962). *Nature (London)* **192**, 55.
Black, C. C., and Mayne, B. C. (1970). *Plant Physiol.* **45**, 738.
Blinks, L. (1960). *Proc. Nat. Acad. Sci. U. S.* **46**, 327.
Boardman, N. K. (1970). *Annu. Rev. Plant Physiol.* **21**, 115.
Boardman, N. K., Anderson, J. M., Kahn, A., Thorne, S. W., and Treffry, T. E.

(1971). In "Anatomy and Biogenesis of Mitochondria and Chloroplasts" (N. K. Boardman, A. W. Linnane, and R. M. Smillie, eds.), pp. 70–84. North-Holland Publ., Amsterdam.
Branton, D. (1968). In "Photophysiology" (A. C. Giese, ed.) Vol. 3, pp. 197–224. Academic Press, New York.
Briantais, J.-M. (1969). Physiol. Veg. **7**, 135.
Briantais, J.-M., and Picaud, M. (1972). FEBS Lett. **20**, 100.
Brown, J. S. (1969). Biophys. J. **9**, 1542.
Brown, J. S. (1971). In "Methods in Enzymology" (A. San Pietro, ed.), Vol. 23, Part A, pp. 477–487. Academic Press, New York.
Donze, M., Haveman, J., and Schiereck, P. (1972). Biochim. Biophys. Acta **256**, 157.
Döring, G., Renger, G., Vater, S., and Witt, H. T. (1969). Z. Naturforsch. B **24**, 1139.
Downton, W. J. S., Berry, J. A., and Tregunna, E. B. (1970). Z. Pflanzenphysiol. **63**, 194.
Duysens, L. N. M. (1963). Nat. Acad. Sci.—Nat. Res. Counc. Publ. **1145**, 1–17.
Duysens, L. N. M., and Sweers, H. E. (1963). In "Studies in Microalgae and Photosynthetic Bacteria" (Jap. Soc. Plant Physiol., eds.), pp. 353–72. Univ. of Tokyo Press, Tokyo.
Duysens, L. N. M., Amesz, J., and Kamp, B. M. (1961). Nature (London) **190**, 510.
Emerson, R. (1957). Science **125**, 746.
Fork, D. C., and Amesz, J. (1969). Annu. Rev. Plant Physiol. **20**, 305.
Fork, D. C., and Heber, U. W. (1968). Plant Physiol. **43**, 606.
Fork, D. C., and Murata, N. (1971). Photochem. Photobiol. **13**, 33.
French, C. S., and Berry, J. A. (1971). Carnegie Inst. Wash., Yearb. **70**, 495.
French, C. S., Myers, J., and McLeod, G. C. (1960). In "Comparative Biochemistry of Photoreactive Systems" (M. B. Allen, ed.), pp. 361–365. Academic Press, New York.
Goedheer, J. C. (1972). Annu. Rev. Plant Physiol. **23**, 87.
Gregory, R. P. F., Raps, S., and Bertsch, W. (1971). Biochim. Biophys. Acta **234**, 330.
Hall, D. O., Edge, H., and Kalina, M. (1971). J. Cell Sci. **9**, 289.
Hall, D. O., Edge, H., Reeves, S. G., Stocking, C. R., and Kalina, M. (1972). Proc. Int. Congr. Photosyn. Res., 2nd, 1971 Vol. 1, p. 701.
Hill, R., and Bendall, F. (1960). Nature (London) **156**, 136.
Huzisige, H., and Yamamoto, Y. (1972). Plant Cell Physiol. **13**, 477.
Huzisige, H., Usiyama, H., Kikuti, T., and Azi, T. (1969). Plant Cell Physiol. **10**, 441.
Izawa, S., and Good, N. E. (1966). Plant Physiol. **41**, 544.
Jacobi, G., and Lehmann, H. (1968). Z. Pflanzenphysiol. **59**, 457.
Kirk, J. T. O., and Tilney-Bassett, R. A. E. (1967). "The Plastids." Freeman, San Francisco, California.
Laetsch, W. M. (1971). In "Photosynthesis and Photorespiration" (M. D. Hatch, C. B. Osmond, and R. O. Slatyer, eds.), pp. 323–349. Wiley (Interscience), New York.
Levine, R. P. (1969). Annu. Rev. Plant Physiol. **20**, 523.
Mühlethaler, K. (1972). Abstr. Int. Congr. Photobiol. 6th, 1972 p. 028.
Murata, N. (1969). Biochim. Biophys. Acta **189**, 171.
Myers, J., and French, C. S. (1960). J. Gen. Physiol. **43**, 723.
Park, R. B., Steinback, K. E., and Sane, P. V. (1971). Biochim. Biophys. Acta **253**, 204.
Punnett, T. (1971). Science **171**, 284.

Radunz, A., Schmid, G. H., and Menke, W. (1971). *Z. Naturforsch.* B **26,** 435.
Reeves, S. G., and Hall, D. O. (1972). *Abstr. Int. Congr. Photobiol., 6th, 1972* p. 246.
Remy, R. (1971). *FEBS Lett.* **13,** 313.
Sane, P. V., Goodchild, D. J., and Park, R. B. (1970). *Biochim. Biophys. Acta* **216,** 162.
San Pietro, A., ed. (1971). "Methods in Enzymology," Vol. 23, Part A. Academic Press, New York.
San Pietro, A., ed. (1972). "Methods in Enzymology," Vol. 24, Part B. Academic Press, New York.
Sauer, K., and Park, R. B. (1965). *Biochemistry* **4,** 2791.
Schwelitz, F. D., Dilley, R. A., and Crane, F. L. (1972). *Plant Physiol.* **50,** 166.
Stubbe, W. (1960). *Z. Bot.* **48,** 191.
Vernon, L. P., and Shaw, E. R. (1969). *Plant Physiol.* **44,** 1645.
Vernon, L. P., Shaw, E. R., Ogawa, T., and Raveed, D. (1971). *Photochem. Photobiol.* **14,** 343.
Vernon, L. P., Klein, S., White, F. G., Shaw, E. R., and Mayne, B. C. (1972). *Proc. Int. Congr. Photosyn. Res., 2nd, 1971* Vol. 1, p. 801.
Warden, J. T., and Bolton, J. R. (1972). *J. Amer. Chem. Soc.* **94,** 4351.
Weaver, E. C., and Weaver, H. (1972). *In* "Photophysiology" (A. C. Giese, ed.), Vol. 7, pp. 1–32. Academic Press, New York.
Weier, T. E., Stocking, C. R., and Shumway, L. K. (1967). *Brookhaven Symp. Biol.* **19,** 353.
Wessels, J. S. C., and Voorn, G. (1972). *Proc. Int. Congr. Photosyn. Res., 2nd, 1971* Vol. 1, p. 833.
Wessels, J. S. C., Van Alphen-Van Waveren, O., and Voorn, G. (1973). *Biochim. Biophys. Acta* **292,** 741.
Wiessner, W., and Amelunxen, F. (1969). *Arch. Mikrobiol.* **67,** 357.

Chapter 4

THE ROLE OF CATION FLUXES IN CHLOROPLAST ACTIVITY

Geoffrey Hind
Biology Department, Brookhaven National Laboratory, Upton, New York

Richard E. McCarty
*Section of Biochemistry, Molecular and Cell Biology,
Cornell University, Ithaca, New York*

1. General Introduction 114
2. The Measurement of Ion Fluxes in Organelle Suspensions 114
   2.1 Introduction 114
   2.2 Electrometer Design 115
   2.3 Kinetic Characteristics of Electrodes 116
   2.4 Electrode Sensitivity, Specificity, and Noise 118
   2.5 Reference Electrodes 120
   2.6 Monitoring Optical Density 121
3. Relation of Cation Fluxes to Energy Transduction 123
   3.1 Introduction 123
   3.2 The Nature and Magnitude of the Electrochemical Gradient in H^+ . 124
   3.3 Is the Magnitude of the H^+ Electrochemical Gradient Sufficient to Drive ATP Formation? 135
   3.4 H^+ Uptake and Postillumination ATP Synthesis 136
   3.5 The Interaction of H^+ Uptake with Phosphorylation and Electron Flow 138
   3.6 Role of Coupling Factors in H^+ Uptake 139
   3.7 Concluding Statement 141
4. Cation Functions in Catalysis and Control 142
   4.1 Background 142
   4.2 Fluorescence Lowering by Cofactor-Mediated Cyclic Electron Flow . 143
   4.3 Regulation of Energy Distribution by Cations 144
   4.4 Enhancement and Divalent Cations 145
   4.5 Coupling of Electron Flow at Low Salt Concentration . . . 148
   4.6 Summary 148
5. Mechanisms for Possible Control of Photosynthesis by Mg Ion . . 148
   5.1 Mg^{2+}-Controlled Enhancement 148
   5.2 Control of Carbon Pathway Enzymes 149
   5.3 Mg Ion Fluxes in Isolated Chloroplasts 150

5.4 Regulation in the Green Cell 152
References 153

1. General Introduction

The discovery of H⁺ uptake in illuminated chloroplasts by Jagendorf and Hind (1963) and the proposal of the chemiosmotic hypothesis of energy conservation by Mitchell (1961, 1966) provided the framework for new experimental approaches in the investigation of chloroplast activities. Certainly, the heaviest emphasis has been placed on the role of H⁺ uptake in the mechanism of photosynthetic phosphorylation. More recently, however, it has been found that divalent cations (especially Mg^{2+}) have profound effects on many chloroplast activities. Although the full importance of ion fluxes in photosynthesis will not be realized for some time, it appears that ion gradients across chloroplast membranes are a form of energy storage and, further, that ion fluxes may contribute to the regulation of photosynthetic activity.

To avoid confusion, it is important to emphasize that the ion fluxes described in this chapter take place across the inner (thylakoid) chloroplast membrane system. The outer chloroplast membranes are readily destroyed during the isolation of chloroplasts unless special precautions are taken. The "chloroplast" used by most workers in the field consists of the inner membrane freed of its outer membranes and stroma.

The literature on ion fluxes in chloroplasts is voluminous and diverse. This article is not a review of the entire field. Rather, we have chosen to emphasize certain aspects of the problem; therefore, many areas, such as the mechanism of H⁺ uptake, anion fluxes, and light-induced volume changes associated with ion fluxes, have not been treated here.

2. The Measurement of Ion Fluxes in Organelle Suspensions

2.1 Introduction

A survey of present understanding of the interaction between ions and chloroplast membranes would not be complete without discussion of advances in ion-specific electrode technology and instrumentation which have recently broadened the scope of ion flux studies and, in particular, of multiparameter analysis as pioneered by Pressman (1967). It is also appropriate to explore here the special problems which result, in photosynthetic systems, from the use of actinic irradiation to drive electron flow and the associated ion movements. This section will otherwise be generally applicable to membrane systems in suspension.

The theory and applications of ion-selective electrodes, and related topics, have been comprehensively reviewed by Buck (1972) and Koryta (1972).

2.2 Electrometer Design

Ion-specific electrodes, including those having glass membranes (for example the pH electrode) are in effect ion-exchange systems as distinct from the reversible redox couples such as Ag/AgCl or Hg/Hg_2SO_4 used as reference half-cells. In contrast to these reference electrodes, membrane electrodes have very high impedance due to the limited availability of charge carriers in the matrix, so that an amplifier used to determine the membrane potential must draw low input current relative to that provided by the membrane.

The vibrating-reed electrometer was the instrument of choice in such applications until recently, when a variety of solid state devices with adequate, if not equal, performance became available. Table I summarizes the relative cost and performance of a varactor bridge electrometer amplifier; a small junction, high impedance input, field effect transistor (FET)* operational amplifier; a laboratory pH meter with FET input, and a research grade vibrating-reed electrometer. Besides affording high reliability, the two amplifiers are economical and simple to incorporate into specially designed circuits such as will be described later. Design flexibility allows the matching of amplifier output to the recorder or converter, and the selection of time constants and gains appropriate to the signal quality. The time constants of laboratory pH meters are too large for kinetic experiments (Izawa and Hind, 1967).

Ion-selective electrodes have impedances in the range 5–800 MΩ (Buck and Krull, 1968) and will need, in our experience, to register changes as small as 0.02 mV. The bias current of an FET operational amplifier such as Analog Devices 42-L is 0.15 pA, so that unless the electrode impedance is known to be below 100 MΩ the varactor bridge amplifier would be a safer choice. In this respect, attention is called to the marked dependence of glass membrane impedances on temperature (Buck and Krull, 1968; Eckfeldt and Perley, 1951; Bates, 1964); for example, Buck and Krull showed that the Radiometer G202B pH electrode had

---

* The nonstandard abbreviations used are: FET, field effect transistor; MOS, metal oxide silicon; ΔpH, the pH differential across chloroplast membranes; 9-AA, 9-aminoacridine; Δψ, the electric potential difference across chloroplast membranes; phenyldicarbaundecaborane (PCB⁻); CF₁, a coupling factor for photophosphorylation; NEM, N-ethylmaleimide; DCMU, 3-(3,4-dichlorophenyl)-1,1-dimethyl urea; PMS, N-methylphenazonium methosulfate; CCCP, carbonyl cyanide m-chlorophenylhydrazone; DAD, diaminodurene; M^{2+}, divalent cation; M^+, monovalent cation; DCIP, 2,6-dichlorophenolindophenol.

TABLE I
Comparative Cost and Performance of Electrometer Devices[a]

Device	Input impedance	Bias current (pA)	Frequency response (Hz)	Cost
Vibrating reed electrometer	10^{16} to 10^{17}	<0.005	10–100	High
Varactor bridge operational amplifier	10^{14}	0.01	10	Low
FET-input operational amplifier	10^{13}	0.1–0.5	10^3 to 10^5	Low
FET-input pH meter	10^{12} to 10^{13}	—	<1	Medium

[a] Frequency responses are the average range of maximum values at full power. Even with additional necessary circuitry and power supplies, the operational amplifiers are lowest in cost.

membrane resistances of 567, 235, and 88 MΩ at 15, 25, and 35°C, respectively. Loss of sensitivity with decreasing temperature may be experienced under marginal operating conditions. Circuits for the determination of electrode resistance and impedance are available (Brand and Rechnitz, 1969; Buck and Krull, 1968; Krebs, 1972).

Electrometer amplifiers with metal oxide silicon (MOS) FET input stages are available commercially, and theoretically they combine the extremely low bias currents of the varactor bridge with the fast response and freedom from offset voltage drift which are typical of FET inputs. Warm-up drifts are more severe with MOS FET stages, however, and the gate itself is prone to destruction by voltage transients. Protection of the gate with zener diodes is practiced, but leakage paths are thereby introduced, and the input impedance may fall to 10^{12} Ω.

Figure 1 shows an electrometer configuration in use at Brookhaven National Laboratory. Five electrode channels are monitored with an 8-channel time-sharing recorder (Brush 816 Multipoint Recorder, Gould Inc., Cleveland, Ohio), which has full scale sensitivity of ±2.5 V. The highest amplifier gain corresponds to 1.0 mV full scale. In three channels, the first stage is an Analog Devices 42L FET input operational amplifier, but in the others 311J varactor bridge electrometers are used in conjunction with glass electrodes.

2.3 Kinetic Characteristics of Electrodes

The membrane resistance and capacitance determine the electrode response time, which may be very brief. pH and cation-sensitive glass electrodes can respond within 2 msec (Buck and Krull, 1968; Johansson and Norberg, 1968), though values between 25 and 100 msec are more

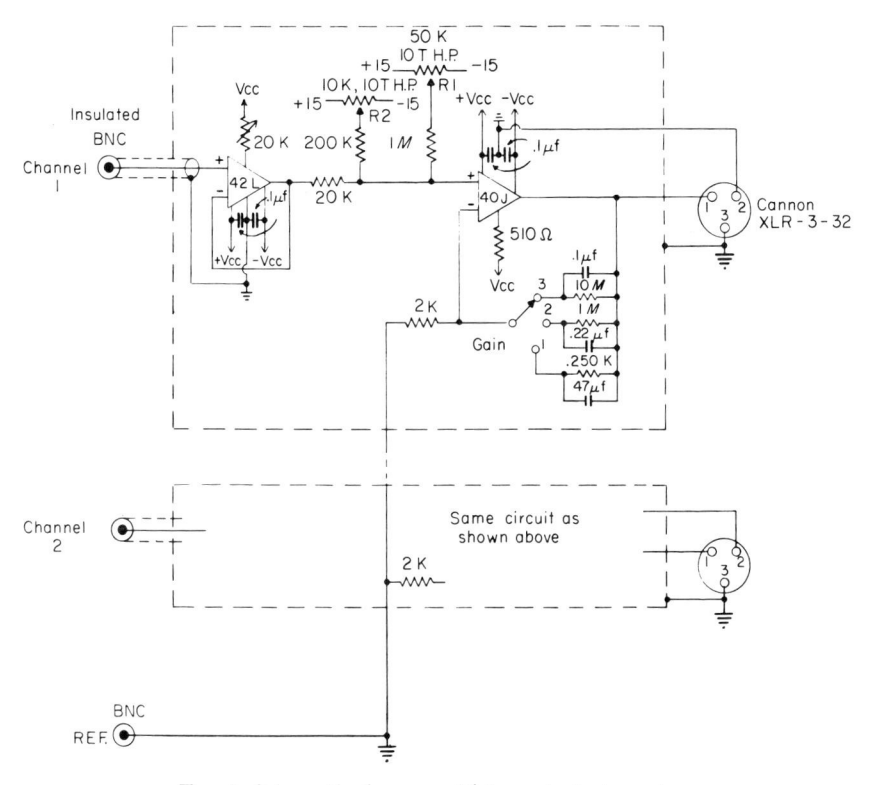

FIG. 1. Schematic for a multichannel electrometer.

typical at 25°C. Liquid ion-exchange electrodes, on the other hand, may require several seconds to attain a stable potential, so for these a continuous flow technique is necessary for fast kinetic studies (Fleet and Rechnitz, 1970; Izawa and Hind, 1967; Thompson and Rechnitz, 1972). Stable readings with liquid membranes are reached more rapidly as the activity of the specific ion rises (Lal and Christian, 1970). For example, the Orion K^+ electrode settles within 1 minute in 10 mM KCl but requires over 10 minutes in 0.1 mM salt.

The response of both glass and liquid membranes can be slowed by interfering ions (Rechnitz, 1970; Rechnitz and Kugler, 1967). Thus, an Na^+-selective glass electrode came to equilibrium, after an addition of Na^+, about 100 times more slowly in the presence of a 5-fold excess of K^+ than in its absence. Similarly, a slight excess of Mg^{2+} increased the response time of the Ca^{2+} liquid membrane electrode from about 10 seconds to above 30 seconds. Even the former value would be too slow for monitoring fluxes in chloroplast suspensions. In our experience, however, liquid membrane electrodes respond much more rapidly to small activity changes generated within a suspension than would be predicted from the quoted response time, obtained by stepwise addition or dilution.

Associated with the slowing effect of interfering ions is a transient potential spike observed upon addition, for example, of K^+ or NH_4^+ to an Na^+-selective electrode; the potential subsequently equilibrating to the value predicted from the ion selectivity constants. Rechnitz and Kugler (1967) point out that the early time course of the transient is the same as that for addition of the ion to which the electrode is nominally selective. Furthermore, transients arise even upon addition of divalent ions to K^+- or Na^+-selective electrodes. It is important, therefore, to ignore transients and allow ample time for equilibration when calibrating an electrode for the effect of an interfering ion. Transient responses to pH changes have been observed with the Orion Ca^{2+} electrode (Bagg and Vinen, 1972), but in our experience, do not arise from small pH changes in the physiological range.

2.4 Electrode Sensitivity, Specificity, and Noise

2.4.1 Effects of Stirring

Stirring is essential in studies with organelle suspensions. It prevents settling, mixes added reagents, minimizes thermal and illumination gradients, and replenishes the portion of sample in contact with the electrodes. Without stirring, light-induced ion fluxes monitored by the glass electrodes appear to be slower and less extensive, whereas with liquid membrane electrodes the response may be entirely lost.

The noise of glass electrodes is not much increased by stirring the sample, owing perhaps to membrane rigidity and large surface area (1–3 cm^2). With liquid membrane electrodes such as the Orion Ca^{2+} electrode, the sensing element is small and flexible and stirrer noise of 1.0 mV may be experienced. This may be diminished by randomizing turbulence in the following ways. (1) by using a small stirrer rotating rapidly, (2) by attention to sample geometry and electrode placement, (3) by increasing the functional area of the ion exchange membrane. Simple electrode designs which allow this have been described (Ross, 1969; Moore, 1969).

2.4.2 Interaction with the Sample

Chloroplasts and mitochondria produce or consume reductants which may interact with an electrode surface capable of undergoing redox changes. Thus, AgCl coated on Ag wire or suspended in an inert matrix will very quickly become an Ag/Ag^+ redox electrode over and above its intended Cl$^-$ selectivity. Solid state Cl$^-$ electrodes are generally advertised as being "resistant to *oxidizing* agents." There is no reason to interpret the light-induced potential changes noted by Walz and Avron (1969) using chloridized Ag wire in a chloroplast suspension, as anything other than shifts in the redox state of the mediator (pyocyanine).

Chloroplasts directly interfere with some liquid membranes by sticking at the interface and diminishing the electrode response. This effect is related to the polarity of the charge carrier in the organic phase, for the Orion Cl⁻-selective membrane rapidly becomes green, whereas the Orion divalent metal electrode is unaffected in chloroplast suspensions. This observation is in accord with the notion that chloroplasts carry a fixed negative charge (Nobel and Mel, 1966). At the expense of slowing the response time, this problem can be overcome by covering the electrode tip with stretched dialysis membrane (Ross, 1969).

Conversely, interference with the properties of the organelle membrane by the organic phase and its components has to be considered when liquid membrane electrodes are used. It remains to be determined, for example, how much valinomycin reaches a suspension of organelles in contact with K⁺-selective probes, and what effect the ion exchangers used in electrodes selective for other ions might have on energy conversion reactions of mitochondria and chloroplasts. Since contamination by leaking ion exchanger will be most serious in the immediate vicinity of the membrane, and since this is the region of the sample in which the ion activity is determined, the need for thorough stirring must be reiterated.

2.4.3 Lack of Specificity

The glass pH electrode and the liquid membrane K⁺ electrode are unusual in being highly specific for H⁺ and K⁺, respectively. More often, interference by some other ion in an organelle suspension will present the experimenter with the task of somehow correcting the data to reveal the true ion fluxes. Pressman (1967) has outlined the equations which govern electrode behavior in the presence of an interfering ion, and the technique of correcting by solution of the simultaneous equations. While this method lends itself well to digital analysis, analog correction is possible and has been used with success in the circuit shown in Fig. 1. Correction rests on the assumption that the pH electrode is specific for H⁺ within the physiological range, and the fact that H⁺ is the most frequently interfering ion (OH⁻ interferes with the Orion Cl⁻ electrode). A compensation current is drawn from the output of the first or second amplifier in the pH channel, and is injected through a 10-turn potentiometer into the (−) input of the second stage of the channel to be compensated. Each such channel has its own potentiometer, and these are adjusted for each sample type before the experiment so that small amounts of added base or acid produce no net response (a transient may persist). Care should be taken not to change the gain of the pH channel by withdrawing excessive correction currents.

The addition of buffer to decrease pH changes to a small but measurable level is recommended to minimize interference, and to suppress the effects of changes in organelle buffering capacity (Polya and Jagendorf, 1969).

2.4.4 LIGHT ARTIFACTS

Actinic irradiation produces two types of spurious signals which can be eliminated as follows: (1) a fast transient due to photosensitivity of Ag/AgCl internal junctions in many specific ion electrodes—this is avoided by the use of red actinic light; (2) a slow transient due to heating of the sample by absorption of visible and infrared light—this can be prevented by filtering out infrared and by water-jacketing the sample. Detection of these artifacts is possible only in chloroplast samples with fully inhibited and uncoupled electron flow.

2.5 Reference Electrodes

In work with organelle suspensions, electrodes are used not to determine the absolute ion activities, but the small changes over a narrow activity range which accompany energization or relaxation of the biological membrane. Since the junction potential developed at the reference electrode will be essentially constant for a given sample, and will be taken into account during subsequent calibration, it might seem that the junction potential need not be the prime concern in reference electrode design, except insofar as uneven stirring will affect it. However, stirrer noise is introduced primarily at the reference rather than the specific ion electrode, and the considerations discussed in Section 2.4.1 apply. Furthermore, reference electrode filling solutions should for this reason be equitransferent and at least tenfold higher in ionic strength than the sample.

The gradual leakage of electrolyte into the sample from the reference electrode can be a major source of drift in specific ion monitoring. We have been unsuccessful in locating a salt which has adequate conductivity and biological inertness to serve as a bridge between the sample and a calomel electrode. However, satisfactory results are given by a bridge of 150 mM KCl in 2% agar leading to a reservoir of 150 mM KCl into which a Ag/AgCl electrode is dipped. Increasing the surface area of the junction will diminish both stirrer noise and electrode resistance, without introducing serious drift at the K^+ and Cl^- ion specific electrodes.

An alternative approach is to use a high impedance specific ion electrode as the reference. In this way, the problem of salt leakage is avoided, but is replaced by that of finding an ion which undergoes no appreciable activity change in the system under study. The electrode must also be free from interference by ions of interest. Brand and Rechnitz (1970a,b) have suggested a circuit design for this differential technique. They observed that when magnetic stirring was used, a solution ground became necessary. Another differential approach is that of Luzzana et al. (1971) in which two identical (pH) electrodes are placed in two sample vessels connected by a buffer-filled bridge. Ion fluxes are induced in one sample only, and the amplifier detects the difference signal which results. Here again, a solution ground was required.

Differential techniques such as the above, offer advantages if a passive ion is available (and an appropriate electrode) or where, in the latter case, stirrer noise is negligible. The requirement for a solution ground when high impedance reference electrodes are employed is a reminder that this need is satisfied gratuitously by conventional reference half-cells.

2.6 Monitoring Optical Density

The value of ion flux studies can be substantially enhanced if the correlative rate of electron transport and the change in particle volume due to water redistribution are determined. If a closed reaction chamber is available, oxygen production or consumption can be followed polarographically; however, a Clark-type probe is needed so that the sample is electrically insulated from the electrodes. The rate of electron transfer to and from colored substances, such as ferricyanide, can be observed spectrophotometrically, as can changes in turbidity which reflect shrinkage or swelling of the organelle. Almost any spectrophotometer with a modulated measuring beam can be adapted for studying chloroplast reactions, by provision of actinic cross-illumination and appropriate complementary filters for protection of the photomultiplier. However, sample size or geometry may render operation within the sample compartment impossible. Accordingly, the measuring beam and detector must be interfaced remotely to the sample by means of light pipes.

Figure 2 shows an optical densitometer circuit for use in conjunction with a Beckman DU spectrophotometer which, owing to modular design, adapts readily to the interposition of light pipes and a remote sample between the monochromator and photomultiplier. The light is

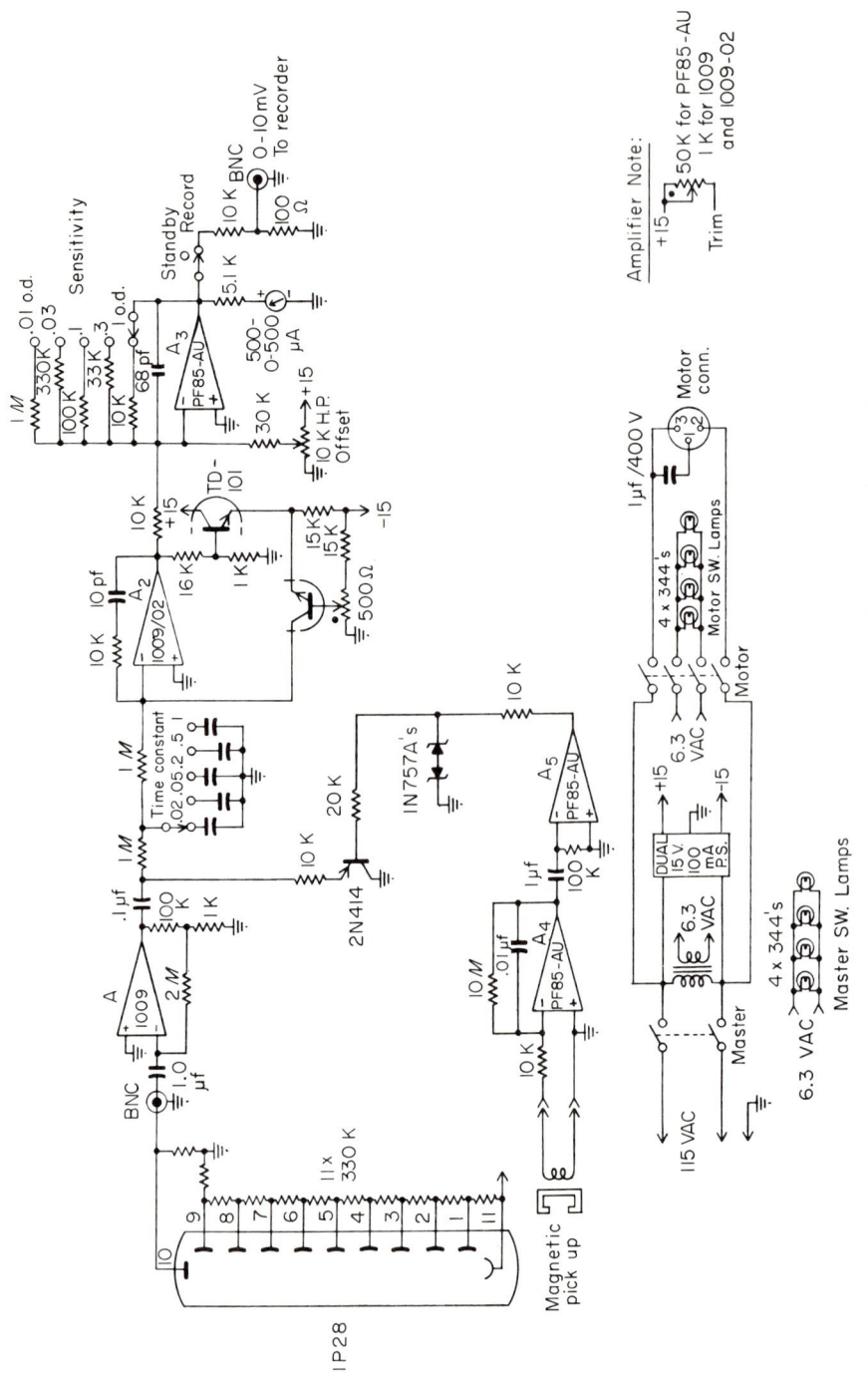

FIG. 2. Schematic for an optical densitometer circuit.

chopped at 180 Hz, on leaving the monochromator, by passage through a vertical hollow cylinder rotating at 60 Hz and having three pairs of opposed vertical slits around its circumference. A magnetic pickup close to the slits supplies a reference signal for the rectifier. This instrument has performed well, with or without light pipes, and with cross-illumination serves as a kinetic spectrophotometer for electron transport rate studies in chloroplasts.

Interfacing fiber optics to a sample cell is not easy, however. French *et al.* (1970) attempted, without success, to excite and collect fluorescence in the sample chamber of a fluorimeter by means of fiber optics. The problem lay principally in collimating the light from the collector bundle before presentation to the analyzing monochromator, since light emerging from fiber optic bundles scatters at a wide solid angle. In the densitometer described above, the collector bundle emits directly into the photomultiplier (through blocking filters) so the problem is confined to the sample region itself. It is manifestly not possible to measure turbidity changes if the measuring beam is initially randomly scattered; however, by means of condenser and field lenses, collimation can be achieved. The sensitivity of the densitometer to turbidity changes can be further increased by placing narrow slits or apertures in the light path between the sample and both the emitting and collecting fiber bundles (Shibata, 1959).

3. Relation of Cation Fluxes to Energy Transduction

3.1 Introduction

The theory that ion movements may play a fundamental role in energy conservation in mitochondria, chloroplasts, and chromatophores has been most explicitly detailed by Mitchell (1961, 1966). Mitchell's chemiosmotic theory has caused at least a minor revolution in research in photophosphorylation. It is now clear that H^+ uptake is closely related to phosphorylation and that large H^+ gradients can be established across chloroplast or chromatophore membranes. Furthermore, the experiments of Jagendorf and his co-workers (Jagendorf and Uribe, 1966) established that a pH gradient can be the driving force for ATP synthesis in chloroplasts. However, it remains to be resolved whether H^+ uptake is a direct consequence of electron flow as suggested by Mitchell or whether H^+ uptake is driven at the expense of a high-energy intermediate. In Mitchell's view, H^+ movement is the primary act in energy conservation and the H^+ electrochemical gradient the driving force for ATP synthesis. Proponents of the chemical hypothesis, however, hold that the primary act in energy conservation is the formation of a high-energy intermediate that can drive either phosphorylation or H⁺ movements.

Since there appears to be no compelling evidence that would entirely rule out either the chemical or chemiosmotic hypothesis, we have chosen not to dwell upon this issue here. Furthermore, the relation of ion fluxes to phosphorylation has been extensively discussed from several viewpoints (Dilley, 1971; Walker and Crofts, 1970; Witt, 1971; Avron and Neumann, 1968). Therefore, this section will be concerned with some of the more recent developments treated in some detail rather than with a broad overview of the field.

3.2 The Nature and Magnitude of the Electrochemical Gradient in H⁺

3.2.1 H⁺ Concentration Gradients

Although Neumann and Jagendorf (1964; Jagendorf and Neumann, 1965) established several years ago that illuminated chloroplasts catalyze a rapid and relatively massive H⁺ uptake, the magnitude of the H⁺ concentration gradient (ΔpH) has only recently been measured. Rumberg and Siggel (1969) were the first to attempt to measure ΔpH. They determined the relaxation time of absorption changes at 703 nm [attributable to chlorophyll a_1^+ (P700)], after turning off the actinic illumination as a function of pH in the presence of a high concentration of gramicidin D. Since gramicidin D seems to induce high permeability to H⁺, it was assumed that the internal pH was equivalent to the external pH. The relaxation time of the 703 nm absorption change was then determined at pH 8 in coupled chloroplasts as a function of the illumination time. The reciprocal of the half time fell with increasing periods of illumination up to 2 seconds. The reciprocal half time in coupled chloroplasts after 2 seconds of illumination at pH 8 was equivalent to that in gramicidin D uncoupled chloroplasts at pH 5.1. Assuming that internal pH controls the rate of electron flow, this would mean that ΔpH in chloroplasts can be 2.9 units.

Rottenberg et al. (1971, 1972) used a microcentrifugation technique to determine the distribution of methylamine-¹⁴C between the internal volume of chloroplasts and the suspending medium. Total water in the chloroplast pellets was estimated by determining the amount of ³H₂O trapped in the pellets whereas water external to the chloroplasts was determined with sorbitol-¹⁴C. From these data, the internal volume could be calculated, allowing the estimation of internal methylamine-¹⁴C concentrations. Since the external methylamine-¹⁴C concentrations could also be determined, the methylamine concentration gradients could be calculated. Gaensslen and McCarty (1971) used a different microcen-

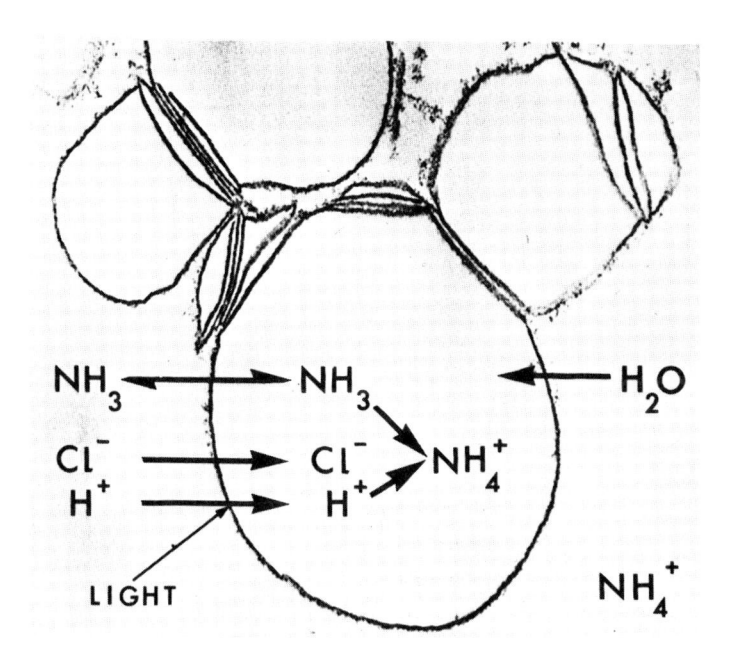

FIG. 3. Light-induced ion fluxes in a chloroplast uncoupled by NH₄Cl. The arrows would be more or less reversed in direction in the ensuing dark, depending on the amount of water which had entered the thylakoid. ×85,000.

trifugation method to study ethylamine uptake. Under ideal conditions, these gradients should be equivalent to those of H⁺. Crofts (1967) showed that amines permeate chloroplasts in the uncharged form. In the acid interior of illuminated chloroplasts, uncharged amines would be rapidly protonated resulting in a decrease in the uncharged amine concentration inside the chloroplasts and, therefore, further amine uptake (Fig. 3). Amine uptake should continue until, at the steady state, the concentration gradient of amine cation approximates that of H⁺. The relationship between the concentration gradient of H⁺ and that of amine cation may, in the case of a monobasic amine, be readily derived. At the steady state, the pH of the internal and external phases of a chloroplast suspension may be defined by the Henderson-Hasselbach equation:

$$\mathrm{pH_{in}} = \mathrm{p}K_a + \log \frac{[A]_{in}}{[A^+]_{in}} \quad (1)$$

and

$$\mathrm{pH_{out}} = \mathrm{p}K_a + \log \frac{[A]_{out}}{[A^+]_{out}} \quad (2)$$

where "in" refers to the internal osmotic compartment of the chloroplasts and "out" to the suspending medium. [A] is the concentration of unprotonated amine and [A⁺], the concentration of amine cation. Assuming that the pK_a of the amine is the same in the two compartments, ΔpH may be expressed

$$pH_{out} - pH_{in} = \Delta pH = \log[A]_{out} - \log[A^+]_{out} - \log[A]_{in} + \log[A^+]_{in} \quad (3)$$

Since chloroplasts seem to be freely permeable to unprotonated amine, $[A]_{in}$ should be equal to $[A]_{out}$. Therefore, Eq. (3) can be reduced to

$$\Delta pH = \log[A^+]_{in} - \log[A^+]_{out} = \log \frac{[A^+]_{in}}{[A^+]_{out}} \quad (4)$$

Since ΔpH is equivalent to $\log[H^+]_{in} - \log[H^+]_{out}$, it follows that

$$\frac{[H^+]_{in}}{[H^+]_{out}} = \frac{[A^+]_{in}}{[A^+]_{out}} \quad (5)$$

It should be emphasized that this will be true only when the permeability of chloroplast membranes to the amine cation is very much less than that to the unprotonated species. Furthermore, a low amine concentration must be used to avoid uncoupling by the amine. If, for example, it is assumed that ΔpH is 3 units, chloroplasts illuminated in the presence of an amine solution which contained 1 mM amine cation would attempt to establish an amine concentration gradient of 1000-fold. Even if the external amine cation concentration fell to 0.5 mM, the internal amine cation concentration would have to be 0.5 M to give a 1000-fold gradient, and it is likely that the chloroplasts would be damaged under these conditions.

ΔpH has also been estimated by determining the uptake of NH_4^+ in illuminated chloroplasts with an electrode sensitive to ammonium ions (Rottenberg and Grunwald, 1972). In addition, the distribution of fluorescent amines has been used in the determination of ΔpH (Schuldiner et al., 1972). Kraayenhof (1970) showed that the quenching of atebrine fluorescence was related to the high energy state. Since atebrine fluorescence quenching in chromatophores was sensitive to reagents, such as nigericin and NH_4Cl, which inhibit H^+ uptake but not phosphorylation, it is apparent that quenching is related to H^+ uptake (Gromet-Elhanan, 1971). Schuldiner and Avron (1971) suggested that the quenching in chloroplasts was the result of atebrine accumulation via the mechanism proposed by Crofts (1967) for amine uptake. In an elegant paper, Deamer et al. (1972) reported that the quenching of the fluorescence of the monobasic amine 9-aminoacridine (9-AA) was directly related to the magnitude of the pH gradient established across the membranes of phospholipid vesicles. It should be pointed out, however, that it is difficult to account for the strong quenching of 9-AA fluorescence merely on the basis of its uptake into chloroplasts. Schuldiner and Avron (1971) suggested that the fluorescence of the accumulated atebrine could be absorbed by chloroplast pigments, thereby resulting in an apparent reduced emission of atebrine fluorescence. Since 9-AA fluorescence is strongly quenched in pigment-free vesicles (Deamer et al., 1972), this explanation does not appear to be too likely. Self-quenching could also occur since the internal concentration of 9-AA in illuminated chloroplasts could be in the millimolar range. Deamer et al. (1972) showed the 9-AA fluorescence was quenched by organic buffers. If there is an energy-dependent change in chloroplast membranes, it is possible that there is an increased interaction between 9-AA and a component of the membrane which causes quenching. This possibility is given some credence by the observations of Heath and Hind (1972), who showed that the apparent binding of the dye ethyl red to chloroplast membranes is affected by illumination. Thus, although 9-AA fluorescence quenching does appear to reflect pH changes, the mechanism of this effect is obscure.

In general, it is quite difficult to distinguish between uptake of a substance and its binding to a membrane. From the swelling which occurs when chloroplasts are illuminated in the presence of high amine concentrations (Izawa, 1965; Gaensslen and McCarty, 1971), it is fair to assume that the amine is accumulated in the internal osmotic compartment of the chloroplasts. However, ΔpH must be determined at relatively low amine concentrations to obviate uncoupling. Therefore, a significant part of the apparent accumulation of the amine cation could be in fact energy-dependent binding to the membrane. If this occurs, ΔpH calculated from the distribution of amines may be too high.

The complexity of the chloroplast may also complicate the determination of ΔpH. For example Kahn (1971) reported evidence that Euglena chloroplasts catalyze both H^+ uptake and release upon illumination. In the presence of high concentrations of amines which inhibited H^+ uptake, an H^+ release was observed in the light. This H^+ release was sensitive to dinitrophenol and, in the absence of amines, dinitrophenol stimulated H^+ uptake. Although these phenomena are not generally observed in chloroplasts from higher plants, it is still possible that chloroplasts contain two compartments. One of these compartments may acidify upon illumination whereas the other may become more basic. The ΔpH between the two compartments could be much greater than the difference between the average internal pH and the external pH.

At external pH values of 8–9, ΔpH in illuminated chloroplasts was found to range from 2.1 to 3.5, depending on the method used (Table

II). The lower values were obtained by the methyl- or ethylamine uptake method. These estimates are probably low since some of the accumulated amine is likely to be lost during centrifugation of the chloroplasts. The higher values of ΔpH were calculated either from NH_4^+ uptake data or 9-AA fluorescence quenching data. By all three methods, ΔpH in illuminated chloroplasts was found to be strikingly dependent on external pH. Whereas H^+ uptake is maximal at about pH 6.5 (Neumann and Jagendorf, 1964), ΔpH was found to be maximal between pH 8 and 9 (Rottenberg et al., 1972; Rottenberg and Grunwald, 1972; Schuldiner et al., 1972) and to decrease with decreasing pH. Over an external pH range of 6–7.5 internal pH was nearly constant (Rottenberg et al., 1972). These results led Rottenberg et al. (1972) to suggest that chloroplasts contain a strong internal buffering agent with a pK_a in the range of 4.5–5.0. Thus, although more H^+ is accumulated at pH 6 than at pH 8, H^+ uptake at the lower pH value was thought to generate a low ΔpH due to strong buffering.

However, the pH dependence of ΔpH may in part be a result of the pH dependence of amine uptake. Until recently, ΔpH had been determined only with amines of high pK_a (9.3–10.7). With amines of high pK_a, the concentration of unprotonated amine decreases 10-fold for every decrease of 1 pH unit. In a 0.1 mM methylamine solution, the concentration used by Rottenberg et al. (1972) to estimate ΔpH, the unprotonated amine concentration at pH 8 would be about 200 nM

TABLE II
ESTIMATES OF ΔpH IN ILLUMINATED CHLOROPLASTS

Method used	$pH_{(out)}$	$pH_{(in)}$	ΔpH	Reference
Methylamine uptake	6.5	5.1	1.4	Rottenberg et al. (1972)
	7.0	5.3	1.7	
	8.0	5.9	2.1	
	8.5	6.1	2.4	
Quenching of 9-amino-	6.5	4.5	2.0	Schuldiner et al. (1972)
acridine fluorescence	7.0	4.4	2.6	
	8.0	4.9	3.1	
	8.5	5.2	3.3	
NH_4^+	6.0	4.3	1.7	Rottenberg and Grunwald (1972)
	7.0	4.5	2.5	
	8.0	4.5	3.5	
	9.0	5.5	3.5	
Aniline uptake	6.5	3.7	2.8	Portis and McCarty (1972)
	7.1	4.0	3.1	
	8.0	5.0	3.0	
Electron flow	8.0	5.1	2.9	Rumberg and Siggel (1969)

whereas at pH 6.5, it would be only 9 nM. At very low concentrations of the unprotonated amine, it may be no longer valid to assume that the permeability of the charged amine is negligible with respect to that of the uncharged amine. It may be calculated, for example, that the internal methylamine cation concentration in chloroplasts illuminated in the presence of 0.1 mM methylamine of pH 6.5 can be over a million times that of the external uncharged amine. Therefore, estimates of ΔpH with amines of high pK may be erroneously low. This idea is supported by the observation (A. R. Portis and R. E. McCarty, unpublished observations, 1972) that ΔpH determined from the distribution of charged aniline was essentially independent of pH over the range of 7–8. Since aniline has a pK_a of 4.6, uncharged aniline is the predominant species in solution and its concentration would change little over the pH range of 7–8. Therefore, it is unlikely that aniline uptake would be limited by low uncharged aniline concentrations and, therefore, aniline distribution may be a more accurate reflection of the pH gradient at low external pH values.

In summary, ΔpH in illuminated chloroplasts, as determined from the distribution of amines or from electron transport measurements, is rather high, probably 3 units or higher. Therefore, it is clear that the H$^+$ concentration gradient must be a major form of energy storage in illuminated chloroplasts.

3.2.2 THE MEMBRANE POTENTIAL

The magnitude of the electric gradient ($\Delta\psi$) associated with H$^+$ uptake in chloroplasts is not as well established as that of ΔpH. The argument in favor of a relatively large $\Delta\psi$ in illuminated chloroplasts at the steady state rests primarily on the interpretation of experiments on the light-induced absorbance increase in the region of 515 nm. Duysens (1954) first discovered this absorption change in illuminated *Chlorella*. This phenomenon has been intensively studied in Witt's laboratory. Since their experiments have been recently reviewed in detail (Witt, 1971), only the essential features of their work will be discussed.

Rumberg *et al.* (1965) and Witt *et al.* (1965) showed that uncouplers such as ethanol or desaspidin abolish the 515 nm shift in chloroplasts and this observation has been confirmed in several laboratories. Although it was originally assumed that the 515 nm change was due to chlorophyll b, it is now clear that carotenoids (Strichartz, 1972) and probably chlorophyll a also contribute to the change. Multilayers of chlorophylls a plus b and carotenoids were found to undergo changes in their absorption when an electric field of 10^6 V/cm was applied

(Schmidt et al., 1971). The spectrum of these changes was similar to that of the rapid, light-induced change in chloroplasts. On the basis of this observation and on the observation that chloroplast-bound rhodamine B undergoes an absorption change on illumination (Emrich et al., 1969), it was proposed that the 515 nm change is an indicator of an electric field across chloroplast membranes. This conclusion is supported by the finding that uncouplers and ionophorous antibiotics such as valinomycin (Grünhagen and Witt, 1970) enhanced the dark decay rate of the flash-induced 515 nm change. By inducing high permeability to H^+ in the case of uncouplers or to K^+ in the case of valinomycin, the ability of the membrane to maintain a potential would be reduced.

The rise time of the 515 nm change is remarkably fast, possibly less than 20 nsec (Wolff et al., 1969). To explain this rapid rise, it was suggested that a primary photochemical reaction takes place across the membrane such that an electric field is established across the membrane. In time, the slower H^+ uptake is thought to occur, possibly via the mechanism proposed by Mitchell (1966) and, eventually, the field generated by the photochemical charge separation would be replaced with an H^+ electrical gradient. Other ions, such as K^+ or Cl^-, would then move in response to this potential.

The rate of the dark decay of the flash-induced 518 nm absorption change is influenced not only by ionophorous antibiotics, but also by phosphorylation (Rumberg and Siggel, 1968). Phosphorylation decreased the half-time of the decay by 2- to 5-fold. Furthermore, the energy transfer inhibitor phlorizin prevented this effect of ADP and P_i. Neumann et al. (1970), however, failed to observe an accelerated decay of the 518 nm shift due to phosphorylation, but they found that high (17 mM) concentrations of divalent cations did speed the decay. Witt (1971) in his review stated that Mg^{2+}, ADP, P_i, or phlorizin alone did not accelerate the decay as long as critical concentrations of these reagents were not exceeded.

In the view of the Berlin group, the 518 nm absorption change is a shift in the absorption spectrum of chloroplast pigments caused by an electric field across chloroplast membranes. This electric field is thought to be generated at least in part by electrogenic H^+ uptake. The decay rate of the 518 nm change is considered to be a reflection of the permeability of the membranes to H^+ and other ions. Based on approximations of the area, thickness and dielectric constant, Schliephake et al. (1968) calculated the capacitance of chloroplast membranes to be about 1 μF. Since the relationship between the 518 nm change and H^+ uptake appeared linear (Reinwald et al., 1968) and since one turnover of the electron transport chain should cause the transport of 2 charges, $\Delta\psi$ per turnover was calculated to be $2e/1~\mu F$ or about 50 mV. In the steady state, this value was estimated to be 100 mV.

The concept that light-induced shifts in the absorption spectrum of photosynthetic pigments is related to the generation of a membrane potential is supported by the observation that K^+ diffusion potentials can elicit similar changes in the dark. Fleischman and Clayton (1968) showed that the light-induced carotenoid band shift in chromatophores was closely related to the high energy state. Jackson and Crofts (1969) established K^+ diffusion potentials in the dark across the membranes of *Rhodopseudomonas spheroides* chromatophores by pulsing chromatophore suspensions with KCl in the presence of valinomycin. The shift in the carotenoid spectrum elicited by KCl addition was quite similar to that of the light-induced shift. Since the internal K^+ concentration was known, it was possible to calculate, from the Nernst equation, the magnitude of the diffusion potential. A linear relationship between the log of the external KCl concentration and the absorption change was found in accordance with the Nernst equation. Using this relationship, they calculated from the magnitude of the light-induced carotenoid shift a potential of 415 mV at the peak and 285 mV at the steady state. The validity of this calculation depends, of course, on the assumption that the light-induced carotenoid shift is caused primarily by a membrane potential resulting from an ion gradient.

Similar experiments with spinach chloroplasts have recently been reported by Strichartz and Chance (1972). A stopped flow instrument was used to mix chloroplasts with various salts in the dark and absorption changes were followed with a dual beam spectrophotometer. On mixing chloroplasts with 0.1 M KCl in the dark, a rise in absorption at 515 nm was observed with a half-time of around 0.6 second. The spectrum of the KCl-induced change was similar to that elicited by light except that the decrease in absorption around 475 nm was more pronounced in the case of the KCl-induced change. Valinomycin enhanced the rate and extent of the KCl-induced change, but did not affect that induced by NaCl. In the presence of valinomycin, a linear relationship between the magnitude of the absorbance change and the log of the KCl concentrations was found. From the light-induced absorption change and this relationship, a membrane potential of 30 mV was calculated, assuming an internal K^+ of 2 mM.

The addition of salts to chloroplasts causes an increase in the emission of delayed light. Barber and Kraan (1970) suggested that diffusion potentials, which would be formed on salt addition, could lower the activation energy barrier for light emission. Valinomycin enhanced the KCl-dependent increase in delayed light emission and

Barber (1972) calibrated this increase with respect to KCl concentration. Using this curve, he calculated that the valinomycin-sensitive portion of delayed light was equivalent to 75–105 mV.

Further evidence that H⁺ uptake is electrogenic has resulted from studies on the effects of ionophorous antibiotics on ion movements in chloroplasts, subchloroplast particles, and chromatophores. The rate and extent of light-dependent H⁺ uptake in chromatophores (Jackson et al., 1968) and subchloroplast particles (McCarty, 1970) are markedly enhanced by valinomycin in the presence of K⁺. This effect may be rationalized in terms of the known ability of valinomycin to promote the rapid permeation of K⁺. In the absence of valinomycin, H⁺ uptake should generate a membrane potential as long as the permeability of the membranes to other ions is low with respect to that of H⁺. This potential would limit H⁺ uptake, and its dissipation would increase the rate and extent of H⁺ flux. Valinomycin, by promoting K⁺ permeability, would allow K⁺ to act as a counterion for H⁺ uptake, resulting in the abolishment of the membrane potential and increased H⁺ uptake.

Nigericin and K⁺, on the other hand, was found by Shavit et al. (1968a) and by Thore et al. (1968) to inhibit H⁺ accumulation in chromatophores. Jackson et al. (1968), showed that nigericin appeared to catalyze an electrically neutral exchange of H⁺ for K⁺ so that the H⁺ gradient would be replaced by a K⁺ gradient. Thus, although H⁺ uptake was inhibited, the K⁺ gradient formed in the presence of nigericin would probably have both the electric and concentration components.

The effects of ionophorous antibiotics on phosphorylation in chromatophores and subchloroplast particles seem to be consistent with the notion that the capacity of the coupling membranes to maintain both an ion concentration gradient as well as an electric gradient must be destroyed to inhibit phosphorylation. For example, although neither valinomycin nor nigericin alone inhibited phosphorylation in chromatophores in the presence of K⁺, marked inhibition was observed in the presence of both antibiotics (Jackson et al., 1968). Since valinomycin would rapidly destroy the K⁺ gradient established in the presence of nigericin, the capacity of the chromatophores to maintain an ion gradient would be minimal.

The case for the existence of a membrane potential in chloroplasts, based on the response of phosphorylation and ion movements to ionophorous antibiotics, is not as strong as that for chromatophores. Antibiotics of the nigericin type are effective uncouplers of phosphorylation in chloroplasts even in the absence of valinomycin (Shavit et al., 1968b). This result could mean that dissipation of ΔpH is sufficient to inhibit phosphorylation in chloroplasts. However, although the K^+ uptake induced by nigericin in chromatophores appears to be electrically neutral, the ratio of K^+ to H^+ in chloroplasts was about 6:10 (Shavit et al., 1970). This result indicates that nigericin-induced K^+ uptake in chloroplasts is not electrically neutral and would at least partially dissipate a membrane potential. Furthermore, Degani and Shavit (1972) showed that low concentrations of nigericin-type ionophores can promote K^+ uptake in chloroplasts without a marked inhibition of phosphorylation. Under these conditions, valinomycin did inhibit phosphorylation and, therefore, a role of a membrane potential in chloroplast phosphorylation was postulated.

Although amines are not effective uncouplers of phosphorylation in chromatophores or subchloroplast particles, they inhibit phosphorylation in chloroplasts (Good, 1960). Since amine uptake via the mechanism proposed by Crofts (1967) should be electrically neutral, it has been assumed that ΔpH is the predominant part of the electrochemical H^+ gradient and that dissipation of ΔpH is sufficient to abolish phosphorylation. However, it should be pointed out that massive amounts of amines are accumulated by chloroplasts illuminated in the presence of uncoupling concentrations of amines (Gaensslen and McCarty, 1971). This uptake causes marked swelling of the chloroplasts (Izawa, 1965; Gaensslen and McCarty, 1971), and it is possible that swollen membranes become leaky to amine cations. If amine cations were to leak out of the chloroplasts, an electric gradient would be dissipated. Thus, a role for a membrane potential in phosphorylation in chloroplasts may not be ruled out simply because amines uncouple phosphorylation.

Attempts have been made to estimate the membrane potential by studying the distribution of ions across chloroplast and chromatophore membranes. The energy-dependent uptake of permeant anions, such as phenyldicarbaundecaborane (PCB^-), in chromatophores has been studied (Skulachev, 1971). Chromatophore suspensions were separated from an outer solution by an artificial phospholipid membrane. PCB^- was present in both the inner (chromatophore) and outer compartments. Uptake of PCB^- in the inner compartment resulted in the formation across the phospholipid membrane of a potential which could be detected with electrodes. The properties of PCB^- uptake are consistent with the notion that uptake is driven by a membrane potential. In illuminated chromatophores, this potential was estimated to be roughly 200 mV, in fair agreement with the value obtained by Jackson and Crofts (1969).

Rottenberg et al. (1972) used a microcentrifugation technique to determine the distribution of $^{36}Cl^-$ and $^{56}Rb^+$ across chloroplast membranes. The potential calculated from the distribution of these ions was about 10 mV more positive in the light than in the dark. This estimate is probably low. The microcentrifugation method apparently underestimates amine uptake since ΔpH determined from amine uptake data obtained by this technique is lower than that obtained by either the NH$_4^+$ uptake or 9-AA fluorescence quenching method. This underestimation may result from loss of accumulated amine during the 3-minute centrifugation. Since a membrane potential should decay more rapidly than an amine concentration gradient, the gradients in concentration of ions that reflect a membrane potential should also decay rapidly. Thus, the microcentrifugation method probably does not give maximal estimates of the ion gradients.

In summary, H$^+$ uptake in chromatophores and chloroplasts probably generates a membrane potential at the steady state (see Table III for summary). The magnitude of this potential is probably less in chloroplasts than in chromatophores, possibly because chloroplasts are more permeable to ions such as Cl$^-$. The absolute magnitude of the membrane potential is open to question. For example, it may not be valid to assume that all the light-induced absorption change of chloro-

TABLE III
Estimates of Δψ in Illuminated Chloroplasts and Chromatophores at the Steady State

Method used	Organelle	Δψ (mV)	Reference
Estimates of capacitance of chloroplast membranes and charges transported	Chloroplasts	About 100	Witt (1971)
Calibration of valinomycin and K$^+$-sensitive delayed light	Chloroplasts	75–105	Barber (1972)
Comparison of the light-induced 515 nm change to that induced by valinomycin and K$^+$ in the dark	Chloroplasts	30	Strichartz and Chance (1972)
Distribution of ^{36}Cl$^-$ and ^{56}Rb$^+$	Chloroplasts	10	Rottenberg et al. (1972)
Distribution of PCB$^-$	Chromatophores	About 200	Skulachev (1971)
Comparison of the light-induced carotenoid shift to that induced by valinomycin and K$^+$ in the dark	Chromatophores	285	Jackson and Crofts (1969)

plast pigments is directly related to a membrane potential. Other possibilities have been discussed by Strichartz and Chance (1972).

In this discussion, the steady state of H^+ uptake has been primarily considered. In the initial period of H^+ uptake, the magnitude of the membrane potential should be higher. In flashing light, ATP formation can take place, but an H^+ concentration gradient is not formed. Presumably, a membrane potential is formed since the 518 nm absorption change is observed in flashing light. Furthermore, valinomycin in the presence of K^+ inhibits phosphorylation in flashing light, but not that in continuous illumination (Junge et al., 1970). Thus, the dissipation of a membrane potential under conditions where an H^+ concentration gradient cannot be formed may be sufficient to inhibit phosphorylation in chloroplasts.

3.3 Is the Magnitude of the H+ Electrochemical Gradient Sufficient to Drive ATP Formation?

Although there is little doubt that the H^+ gradient must be a major form of energy storage in photosynthetic organelles, whether this gradient is sufficient to drive ATP synthesis is uncertain. Not only is it difficult to assess the magnitude of the H^+ gradient but also the magnitude of the potential against which the organelles can synthesize ATP. This potential $\Delta G'$, the phosphate potential, is defined by the following equation:

$$\Delta G' = \Delta G'_o + 1.36 \log \frac{[ATP]}{[ADP][P_i]}$$

where $\Delta G'_o$ is the standard free energy change for the hydrolysis of ATP. Kraayenhof (1969) calculated a $\Delta G'$ of about 15.5 kcal/mole for class I (intact) chloroplasts in the steady state. He used a value of 9.25 kcal/mole for the $\Delta G'_o$ of ATP hydrolysis (Benzinger et al., 1959). However, Rosing and Slater (1972) have recently calculated that the $\Delta G'_o$ for ATP hydrolysis is about 2 kcal/mole less than the earlier estimates. Thus, $\Delta G'$ for ATP synthesis in chloroplasts is probably closer to 13.5 kcal/mole.

There are some problems with the measurement of $\Delta G'$ in chloroplasts. It may not be valid to assume that the steady state of phosphorylation is in true thermodynamic equilibrium. Furthermore, a chloroplast suspension is a multiphase, multicompartment system and the concentration of the reactants at the site of ATP formation may not be the same as those in solution. Yet, for the sake of discussion a phosphate potential of 13.5 kcal/mole will be assumed.

Estimates of ΔpH as high as 3.5 units and of $\Delta \psi$ of 100 mV have been made in illuminated chloroplasts at the steady state. Thus, the

H⁺ gradient could be as high as 310 mV. Even if $\Delta\psi$ is as low as 30 mV, the H⁺ gradient could be equivalent to 240 mV. In illuminated chromatophores the H⁺ gradient was estimated to be from 260 to 340 mV (Crofts et al., 1972). A $\Delta G'$ of 13.5 kcal/mole is equivalent to a potential drop of about 590 mV. Whether a gradient of 240 to 340 mV is sufficient to drive phosphorylation against a potential of 650 mV depends on the H⁺:ATP ratio. Schwartz (1968) determined that about 2H⁺ were required for the formation of 1 ATP. Schröder et al. (1972) found the H⁺:ATP ratio to be 1.5 in the absence of valinomycin and K⁺ and 3 in their presence. They criticized Schwartz' results since he did not have valinomycin present. In the absence of this antibiotic, the rate of H⁺ efflux could be limited by the rate of counterion movements. Since the H⁺:ATP ratio is calculated from the rates of H⁺ efflux and of ATP formation, values of H⁺:ATP in the absence of valinomycin and K⁺ would, in the view of Schröder et al., be erroneously low. Under certain conditions, chloroplasts catalyze an ATP-dependent H⁺ uptake in the dark (see Section 3.6). For each ATP hydrolyzed, 2H⁺ were accumulated (Carmeli, 1970; Gaensslen and McCarty, 1971). Since valinomycin and K⁺ were not added, it is possible that the initial rates of H⁺ uptake were underestimated and, therefore, the H⁺:ATP ratio could be higher. The measurement of H⁺:ATP ratios are further complicated by the fact that ATP and ADP decrease the rate of H⁺ efflux (McCarty et al., 1971; Telfer and Evans, 1972).

If the H⁺:ATP ratio is 3, an H⁺ gradient of 240 to 340 mV would be equivalent to a total potential drop of 720–1020 mV, which is in excess of the 650 mV phosphate potential. Even with an H⁺:ATP ratio of 2, the upper estimate of the H⁺ gradient is close to the phosphate potential. Since there are uncertainties involved in the estimation of both $\Delta G'$ and the H⁺ gradient, the only conclusion that can reasonably be reached is that the magnitude of the H⁺ gradient is sufficiently close to that of $\Delta G'$ to warrant further measurements.

3.4 H⁺ Uptake and Postillumination ATP Synthesis

A close correlation between H⁺ uptake and the ability of chloroplasts to form ATP in the dark after illumination (Hind and Jagendorf, 1963) has become more evident. A number of reagents or treatments which affect H⁺ uptake have been found to affect postillumination ATP synthesis in a similar way.

The rate and extent of H⁺ uptake in *R. rubrum* chromatophores (Jackson et al., 1968) and digitonin subchloroplast particles (McCarty, 1970) are stimulated by valinomycin and K⁺. Although these reagents have no effect on phosphorylation in continuous light, they enhanced postillumination ATP synthesis (McCarty, 1970). Furthermore, the rate of the formation of the intermediate (X_E) which is responsible for ATP formation in the dark was accelerated by valinomycin and K^+, and the half-time for H^+ uptake in the presence of valinomycin was similar to that of X_E formation.

Nelson et al. (1971) reported that pyridine (pK_a 5.3) greatly stimulates H^+ uptake in chloroplasts. The mechanism of the pyridine stimulation of H^+ uptake is probably similar to that of the inhibition of H^+ uptake by amines of high pK_a (Crofts, 1967) (see Fig. 3). However, because of the low pK_a of pyridine, pyridine uptake at pH values above about 6 does not result in the liberation of H^+ into the medium as is the case for amines of high pK_a. Pyridine, therefore, acts essentially as an internal buffering agent, thereby increasing the amount of H^+ taken up from the suspending medium. At concentrations which stimulate H^+ uptake, but do not seriously inhibit phosphorylation in continuous light, pyridine also enhanced postillumination ATP synthesis. This stimulation is probably not caused merely by the increased H^+ uptake since it would be expected that the pyridine cation gradient is equivalent to that of H^+. More likely, the effect results from the fact that pyridine approximately halves the rate of collapse of the pH gradient in the dark. This increased stability of the high energy state should allow enhanced ATP formation in the dark.

Remarkably low concentrations of ATP (1–10 μM) enhance the extent of H^+ uptake in chloroplasts and slow the rate of H^+ efflux in the dark (McCarty et al., 1971; Telfer and Evans, 1972). Postillumination ATP formation was also stimulated by ATP (McCarty et al., 1971) when the ATP was present in the illumination stage reaction mixture.

Chloroplasts suspended in 10 mN sodium polygalacturonic acid are largely uncoupled (Cohen and Jagendorf, 1972). The addition of salts of Cl^-, NO_3^-, and SO_4^{2-} restored H^+ uptake, phosphorylation, and postillumination ATP synthesis, whereas salts of less permeable anions and sucrose were less effective. A close correlation between the restoration of H^+ uptake and of postillumination ATP synthesis was found.

These and previous observations (Jagendorf and Uribe, 1966; Galmiche et al., 1967; Izawa, 1970) provide evidence for the notion that X_E may be identical to ΔpH. This possibility is also supported by the finding that whereas postillumination ATP synthesis in subchloroplast particles is very sensitive to NH_4Cl, phosphorylation in continuous light is not (McCarty, 1968). Thus, it is probable that a light-induced pH gradient as well as an artificially formed pH gradient (Jagendorf and Uribe, 1966) can be the driving force for ATP formation in the dark.

3.5 The Interaction of H⁺ Uptake with Phosphorylation and Electron Flow

If phosphorylation and H⁺ uptake were to compete for a common energy source, phosphorylation should decrease the extent of the H⁺ gradient. Although Crofts (1968) clearly showed a competition between NH_4^+ uptake and phosphorylation, it has been more difficult to show that phosphorylation inhibits H⁺ uptake itself. Karlish and Avron (1967) reported that the simultaneous presence of ADP, P_i, and Mg^{2+} stimulates H⁺ uptake in lettuce chloroplasts and suggested that this stimulation was caused by transport of an $ADP-P_i-Mg^{2+}$ complex. Even in the presence of pyridine which greatly stimulated H⁺ uptake, phosphorylation failed to reduce the extent of H⁺ uptake (Nelson et al., 1971). The failure of phosphorylation to inhibit the extent of H⁺ uptake is difficult to understand. Karlish and Avron (1967) suggested that the energy required to drive H⁺ uptake may be only a small fraction of the total energy available in the high energy state. In view of the recent estimations of the magnitude of the H⁺ gradient, this possibility seems remote.

However, Dilley and Shavit (1968) showed that the addition of either ADP or Mg^{2+} to a spinach chloroplast suspension, which contained all the other reagents necessary for phosphorylation, stimulated electron flow under the low salt conditions used by Karlish and Avron. Mg^{2+} also seemed to reduce the permeability of chloroplast membranes to H⁺. Dilley and Shavit further reported that the rate and extent of the postillumination pH decay was reduced by phosphorylation under high salt conditions (0.1 M KCl). Similar results were obtained by Schwartz (1968) and by Schröder et al. (1972).

In the experiments of Dilley and Shavit (1968), the addition of ADP to chloroplasts supplemented with arsenate and Mg^{2+} stimulated the extent of H⁺ uptake even in the presence of 0.1 M KCl. It was assumed that this stimulation was caused by an increased rate of electron flow in the presence of ADP, arsenate, and Mg^{2+}. However, since arsenate appears to interact with the phosphorylating system in a manner similar to P_i, it is difficult to understand why the extent of H⁺ uptake was not reduced by the presumed arsenolation of ADP.

The failure to observe an inhibition of the extent of H⁺ uptake by phosphorylation may be due to the fact that ATP, at very low concentrations, stimulates H⁺ uptake (McCarty et al., 1971; Telfer and Evans, 1972). Since ATP inhibits electron flow (Avron et al., 1958; Izawa and Good, 1969), this stimulation cannot be explained by an enhancement of the rate of electron transport. ADP was as effective as ATP, but UTP, ITP, GTP, and AMP had little effect on either H⁺ uptake or electron flow. Neither arsenate nor hexokinase and glucose markedly affected the ADP stimulation of H⁺ uptake. However, ADP actually inhibited H⁺ uptake when arsenate and hexokinase and glucose were present simultaneously (McCarty et al., 1971). Under these conditions, there would be little or no ATP in the reaction mixture. Thus, an inhibition of the extent of H⁺ uptake may be observed if the stimulatory effect of ATP on H⁺ uptake is minimized. Much higher concentrations of ATP (1–10 mM) were found by Shavit and Herscovici (1970) to inhibit phosphorylation and electron flow. ATP at these concentrations appears to act as an energy transfer type inhibitor.

Phosphorylation increases the rate of electron flow in chloroplasts. The mechanism of this phenomenon is as obscure as that of photophosphorylation itself, but it can be interpreted in terms of either the chemiosmotic or chemical hypothesis. More recently it has been proposed that the internal pH may control the rate of electron flow (Rottenberg et al., 1971, 1972). The pH optimum of noncyclic electron flow in chloroplasts is about 8.5. In chloroplasts uncoupled by many different treatments, the optimum shifts toward more acid pH (Good et al., 1966). A plot of the rate of ferricyanide reduction against internal pH showed an optimum rate of reduction at about pH 5. Thus, assuming a ΔpH of 3 units, the internal pH of chloroplasts illuminated at an external pH of 7 would be 4, and this pH would be too low to allow maximum rates of electron flow. At an external pH of 8, close to the optimum for electron flow, the internal pH would be 5. In uncoupled chloroplasts ΔpH is reduced and therefore, lower external pH values are required to reach an internal pH of 5.

High concentrations of uncouplers strongly inhibit electron flow in the pH range of 8–9 (Good et al., 1966). This inhibition may be rationalized in terms of the concept that internal pH controls electron flow since uncouplers would increase the internal pH above the pH 5 optimum for electron flow by reducing ΔpH.

It should be pointed out that the stimulation of electron flow by phosphorylation cannot be explained solely by the concept that internal pH controls the rate of electron flow. Phosphorylation, like an uncoupler, should decrease ΔpH and, therefore, shift the optimum of electron flow to a more acid pH. However, this is not the case and, therefore, the stimulation of electron flow by phosphorylation must be explained in other ways.

3.6 Role of Coupling Factors in H⁺ Uptake

Although the coupling factor of phosphorylation in chloroplasts (CF$_1$) appears to play no direct role in light-dependent H⁺ uptake

(McCarty and Racker, 1966), recent evidence suggests that the interaction of CF_1 with adenine nucleotides may modify H^+ uptake. H^+ uptake is stimulated by low concentrations of ATP (McCarty et al., 1971; Telfer and Evans, 1972). Although H^+ uptake in the absence of ATP is resistant to inhibition by an antiserum to CF_1, the ATP-stimulated portion of H^+ uptake is quite sensitive (McCarty et al., 1971). Furthermore, antibodies have been prepared against the purified subunits of CF_1. The ATP-stimulated part of H^+ uptake is specifically inhibited by an antiserum to the largest subunit at amounts that inhibit phosphorylation only slightly (Nelson et al., 1973).

A mutant of *Chlamydomonas reinhardi* has been isolated which apparently has a modified CF_1 (Sato et al., 1971). Although the mutant chloroplasts could not carry out photophosphorylation, the rates and extents of H^+ uptake in these chloroplasts were quite similar to those in chloroplasts from the wild-type alga. However, although the combination ADP, Mg^{2+} and arsenate enhanced H^+ uptake in chloroplasts from the wild-type alga, no stimulation was observed in the mutant chloroplasts. These results support the notion that CF_1 is not directly involved in H^+ uptake, but does play a role in the H^+ uptake which is enhanced by ATP or ADP.

Baccarini-Melandri et al. (1970) removed a coupling factor from *Rhodopseudomonas capsulata* chromatophores by sonication of the chromatophores in EDTA solutions. A cold labile, Mg^{2+}-dependent ATPase activity was solubilized by this procedure. Although phosphorylation was inhibited by the removal of the factor, H^+ uptake was unaffected. Thus, this coupling factor does not appear to be required for H^+ uptake.

CF_1 may have a high affinity site for ATP. Since the efflux rate of H^+ from chloroplasts is slowed by ATP, it is possible that ATP binding to CF_1 may regulate the flow of H^+ through either CF_1 or some other part of the membrane. It was postulated (McCarty et al., 1971) that ATP modifies the conformation of chloroplast-bound CF_1. Ryrie and Jagendorf (1971) showed that when chloroplasts are briefly illuminated in the presence of ^3H_2O, about 100 nmoles of ^3H are found in 1 nmole of CF_1 whereas almost none was found in the CF_1 of chloroplasts incubated in the dark with ^3H_2O. Thus, it appears that CF_1 undergoes a conformational change in the light which allows the incorporation of ^3H_2O. Furthermore, the reaction of bound CF_1 with N-ethylmaleimide is also affected by light. Whereas N-ethylmaleimide (NEM) reacts with chloroplast-bound CF_1 in the light to produce a partial inhibition of phosphorylation, the incubation of chloroplasts with NEM in the dark has no effect on phosphorylation (McCarty et al., 1972). Since low concentrations of ATP diminish the inhibition of phosphorylation by light and NEM, it is possible that ATP alters the conformation of CF_1. Obviously, more work must be done before the full importance of conformational changes in CF_1 on H^+ uptake or phosphorylation can be realized.

The uptake of NH_4^+ (Crofts, 1968), amines (Gaensslen and McCarty, 1971), or H^+ (Carmeli, 1970) can be coupled to ATP hydrolysis in darkened chloroplasts. Although chloroplasts normally exhibit little ATP hydrolysis, a Mg^{2+}-dependent ATPase may be activated by illumination of chloroplasts in the presence of a sulfhydryl compound (Petrack and Lipmann, 1961; Petrack et al., 1965). Once the ATPase is activated, it catalyzes ATP hydrolysis in the dark. Since this ATPase is sensitive to a CF_1 antiserum (McCarty and Racker, 1966) as well as energy-transfer inhibitors (McCarty and Racker, 1968), it is apparent that CF_1 participates in this reaction. Furthermore, light-dependent ethylamine (Gaensslen and McCarty, 1971) or NH_4^+ (Crofts, 1968) uptake is not inhibited by the energy transfer inhibitor Dio-9 whereas the ATP-dependent uptake of these compounds is quite sensitive. Therefore, CF_1 is probably required for ATP-driven ion uptake in chloroplasts. More than likely ATP hydrolysis by CF_1 provides the energy for H^+ uptake, but the properties of the ATP-driven H^+ uptake are remarkably similar to those proposed by Mitchell (1966) for a proton translocating ATPase.

3.7 Concluding Statement

From the recent measurements of the light-induced H^+ gradient in both chromatophores and chloroplasts, there is little doubt that major amounts of energy can be stored in the form of H^+ gradients. It is also becoming increasingly apparent that, at least in chloroplasts, ATP synthesis can occur at the expense of an H^+ gradient. However, it is not clear whether the H^+ gradient is a primary or secondary form of energy storage. Furthermore, the mechanism of H^+ uptake remains obscure.

It should also be emphasized that different roles for H^+ uptake in phosphorylation have been proposed. In Dilley's interpretation (1971), for example, H^+ uptake is thought to result in the protonation of a macromolecule which must be in the protonated form to allow ATP formation. In his view, therefore, H^+ uptake is involved in the activation of a part of the phosphorylation device, not as the driving force for ATP formation. This proposal is certainly not unreasonable.

Thus, although it is generally agreed that H^+ uptake is intimately involved in the phosphorylation mechanism, the manner in which H^+ uptake is related to ATP synthesis remains controversial. Although many possibilities may be proposed, any future hypotheses for the role of ion uptake in phosphorylation must take into account that the magnitude of the H⁺ gradient is rather large.

4. Cation Functions in Catalysis and Control

4.1 Background

In a recent contribution to this series, Govindjee and Papageorgiou (1971) comprehensively reviewed interpretations of the slow rise and fall in fluorescence of chlorophyll a, which is observed during the first few minutes of illumination of whole algal cells (see also Myers, 1971). Briefly, this induction phase does not result from a change in the redox state of the fluorescence quencher, since it is independent of the oxygen evolution rate (which rises as the fluorescence rises) and can be observed both in the presence and absence of the inhibitor of oxygen evolution, DCMU [3-(3,4-dichlorophenyl)-1,1-dimethylurea] (Govindjee et al., 1966; Duysens and Talens, 1969). It is also found in algal mutants lacking photosystem II activity, but not in those lacking photosystem I (Bannister and Rice, 1968). Along with data from studies with uncoupling agents (Duysens and Talens, 1969; Papageorgiou and Govindjee, 1968), this latter finding led to the belief that endogenous cyclic photophosphorylation influences in some way the fluorescence yield of system II.

Slow *chromatic* transients in the fluorescence yield and in the relative distribution of fluorescence emission between the photosystems were also observed in whole cells. Bonaventura and Myers (1969) and Murata (1969a) developed the concept of two states of system II, interconvertible by light, which are characterized by their fluorescence yield. In state 1, obtained after prolonged darkness or by illumination with light which activates only system I, oxygen evolution rates are maximal and fluorescence (from system II) is high. State 2, reached by prolonged illumination with system II light, is recognized by its low fluorescence yield and oxygen evolution rate coupled with enhanced system I fluorescence emission (F735) at 77°K. It is clear that redistribution of excitation energy, rather than reversible activation of system II units, accounts for the state 2 to 1 transition (Bonaventura and Myers, 1969; Duysens, 1972).

Interpretation, on the basis of the above, of whole cell data showing slow fluorescence induction is difficult however, doubtless because of the complex equilibration between processes in the thylakoid, stroma, and cytoplasm which must occur during establishment of steady-state photosynthesis. Neither is cross correlation of data obtained from widely separated taxa always possible. Thus, a strong incentive has arisen to demonstrate transitions of state in isolated chloroplasts.

4.2 Fluorescence Lowering by Cofactor-Mediated Cyclic Electron Flow

Murata and co-workers, pursuing the notion that *in vivo* cyclic phosphorylation drives the state 2 to 1 transition, set up cyclic electron flow in isolated and swollen chloroplasts by addition of the mediator, N-methylphenazonium methosulfate (PMS). In the presence of DCMU, PMS promoted a rapid ($t_{0.5} = 2$ seconds) *decline* in fluorescence yield at high light intensity (Murata and Sugahara, 1969). Like the decline accompanying the state 1 to 2 transition in *Chlorella* (Papageorgiou and Govindjee, 1968), this effect was reversed by the uncoupler CCCP (carbonylcyanide m-chlorophenylhydrazone). Another cofactor of cyclic phosphorylation, diaminodurene (DAD) gave results similar to those with PMS, in isolated chloroplasts (Wraight and Crofts, 1970) and both responses could be reversed by a variety of uncoupling agents.

Glutaraldehyde-fixed chloroplasts, which retain the ability to develop pH gradients but not to phosphorylate in the light, do not exhibit PMS-induced lowering of the fluorescence quantum yield. Washing with EDTA also inhibits the lowering (Mohanty et al., 1972). Thus the lowering is not simply a reflection of the presence of a transmembrane pH gradient (Wraight and Crofts, 1970). The sensitivity to EDTA, and high pH optimum (over 8.0), suggest rather that the coupling factor, CF_1, is involved (McCarty and Racker, 1966). A large protein partially buried in the membrane, CF_1 could be instrumental in promoting a change in membrane conformation and hence, perhaps, the transitions of state (Murata, 1969a).

The lowering of fluorescence by PMS or DAD seems to contradict observations with whole cells, in which activation of photosystem I *increases* the fluorescence yield. An additional difference between the PMS effect and the *in vivo* changes is seen upon measuring fluorescence emission spectra of steady states frozen at 77°K. States 1 and 2 show an inverse correlation of the emissions at 684 and 735 nm at low temperature, that is, from photosystem II and I, respectively. In contrast, PMS lowers fluorescence at all wavelengths (Murata and Sugahara, 1969; Mohanty et al., 1972). There is hence as much uncertainty attached to the relation of the PMS and DAD effects to the *in vivo* state transitions, as there is regarding the relation of *in vitro* to *in vivo* cyclic phosphorylation. Duysens (1972) concluded that the contrary effects of activating system I in chloroplasts and whole cells were not reconcilable by a single, or simple mechanism.

4.3 Regulation of Energy Distribution by Cations

Seemingly more comparable to the *in vivo* fluorescence changes which accompany state transition, is the large increase in yield obtained upon restoration of Mg^{2+} to chloroplasts isolated in low ionic strength media devoid of divalent cation (M^{2+}) (Homann, 1969; Murata, 1969b). In this case, the frozen steady states show an increase, due to Mg^{2+}, in emission from system II at the apparent expense of system I, as observed in the state 2 to 1 transition. The increase in fluorescence yield was half saturated at about 0.5 mM Mg^{2+} (Homann, 1969; Murata *et al.*, 1970), but other divalent metal cations were almost as effective. Furthermore, monovalent cations at about 100-fold higher concentration produced the same effect.

In the intact chloroplast, the ion concentrations are about 100 mM K^+ and 15 mM each of Ca^{2+} and Mg^{2+} (Nobel, 1969) so that the high fluorescent condition must already exist *in vivo*. It follows that the large fluorescence changes resulting from addition of salts to "low salt" chloroplasts are due to the restoration of a more normal ionic environment to abnormal thylakoids.

Izawa and Good (1966a,b) have provided visible proof of the abnormality of low-salt chloroplasts. They showed that these have no grana/stroma lamellar differentiation and are swollen. Addition of salts, and in particular divalent cations, promoted contraction, increased static light scattering and dynamic light scattering changes during electron flow, and most surprisingly induced reappearance of grana. Murakami and Packer (1971) established a correlation between loss of grana by dedifferentiation, and decrease in fluorescence yield from system II. Furthermore, Ohki *et al.* (1971) showed that physical separation of grana and stroma lamellae by detergents or the French pressure cell depended upon the presence of salts, as the foregoing would indicate, and Murata (1971) established that only the resulting grana preparations (after dilution to decrease the ionic strength) showed the salt-induced fluorescence rise.

The notion emerging from these observations is that redifferentiation of grana might lead to the isolation of some system I units in the nongrana regions (Park and Sane, 1971) and thus diminish the possibility of energy spillover from system II to system I within the grana. The fluorescence yield would rise in consequence. Murata (1971) rejected this interpretation on two grounds: (1) a lack of parallelism between fluorescence yield changes (in the presence of PMS) and light-scattering changes induced by acetate or methylamine; (2) the retention of differentiation, as determined by fractionation studies, in thylakoids swollen by ammonia uncoupling (Ohki et al., 1971). Light-scattering changes, however, can indicate a variety of morphological events (Izawa and Good, 1966a,b; Murakami and Packer, 1970), and Izawa and Good showed that simple amines serve also as salts in reversing the low salt condition. When thylakoids are swollen by illumination in the presence of amines, the grana lamellae are the most resistant to disruption (Fig. 3) and fractionation might even be facilitated.

The evidence seems in favor of a correlation between the salt-induced fluorescence rise and the structural redifferentiation of the membranes into grana and stroma lamellae. Those membrane properties that render possible the reversible self-assembly of grana merit extensive study. What is seen as a gross structural response to a large, and presumably quite unnatural, change in cation activity clearly may have a bearing on the more subtle changes in the membrane which involve cations and which are discussed below.

4.4 Enhancement and Divalent Cations

Murata (1969b) found that Mg^{2+} increased not only the fluorescence of low salt chloroplasts, but also the quantum yield of DCIP (dichlorophenolindophenol) reduction by system II. This effect was more pronounced when actinic light centered at 695 nm rather than 480 nm was used, a finding which implies that Mg^{2+} might increase spillover of excitation energy from system I to system II pigment complexes (Sun and Sauer, 1972). Again in parallel with low-temperature fluorescence yield data, the quantum yield of the system I mediated electron transfer from reduced DCIP to NADP, in 480 nm light, was diminished by Mg^{2+}. Murata in fact concluded that Mg^{2+} served to *suppress* spillover in the reverse (but commonly accepted) direction, from system II to system I. If, however, spillover distributes excitation energy in favor of the photosystem that is rate limiting, conditions that maximize spillover should also maximize the rate of electron flow in NADP reduction with water as the electron donor. Since Mg^{2+} indeed does increase the overall quantum yield, extrapolated to zero light intensity, for this reaction (Avron and Ben-Hayyim, 1969, Rurainski et al., 1971) Mg^{2+} was considered by Sun and Sauer (1972) to increase spillover from system I to II, rather than the converse.

These same authors made the important discovery that enhancement is promoted by Mg^{2+} (or Mn^{2+}), but not by monovalent cations *at any concentration*. [Enhancement of a photosynthetic process is defined as the combination of far-red illumination with light of lower wavelength to give a rate greater than the sum of the rates with the two light beams presented separately (Emerson, 1957).] The data, shown in Figs. 4 and 5,

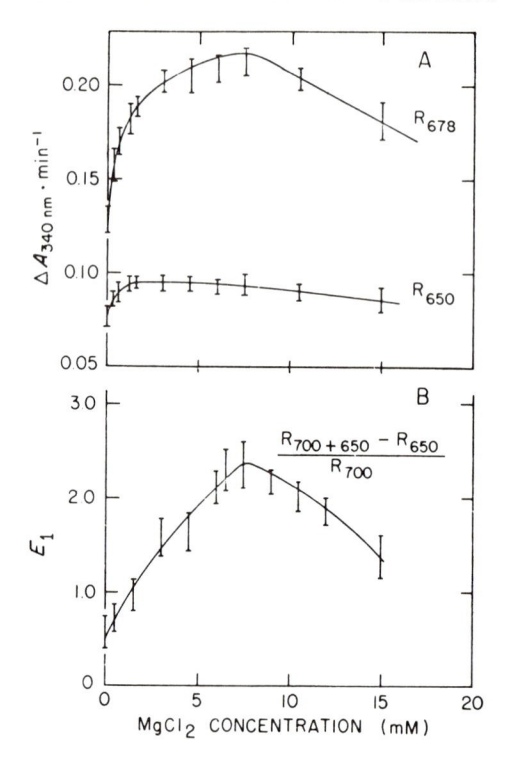

FIG. 4. Effect of the MgCl₂ concentration on the rate (A, upper curves) and on the enhancement E_1 (B, lower curve) of NADP reduction (from water) by broken chloroplasts. [From Sun and Sauer (1972) by kind permission.]

also demonstrate a salt-induced increase in electron flow which is not M^{2+}(divalent cation)-specific. Enhancement is therefore demarcated from this salt effect and from the fluorescence yield phenomenon in either of which M^+ (monovalent cation) can, though at a relatively high concentration, supplant M^{2+}. Sinclair (1972), in contrast, obtained enhancement by KCl in similar studies.

There are slight but significant differences too, in the half-saturating M^{2+} concentrations for the various effects. The value is about 0.5 mM for the rise in fluorescence yield (Murata et al., 1970), 3–10 mM for the rise in light scattering (Murakami and Packer, 1971), and 4 mM for the enhancement effect (Sun and Sauer, 1972; Sinclair, 1972). Enhancement is observed only above 2 mM M^{2+}, at which level the fluorescence rise is saturated. It may be that this specific M^{2+} effect is also associated with a light-scattering change, which would explain why NaCl fails at any concentration to increase light scattering to the extent attained in the presence of Mg^{2+} (Murakami and Packer, 1971). Possibly Mg^{2+}, like

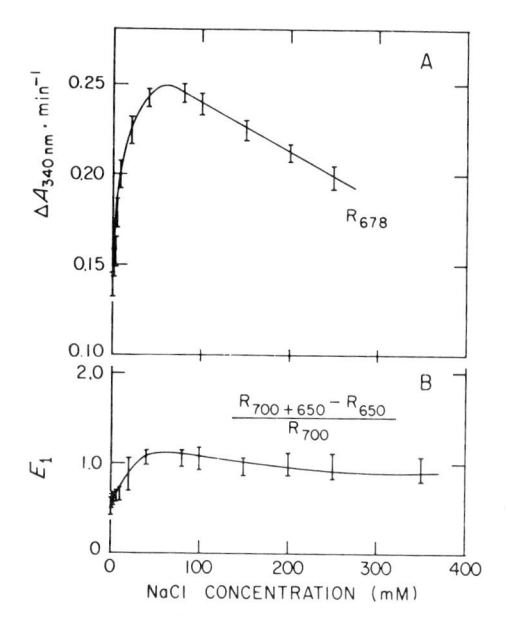

FIG. 5. Effect of the NaCl concentration on the rate (A, upper curve) and on the enhancement E_1 (B, lower curve) of NADP reduction (from water) by broken chloroplasts. [From Sun and Sauer (1972) by kind permission.]

illumination or acidification, changes the thickness of the thylakoid membranes (Murakami and Packer, 1970; Nobel, 1968).

It is useful to consider whether promotion of enhancement by M^{2+} might be related to an effect at the coupling site (CF_1), as was proposed earlier (Section 4.2) for the effects of PMS and DAD on fluorescence. Hauska and Sane (1972) demonstrated that the stroma lamellae are enriched in CF_1, as well as in cyclic phosphorylation activity, relative to the grana lamellae; yet enhancement is possible only in the grana. If CF_1 conformation directly influences enhancement, the minor, grana CF_1 component is functionally distinct from the stroma CF_1. As an alternative, it is difficult to conceive how spillover between the photosystems in the grana could be modified through reactions occurring in the stroma lamellae. Rurainski et al. (1971) found, however, that a decline in P700 turnover accompanied the increase by Mg^{2+} in the rate of NADP reduction from water. They concluded that P700 is not involved in this noncyclic electron flow pathway. If P700 is located principally on the stroma and oxygen evolution on the grana lamellae, a possibility discussed in the review of Park and Sane (1971), tight coupling between the stroma and grana lamellar electron flow systems would be indicated, with Mg ion playing a central role. Punnett (1971) has suggested that control of enhancement *in vivo* occurs by reversible de-differentiation of grana into stroma lamellae, but further anatomical and physiological studies are needed for confirmation.

4.5 Coupling of Electron Flow at Low Salt Concentration

In salt-free chloroplasts, M^{2+} at 0.006 mM or M^+ at 0.6 mM half-saturate a decrease in electron flow rate from a high, uncoupled to a low, coupled level (Gross *et al.*, 1969). Actual restoration of phosphorylation, however, was dependent on a second (Mg^{2+}) function, half-saturated at much higher concentration and presumably operating at the terminal steps of ATP synthesis. The preparations used conventionally as low-salt chloroplasts contain enough salt (10 mM NaCl) to saturate the above low concentration requirement for coupling. The saturation level is so far below the ionic strength of the aqueous compartments of chloroplasts *in vivo*, that regulation by this effect is inconceivable.

4.6 Summary

Isolated thylakoids from higher plants exhibit no fewer than four phenomenologically distinct types of salt requirement: (1) M^+ or M^{2+} induced coupling of electron flow; (2) M^+ or M^{2+} dependent maintenance of differentiation into grana and stroma lamellae; (3) Mg^{2+} dependent phosphorylation of ADP during coupled electron flow; (4) M^{2+} regulated spillover of excitation energy between the photosystems.

To evaluate further the potential association of any of these phenomena with a regulatory mechanism we must explore next the interaction of the thylakoids with the stroma in the intact chloroplast, and this in turn with the cytoplasm.

5. Mechanisms for Possible Control of Photosynthesis by Mg Ion

5.1 Mg^{2+}-Controlled Enhancement

For the enhancement of NADP reduction by Mg^{2+} to serve in a regulatory fashion *in vivo*, the activity of Mg^{2+} in the stroma would need to vary over the range between about 2 and 8 mM. Increased stroma Mg could result from active ion transport out of the thylakoids, from the cytoplasm, or by release of Mg^{2+} from a complex. In this latter regard, Duval and Duranton (1972) demonstrated the presence in thylakoids of a chlorophyll-free protein fraction with a high affinity for Mg^{2+}. Such proteins occur not only in etioplasts, but possibly also in membranes of mitochondria, sarcosomes, and bacteria (Binet and Volfin, 1971).

Lin and Nobel (1971) found that intact chloroplasts isolated rapidly from illuminated leaves contained more total Mg^{2+} than parallel isolates from unilluminated material, and in addition, they fixed CO_2 at a higher rate. Reasonable assumptions about stroma volume indicated a net gain of between 8 and 17 mM Mg^{2+} in the light. The actual change in activity of the ion would depend on the fraction becoming bound, and on the concentrating effect due to light-driven contraction of the chloroplast (Nobel, 1969). The values are, however, appropriate to a first approximation for the control of enhancement. Other possible control points exist for regulation by Mg^{2+} activity fluctuations of this order, and these will now be discussed.

5.2 Control of Carbon Pathway Enzymes

In the stroma, as in the thylakoid, Mg plays an essential role in maintaining enzyme activities. Numerous possibilities for control by Mg are afforded, and current research centers upon identification of the regulatory loci. One such site, at the carboxylation step, seems to be well established and will be dealt with at length; but it would be premature to rule out other sites that have been proposed, and these will be discussed first.

Bassham et al. (1970) described a regulation of metabolite transfer across the chloroplast envelope (probably at the inner membrane (Heldt and Sauer, 1971)) which was governed by Mg ion and endogenous fructose-1,6-diphosphatase. Enzyme activity was determined as an inhibition of CO_2 fixation due to facilitated diffusion of Calvin cycle intermediates from the chloroplasts. The effect of Mg^{2+} was dependent both on the enzyme and pyrophosphate levels, but half-saturation was experienced over the range 1 to 4 mM Mg^{2+}.

In C_4 (dicarboxylic acid pathway) plants, fructose-1,6-diphosphatase is located predominantly in the cytoplasm rather than the chloroplast. On the other hand, in C_4 plants, chloroplast pyrophosphatase levels are unusually high, serving to remove pyrophosphate generated as a byproduct of the phosphopyruvate synthetase reaction, and so driving the following reaction to the right:

$$\text{Pyruvate} + \text{ATP} + P_i = \text{phosphopyruvate} + \text{ATP} + PP_i$$

This pyrophosphatase is specifically activated by Mg ion, with half saturation occurring at about 2 mM (Simmons and Butler, 1969).

Jensen (1971) studied the Mg^{2+}-dependence of phosphoribulokinase in spinach chloroplasts and found half saturation at 6 mM $MgCl_2$. The influence of Mg on metabolite transport, discussed above, was eliminated by osmotically rupturing the chloroplasts in the reaction medium. Finding lower sensitivity of ribulose 1,5-diphosphate (RuDP) carboxylase to Mg ion, Jensen suggested that the kinase regulates CO_2 fixation *in vivo* in response to changes in stroma Mg^{2+} activity. Subsequently, however, Walker (1972) and Jensen and Bahr (1972) observed a steep gradient for activation of RuDP carboxylase by Mg ion in ruptured chloroplasts. Using a reconstituted chloroplast system (Stokes *et al.*, 1972), Walker (1972) was able to demonstrate the "turning on" of CO_2 fixation during illumination upon increasing the $MgCl_2$ concentration from 1 to 5 mM. Maximal rates of photosynthesis developed only after a few minutes' lag, possibly owing to the need to establish optimal pool sizes of Calvin cycle intermediates.

In agreement with data of Sugiyama *et al.* (1968) based on studies with the isolated carboxylase, Walker found that increasing the $MgCl_2$ from 1 to 5 mM decreased the apparent K_m for CO_2 almost 8-fold in the reconstituted system. Of paramount significance is Walker's finding that, at 1 mM $MgCl_2$, CO_2 fixation may be initiated either by adding Mg^{2+} or by increasing the bicarbonate concentration. This latter implies that the carboxylase alone is the site of effective rate limitation and excludes at least for these particular conditions, a regulatory role for phosphoribulokinase.

5.3 Mg Ion Fluxes in Isolated Chloroplasts

The chloroplast clearly does not lack reactions for which Mg^{2+} is a potential regulator. Looking at the other side of the coin, we must now inquire whether the means exist to bring about a change in Mg ion distribution or content in response to illumination. Dilley and Vernon showed, in 1965, that thylakoids lose K^+ and Mg^{2+} to the suspending medium upon illumination. The analytical method, however, did not lend itself to kinetic analysis. Hind *et al.* (1973), using multiparameter analysis of ion fluxes according to the methods described in this chapter, found that a rapid and extensive loss of Mg from the thylakoids occurs during illumination at pH values ranging between 6.6 and 8.3. This response was smaller than the light-induced H^+ uptake in most cases, and was fully reversed in the dark. A balance sheet for all the ions moving in the suspension showed that the fluxes maintained approximate electroneutrality.

Hind *et al.* (1973) studied the effect of the uncoupler CCCP on the Mg^{2+} flux and observed complete inhibition, in parallel with inhibition of the other light-driven fluxes. Figure 6 shows that, in contrast, ammonia does not eliminate the response. The other ion fluxes have been added to the micrograph in Fig. 3 in support of the scheme first presented by Crofts

4. CATION FLUXES IN CHLOROPLASTS

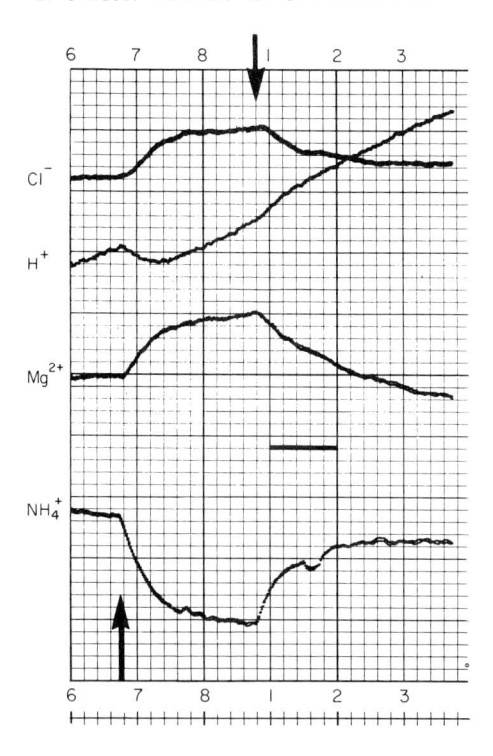

FIG. 6. Light-dependent ion fluxes in spinach chloroplasts uncoupled by NH₄Cl. The reaction medium contained (mM): Tris-HEPES buffer, pH 8.2, 1.0; MgCl₂, 2.0; NH₄Cl, 3.3; pyocyanine, 0.003 and chlorophyll, 0.06 mg/ml. Upward arrow, light on; downward arrow, light off. Time marker = 25 seconds. The maximal trace deflections give, in nanoequivalents per milligram of chlorophyll: Cl⁻ 82 influx, NH₄⁺ 563 influx, Mg²⁺ 533 efflux, H⁺ 31 influx.

(1967). However, it is clear that at the low Cl⁻ levels of Fig. 6, NH₄⁺ uptake is balanced by Mg²⁺ efflux, not by Cl⁻ influx.

The finding of a large reversible light-driven Mg²⁺ flux implies that in whole chloroplasts, the stroma Mg²⁺ activity rises in the light as demanded by the enzyme- or enhancement-mediated control mechanisms discussed earlier. This would be supplemented, probably at a slower pace, by active importation of Mg²⁺ from the cytoplasm (Lin and Nobel, 1971). If the assumptions of Stokes et al. (1972) regarding chloroplast volume and chlorophyll content are made, the light-induced Mg²⁺ flux of Hind et al. (1973) would yield an increase of at least 10 mM in stroma Mg concentration. Jensen and Bassham (1968) and Walker (1972) have noted that an increase in stroma Mg²⁺ and decrease in H⁺ activity could function in synergism to activate RDP carboxylase. The appeal of this concept has not been lost by the finding of large Mg ion fluxes since, just as H^+ changes may be buffered by stroma proteins, so may Mg^{2+} activity changes be diminished by chelation, as noted in Section 5.1.

5.4 Regulation in the Green Cell

It would be remiss, in a treatment of regulatory mechanisms, not to discuss the possible utility of such schemes to the organism. In particular, the inquisitive will wish to know why the chloroplast should be concerned to deactivate the Calvin cycle in the dark, since this would occur in any case as soon as the ATP and NADPH pools were depleted.

Experience with intact plastids and the reconstituted system (Walker, 1972) informs us that in the absence of a "primer" such as ribose 5-phosphate, very long lags will occur in the CO_2 fixation rate until the acceptor pools are recharged; on the other hand, whole plants show only brief induction periods after hours of darkness, and photosynthesis must remain more or less "primed." This desirable state could best be maintained by decreasing the activity of the carboxylase rapidly enough to limit depletion of the pentose phosphate pools during a light/dark transition. The ability to attain maximal rates of photosynthesis very soon after a dark/light transition would confer obvious selective advantages.

Tolbert and co-workers (Andrews et al., 1973) have produced good evidence that photorespiration results from oxidative splitting of ribulose diphosphate, catalyzed by an oxygenase function of RDP carboxylase. This was as dependent on Mg ion as was the carboxylase activity. The notion that both photosynthesis and photorespiration are regulated by Mg ion, at the same locus, has considerable appeal. Certainly the control of either process without the other would be pointless if, as assumed above, a prime function of such regulation is the maintenance of pentose phosphate pools during periods of weak light or darkness.

The state 1 to 2 transition is not observed in systems simpler than the whole cell. Vredenberg (1970) studied fluorescence induction in *Nitella* and found that upon breaking the cells, the slow fluorescence changes due to the state transitions were abolished. In intact cells, there was a close association between the potential across the plasmalemma and the tonoplast, and the slow fluorescence change. This result emphasizes that the transitions in pigment state are responsive to ion transport processes across cytoplasmic, nongreen membranes.

It seems probable that the next decade will see a disappearance of the division of photosynthesis research into "carbon pathway," "electron transport," "photochemistry," and so on, as the barriers to progress transcend these divisions and demand none but a holistic approach. Only at the whole cell level will the fullest understanding of regulatory processes in photosynthesis be gained.

Acknowledgments

The authors are indebted to Walter J. Geisbusch for Fig. 3 and Elsevier Publishing Company for permission to reproduce illustrations. This work was supported by the U. S. Atomic Energy Commission and by a Research Grant from the National Science Foundation (P1B 1646) to R. E. McCarty. R. E. McCarty is a Career Development Awardee of the National Institutes of Health (GM-14, 877-04).

References

Andrews, T. J., Lorimer, G. H., and Tolbert, N. E. (1973). *Biochemistry* 12, 11.
Avron, M., and Ben-Hayyim, G. (1969). *Progr. Photosyn. Res.* 2, 1185.
Avron, M., and Neumann, J. (1968). *Annu. Rev. Plant Physiol.* 19, 137.
Avron, M., Krogmann, D. W., and Jagendorf, A. T. (1958). *Biochim. Biophys. Acta* 30, 144.
Baccarini-Melandri, A., Gest, H., and San Pietro, A. (1970). *J. Biol. Chem.* 245, 1224.
Bagg, J., and Vinen, R. (1972). *Anal. Chem.* 44, 1773.
Bannister, T. T., and Rice, G. (1968). *Biochim. Biophys. Acta* 162, 555.
Barber, J. (1972). *FEBS Lett.* 9, 313.
Barber, J., and Kraan, G. P. B. (1970). *Biochim. Biophys. Acta* 197, 49.
Bassham, J. A., El-Badry, A. M., Kirk, M. R., Ottenheym, H. C. J., and Springer-Lederer, H. (1970). *Biochim. Biophys. Acta* 223, 261.
Bates, R. M. (1964). "Determination of pH: Theory and Practice." Wiley, New York.
Benzinger, T., Kitzinger, C., Hems, R., and Burton, K. (1959). *Biochem. J.* 71, 400.
Binet, A., and Volfin, P. (1971). *FEBS Lett.* 17, 197.
Bonaventura, C., and Myers, J. (1969). *Biochim. Biophys. Acta* 189, 366.
Brand, M. J. D., and Rechnitz, G. A. (1969). *Anal. Chem.* 41, 1788.
Brand, M. J. D., and Rechnitz, G. A. (1970a). *Anal. Chem.* 42, 617.
Brand, M. J. D., and Rechnitz, G. A. (1970b). *Anal. Chem.* 42, 1659.
Buck, R. P. (1972). *Anal. Chem.* 44, 270R.
Buck, R. P., and Krull, I. (1968). *J. Electroanal. Chem.* 18, 387.
Carmeli, C. (1970). *FEBS Lett.* 7, 297.
Cohen, W. S., and Jagendorf, A. T. (1972). *Arch. Biochem. Biophys.* 150, 235.
Crofts, A. R. (1967). *J. Biol. Chem.* 242, 3352.
Crofts, A. R. (1968). In "Regulatory Functions of Biological Membranes" (S. J. Jarnefeldt, ed.), p. 247. Amer. Elsevier, New York.
Crofts, A. R., Jackson, J. B., Evans, E. H., and Cogdell, R. J. (1972). *Proc. Int. Congr. Photosyn. Res., 2nd, 1971* p. 873.
Deamer, D. W., Prince, R. C., and Crofts, A. R. (1972). *Biochim. Biophys. Acta* 274, 323.
Degani, H., and Shavit, N. (1972). *Arch. Biochem. Biophys.* 152, 339.
Dilley, R. A. (1971). *Curr. Top. Bioenerg.* 4, 237.
Dilley, R. A., and Shavit, N. (1968). *Biochim. Biophys. Acta* 162, 86.
Dilley, R. A., and Vernon, L. P. (1965). *Arch. Biochem. Biophys.* 111, 365.
Duval, D., and Duranton, J. (1972). *Biochim. Biophys. Acta* 274, 240.
Duysens, L. N. M. (1954). *Science* 120, 353.
Duysens, L. N. M. (1972). *Biophys. J.* 12, 858.
Duysens, L. N. M., and Talens, A. (1969). *Progr. Photosyn. Res.* 2, 1073.
Eckfeldt, E. L., and Perley, G. A. (1951). *J. Electrochem. Soc.* 98, 37.

Emerson, R. (1957). *Science* **125**, 146.
Emrich, H. M., Junge, W., and Witt, H. T. (1969). *Naturwissenschaften* **56**, 514.
Fleet, B., and Rechnitz, G. A. (1970). *Anal. Chem.* **42**, 690.
Fleischman, D. E., and Clayton, R. K. (1968). *Photochem. Photobiol.* **8**, 287.
French, C. S., Hart, R. W., Murata, N., and Wraight, C. A. (1970). *Carnegie Inst. Wash., Yearb.* **68**, 607.
Gaensslen, R. E., and McCarty, R. E. (1971). *Arch. Biochem. Biophys.* **147**, 55.
Galmiche, J. M., Girault, G., Tyszkiewicz, E., and Fiat, R. (1967). *C. R. Acad. Sci.* **265**, 374.
Good, N. E. (1960). *Biochim. Biophys. Acta* **40**, 502.
Good, N. E., Izawa, S., and Hind, G. (1966). *Curr. Top. Bioenerg.* **1**, 75.
Govindjee, and Papageorgiou, G. (1971). *In* "Photophysiology" (A. C. Giese, ed.), Vol. 6, pp. 1–46. Academic Press, New York.
Govindjee, Munday, J. C., Jr., and Papageorgiou, G. (1966). *Brookhaven Symp. Biol.* **19**, 434.
Gromet-Elhanan, Z. (1971). *Eur. J. Biochem.* **25**, 84.
Gross, E., Dilley, R. A., and San Pietro, A. (1969). *Arch. Biochem. Biophys.* **134**, 450.
Grünhagen, H. H., and Witt, H. T. (1970). *Z. Naturforsch. B* **25**, 373.
Hauska, G. A., and Sane, P. V. (1972). *Z. Naturforsch. B* **27**, 938.
Heath, R. L., and Hind, G. (1972). *J. Biol. Chem.* **247**, 2917.
Heldt, H. W., and Sauer, F. (1971). *Biochim. Biophys. Acta* **234**, 83.
Hind, G., and Jagendorf, A. T. (1963). *Proc. Nat. Acad. Sci. U. S.* **49**, 715.
Hind, G., Nakatani, H. Y., and Izawa, S. (1973). *Proc. Nat. Acad. Sci. U. S.* submitted.
Homann, P. H. (1969). *Plant Physiol.* **44**, 932.
Izawa, S. (1965). *Biochim. Biophys. Acta* **102**, 373.
Izawa, S. (1970). *Biochim. Biophys. Acta* **223**, 165.
Izawa, S., and Good, N. E. (1966a). *Plant Physiol.* **41**, 533.
Izawa, S., and Good, N. E. (1966b). *Plant Physiol.* **41**, 544.
Izawa, S., and Good, N. E. (1969). *Progr. Photosyn. Res.* **3**, 1288.
Izawa, S., and Hind, G. (1967). *Biochim. Biophys. Acta* **143**, 377.
Jackson, J. B., and Crofts, A. R. (1969). *FEBS Lett.* **4**, 185.
Jackson, J. B., Crofts, A. R., and von Stedingk, L.-V. (1968). *Eur. J. Biochem.* **6**, 41.
Jagendorf, A. T., and Hind, G. (1963). *Nat. Acad. Sci.—Nat. Res. Council Publ.* **1145**, 599.
Jagendorf, A. T., and Neumann, J. (1965). *J. Biol. Chem.* **240**, 3210.
Jagendorf, A. T., and Uribe, E. (1966). *Brookhaven Symp. Biol.* **19**, 215.
Jensen, R. G. (1971). *Biochim. Biophys. Acta* **234**, 360.
Jensen, R. G., and Bahr, J. T. (1972). *Proc. Int. Congr. Photosyn. Res., 2nd, 1971* p. 1787.
Jensen, R. G., and Bassham, J. A. (1968). *Biochim. Biophys. Acta* **153**, 227.
Johannson, G., and Norberg, K. (1968). *J. Electroanal. Chem.* **18**, 239.
Junge, W., Rumberg, B., and Schröder, H. (1970). *Eur. J. Biochem.* **14**, 575.
Kahn, J. S. (1971). *Biochim. Biophys. Acta* **245**, 144.
Karlish, S. J. D., and Avron, M. (1967). *Nature (London)* **216**, 1107.
Koryta, J. (1972). *Anal. Chem.* **44**, 270R.
Kraayenhof, R. (1969). *Biochim. Biophys. Acta* **180**, 213.
Kraayenhof, R. (1970). *FEBS Lett.* **6**, 161.

Krebs, W. M. (1972). *Anal. Chem.* **44**, 187.
Lal, S., and Christian, G. D. (1970). *Anal. Lett.* **3**, 11.
Lin, D. C., and Nobel, P. S. (1971). *Arch. Biochem. Biophys.* **145**, 622.
Luzzana, M., Perrella, M., and Rossi-Bernardi, L. (1971). *Anal. Biochem.* **43**, 556
McCarty, R. E. (1968). *Biochem. Biophys. Res. Commun.* **32**, 37.
McCarty, R. E. (1970). *FEBS Lett.* **9**, 313.
McCarty, R. E., and Racker, E. (1966). *Brookhaven Symp. Biol.* **19**, 202.
McCarty, R. E., and Racker, E. (1968). *J. Biol. Chem.* **243**, 129.
McCarty, R. E., Fuhrman, J. S., and Tsuchiya, Y. (1971). *Proc. Nat. Acad. Sci. U. S.* **68**, 2522.
McCarty, R. E., Pittman, P. R., and Tsuchiya, Y. (1972). *J. Biol. Chem.* **247**, 3048.
Mitchell, P. (1961). *Nature (London)* **191**, 114.
Mitchell, P. (1966). *Biol. Rev. Cambridge Phil. Soc.* **41**, 445.
Mohanty, P., Braun, B. Z., and Govindjee. (1972). *Biochim. Biophys. Acta* **292**, 459.
Moore, E. W. (1969). *Nat. Bur. Stand. (U. S.), Spec. Publ.* **314**, 215–285.
Murakami, S., and Packer, L. (1970). *J. Cell Biol.* **47**, 332.
Murakami, S., and Packer, L. (1971). *Arch. Biochem. Biophys.* **146**, 337.
Murata, N. (1969a). *Biochim. Biophys. Acta* **172**, 242.
Murata, N. (1969b). *Biochim. Biophys. Acta* **189**, 171.
Murata, N. (1971). *Biochim. Biophys. Acta* **245**, 365.
Murata, N., and Sugahara, K. (1969). *Biochim. Biophys. Acta* **189**, 182.
Murata, N., Tashiro, H., and Takamiya, A. (1970). *Biochim. Biophys. Acta* **197**, 250.
Myers, J. (1971). *Annu. Rev. Plant Physiol.* **22**, 289.
Nelson, N., Nelson, H., Naim, V., and Neumann, J. (1971). *Arch. Biochem. Biophys.* **145**, 263.
Nelson, N., Deters, D., Nelson, H., and Racker, E. (1973). *J. Biol. Chem.* **248**, 2049.
Neumann, J., and Jagendorf, A. T. (1964). *Arch. Biochem. Biophys.* **107**, 109.
Neumann, J., Ke, B., and Dilley, R. A. (1970). *Plant Physiol.* **46**, 86.
Nobel, P. S. (1968). *Biochim. Biophys. Acta* **153**, 170.
Nobel, P. S. (1969). *Biochim. Biophys. Acta* **172**, 134.
Nobel, P. S., and Mel, H. C. (1966). *Arch. Biochem. Biophys.* **113**, 695.
Ohki, R., Kuneida, R., and Takamiya, A. (1971). *Biochim. Biophys. Acta* **226**, 144.
Papageorgiou, G., and Govindjee. (1968). *Biophys. J.* **8**, 1316.
Park, R. B., and Sane, P. V. (1971). *Annu. Rev. Plant Physiol.* **22**, 395.
Petrack, B., and Lipmann, F. (1961). *In* "Light and Life" (W. D. McElroy and H. B. Glass, eds.), p. 621. Johns Hopkins Press, Baltimore, Maryland.
Petrack, B., Craston, A., Sheppy, F., and Farron, F. (1965). *J. Biol. Chem.* **240**, 906 (1965).
Polya, G. M., and Jagendorf, A. T. (1969). *Biochem. Biophys. Res. Commun.* **36**, 396.
Portis, A. R., Jr., and McCarty, R. E. (1973). *Arch. Biochem. Biophys.* **156**, 621.
Pressman, B. C. (1967). *In* "Methods in Enzymology" (R. W. Estabrook and M. E. Pullman, eds.), Vol. 10, pp. 714–726. Academic Press, New York.
Punnett, T. (1971). *Science* **171**, 284.
Rechnitz, G. A. (1970). *Accounts Chem. Res.* **3**, 69.
Rechnitz, G. A., and Kugler, G. C. (1967). *Anal. Chem.* **39**, 1682.
Reinwald, E., Siggel, U., and Rumberg, B. (1968). *Z. Naturforsch. B* **23**, 1616.
Rosing, J., and Slater, E. C. (1972). *Biochim. Biophys. Acta* **267**, 275.
Ross, J. W. (1969). *Nat. Bur. Stand. (U. S.), Spec. Publ.* **314**, 57–88.

Rottenberg, H., and Grunwald, T. (1972). *Eur. J. Biochem.* **25**, 71.
Rottenberg, H., Grunwald, T., and Avron, M. (1971). *FEBS Lett.* **13**, 41.
Rottenberg, H., Grunwald, T., and Avron, M. (1972). *Eur. J. Biochem.* **25**, 54.
Rumberg, B., and Siggel, U. (1968). *Z. Naturforsch. B* **23**, 239.
Rumberg, B., and Siggel, U. (1969). *Naturwissenschaften* **56**, 130.
Rumberg, B., Schmidt-Mende, P., Skerra, B., Vater, J., Weikard, J., and Witt, H. T. (1965). *Z. Naturforsch. B* **20**, 1086.
Rurainski, H. J., Randles, J., and Hoch, G. (1971). *FEBS Lett.* **13**, 98.
Ryrie, I. J., and Jagendorf, A. T. (1971). *J. Biol. Chem.* **246**, 3771.
Sato, V. L., Levine, R. P., and Neumann, J. (1971). *Biochim. Biophys. Acta* **253**, 437.
Schliephake, W., Junge, W., and Witt, H. T. (1968). *Z. Naturforsch. B* **23**, 1571.
Schmidt, S., Reich, R., and Witt, H. T. (1971). *Naturwissenschaften* **8**, 414.
Schröder, H., Muhle, M., and Rumberg, B. (1972). *Proc. Int. Congr. Photosyn. Res., 2nd, 1971* p. 919.
Schuldiner, S., and Avron, M. (1971). *FEBS Lett.* **14**, 227.
Schuldiner, S., Rottenberg, H., and Avron, M. (1972). *Eur. J. Biochem.* **25**, 64.
Schwartz, M. (1968). *Nature (London)* **219**, 915.
Shavit, N., and Herscovici, A. (1970). *FEBS Lett.* **11**, 125.
Shavit, N., Thore, A., Keister, D. L., and San Pietro, A. (1968a). *Proc. Nat. Acad. Sci. U. S.* **59**, 917.
Shavit, N., Dilley, R. A., and San Pietro, A. (1968b). *Biochemistry* **7**, 2356.
Shavit, N., Degani, H., and San Pietro, A. (1970). *Biochim. Biophys. Acta* **216**, 208.
Shibata, K. (1959). *Methods Biochem. Anal.* **7**, 77–109.
Simmons, S., and Butler, L. G. (1969). *Biochim. Biophys. Acta* **172**, 150.
Sinclair, J. (1972). *Plant Physiol.* **50**, 778.
Skulachev, V. P. (1971). *Curr. Top. Bioenerg.* **4**, 127.
Stokes, D. M., Walker, D. A., and McCormick, A. V. (1972). *Proc. Int. Congr. Photosyn. Res., 2nd, 1971* p. 1779.
Strichartz, G. R. (1972). *Plant Physiol.* **49**, 272.
Strichartz, G. R., and Chance, B. (1972). *Biochim. Biophys. Acta* **256**, 71.
Sugiyama, T., Nakayama, N., and Akazawa, T. (1968). *Arch. Biochem. Biophys.* **126**, 737.
Sun, A. S. K., and Sauer, K. (1972). *Biochim. Biophys. Acta* **256**, 409.
Telfer, A., and Evans, M. C. W. (1972). *Biochim. Biophys. Acta* **256**, 625.
Thompson, H. I., and Rechnitz, G. A. (1972). *Anal. Chem.* **44**, 300.
Thore, A., Keister, D. L., Shavit, N., and San Pietro, A. (1968). *Biochemistry* **7**, 3499.
Vredenberg, W. J. (1970). *Biochim. Biophys. Acta* **223**, 230.
Walker, D. A. (1972). *Proc. Int. Congr. Photosyn. Res., 2nd, 1971* p. 1773.
Walker, D. A., and Crofts, A. R. (1970). *Annu. Rev. Biochem.* **39**, 389.
Walz, D., and Avron, M. (1969). *In* "Electron Transport and Energy Conservation" (J. M. Tager, ed.), pp. 463–468. Adriatica Editrice, Bari, Italy.
Witt, H. T. (1971). *Quart. Rev. Biophys.* **4**, 365.
Witt, H. T., Rumberg, B., Schmidt-Mende, P., Siggel, U., Skerra, B., Vater, J., and Weikard, J. (1965). *Angew. Chem., Int. Ed. Engl.* **4**, 799.
Wolff, C., Buchwald, H. E., Rüppel, H., Witt, K., and Witt, H. T. (1969). *Z. Naturforsch. B* **24**, 1038.
Wraight, C. A., and Crofts, A. R. (1970). *Eur. J. Biochem.* **17**, 319.

Chapter 5

NITROGEN FIXATION IN PHOTOSYNTHETIC BACTERIA*

Donald L. Keister and Darrell E. Fleischman
Charles F. Kettering Research Laboratory, Yellow Springs, Ohio

1. Introduction 157
   1.1 Distribution of Photosynthetic Bacteria 158
   1.2 Distribution of Nitrogenase 159
2. Relationship of Hydrogenase and Nitrogenase 159
3. Role of Nitrogen Fixation in Metabolism 162
4. Control of Nitrogenase 164
   4.1 Control of Nitrogenase Synthesis 164
   4.2 Control of Nitrogenase Activity 166
5. Enzymology of Nitrogen Fixation 168
   5.1 General Properties of the Nitrogenase System 168
   5.2 Nitrogen Fixation in Specific Cell-Free Extracts 175
6. Ecological Significance of Nitrogen Fixation in Photosynthetic Bacteria . 178
7. Concluding Remarks 179
   References 179

1. Introduction

The purple and green photosynthetic bacteria are typical aquatic organisms that inhabit anaerobic and semiaerobic marine and freshwater environments. Within these two small groups of microorganisms, a diversity of morphological forms and a wide variety of physiological characteristics, especially in adaptation to the use of chemical energy sources, is represented. Thus these bacteria have been classified on the basis of their photosynthetic capacity. As is usual with bacteria, it is their diversity that makes them interesting as a group. Even the structure and mechanisms of their photochemical systems are quite diverse. The photosynthetic bacteria are among the metabolically most primitive currently living forms. Therefore, studies on the nature of nitrogen

*Contribution No. 487 from the Charles F. Kettering Research Laboratory.

fixation and photosynthesis and how these processes interact with other biochemical events is important to further our understanding of the origin and mechanism of these reactions.

Many reviews and books are available to the researcher or student who is concerned with either nitrogen fixation or photosynthetic bacteria. Some of the very useful recent ones on photosynthetic bacteria are by Gest (1972), Oelze and Drews (1972), Baltscheffsky et al. (1971), Frenkel (1970), Vernon (1968), Pfennig (1967), and Kondrat'eva (1965). Stanier in 1961 gave a very lucid account of the development of a unitary concept of photosynthesis. Recent reviews and books on nitrogen fixation include those by Dalton and Mortenson (1972), Allen (1972), Benemann and Valentine (1971, 1972), Postgate (1971), Burris (1971), Bergersen (1971), Hardy et al. (1971), and Stewart (1966) and two complete journal issues, *Plant and Soil* (1971), a special volume edited by Lie and Mulder, and *Proceedings of the Royal Society, Series B*, Volume 172, No. 1029 (1969). None of these have been devoted to nitrogen fixation in photosynthetic bacteria, and this article will describe the comparative nitrogen fixation in photosynthetic bacteria and in other microorganisms.

1.1 Distribution of Photosynthetic Bacteria

Photosynthetic bacteria are classified into three major groups: the purple sulfur bacteria (Thiorhodaceae), the green sulfur bacteria (Chlorobacteriaceae), the purple and brown nonsulfur bacteria (Athiorhodaceae).

The photosynthetic bacteria do not evolve oxygen and depend on the presence of external electron donors such as reduced sulfur and/or organic compounds. The purple and green sulfur bacteria are strict anaerobes, photoautotrophic, and oxidize several inorganic sulfur compounds in the process of photosynthesis. Many species of the Thiorhodaceae are potentially mixotrophic, capable of photoassimilating organic compounds. The Chlorobacteriaceae are considered to be obligate photoautotrophic anaerobes; but one species (*Chloropseudomonas ethylicum*) is able to photoassimilate some organic compounds.

The Athiorhodaceae are photoheterotrophs that require organic compounds for their growth, and some species are facultative aerobes that can grow in the dark.

These organisms are widely distributed in nature (Kondrat'eva, 1965). They thrive in moist and muddy soils, ponds, rivers, lakes, sulfur springs, and marine environments. The purple and green sulfur bacteria are often found in great abundance and occur as blooms in lakes, ponds, and mud. They prefer mineral waters and rich organic and sulfide-containing waters, especially areas of low oxygen content. They can grow under low light conditions, and for this reason are found in lakes down to a depth of 35 meters (Pfennig, 1967). The nonsulfur bacteria are ubiquitous but are seldom found in visible masses although exceptions have been noted (Jones, 1956; Collins, 1958).

1.2 Distribution of Nitrogenase

Studies on nitrogen fixation in the late 1930's (Wilson, 1940) had demonstrated that molecular hydrogen inhibited nitrogen fixation in all aerobic organisms investigated, including red clover, *Azotobacter* and *Nostoc*. These studies led to the idea that hydrogenase possibly could be used as a marker enzyme for nitrogenase. Development of the sensitive $^{15}N_2$ assay for nitrogen fixation (Burris and Miller, 1941) led to the testing of other organisms known to contain hydrogenase such as *Escherichia coli, Scenedesmus,* and *Proteus vulgaris,* but disappointing negative results were obtained. In 1949, Gest and Kamen were studying the photoevolution of H_2 by *Rhodospirillum rubrum* and noted the N_2 inhibited H_2 evolution. Since this inhibition was analogous to the inhibition of N_2 uptake by H_2 in other nitrogen-fixing systems, they predicted that *R. rubrum* would fix N_2. Subsequent $^{15}N_2$ assays substantiated this prediction (Kamen and Gest, 1949). This was the first time the isotopic tracer technique was used successfully to demonstrate fixation in an organism not previously known to fix N_2, and this observation was subsequently confirmed by the Kjeldahl procedure (Lindstrom et al., 1949). Many other photosynthetic bacteria were tested shortly thereafter, and N_2 fixation appeared to be a general characteristic of this group of organisms (Lindstrom et al., 1950, 1951).

These successful results and the knowledge that these facultative anaerobes fixed N_2 only anaerobically led to the demonstration or confirmation of nitrogen fixation in several other organisms which contained a hydrogenase including *Klebsiella pneumoniae* (*Aerobacter aerogenes;* Hamilton and Wilson, 1955), *Bacillus polymyxa* (Grau and Wilson, 1962), *Methanobacterium omelianskii* (Pine and Barker, 1954), *Mycobacterium flavum* (Fedorov and Kalininskaya, 1961), and *Desulfovibrio* (Sisler and ZoBell, 1951).

2. Relationship of Hydrogenase and Nitrogenase

A correlation between hydrogenase, the enzyme which reversibly activates molecular hydrogen*

---

* The discovery that the nitrogenase enzyme in cell-free systems can catalyze an ATP-dependent hydrogen evolution makes important a differentiation between

$$H_2 \rightleftharpoons 2H^+ + 2e^-$$

and nitrogenase* began with the observation that hydrogen inhibited symbiotic nitrogen fixation (Wilson and Umbreit, 1937). It was some years later before Hoch *et al.* (1957) demonstrated the presence of hydrogenase in root nodules, and eventually it was shown to be in the bacteroids (Dixon, 1968). Since the initial observation, it has been established that hydrogen inhibits nitrogen fixation in a wide variety of nitrogen-fixing systems and that all nitrogen-fixing organisms contain a hydrogen-activating enzyme. As mentioned above, it was this correlation that led Kamen and Gest (1949) to the discovery that *R. rubrum* is a nitrogen-fixing bacterium. The converse relationship between hydrogenase and nitrogenase is not true, and several microorganisms, such as *E. coli, Scenedesmus,* and *Hydrogenomonas*, which do contain an active hydrogenase do not fix nitrogen. The correlation between nitrogenase and hydrogenase was furthered by the observation that hydrogenase activity was highest in cells grown on molecular nitrogen and like nitrogenase was greatly diminished in ammonia-grown cells (Lee and Wilson, 1943).

With these facts in mind, let us examine the studies on hydrogen evolution in photosynthetic bacteria. Gest and Kamen (1949) first observed the light-dependent formation of molecular hydrogen when *R. rubrum* cells were grown on citric acid cycle compounds, such as malate, fumarate, acetate, and succinate, with an amino acid as the nitrogen source. Cells grown on ammonium salts do not produce hydrogen until the ammonium nitrogen is exhausted from the medium. After exhaustion of the NH_3 a lag period in growth occurs during which the hydrogen-evolving enzymes are rapidly synthesized, and the cells acquire the capacity to photoevolve hydrogen. Ammonia has been shown to give rise to repression of this system (Ormerod *et al.*, 1961). Resting cells grown on glutamate or molecular nitrogen had the capacity to evolve hydrogen at the expense of external organic electron donors at a significant rate. The stoichiometry of H_2 and CO_2 production indi-

---

this reaction and the conventional non-ATP-dependent Knallgas reaction catalyzed by hydrogenase. Cell-free extracts of *A. vinelandii* and other microorganisms evolve H_2 in an ATP-dependent reaction that has the same requirements as N_2 reduction (Bulen *et al.*, 1965a; Burns and Bulen, 1966). The hydrogenase catalyzed H_2-evolution is inhibited by CO whereas the ATP-dependent H_2 evolution catalyzed by the nitrogenase in cell-free extracts is not inhibited. Intact cells do not normally catalyze a CO-insensitive H_2 evolution. (See Section 5.1.4.)

* Nitrogenase is the term used to operationally describe the active entity in N_2 reduction and is now known to consist of two proteins. To borrow the definition of Burris (1971): "Nitrogenase ... refers to the two proteins acting together."

cated that resting cells completely decomposed the organic acids, acetate, succinate, fumarate, and malate and that the decomposition was inhibited by fluoroacetate. This observation coupled with Eisenberg's (1953) observation that *R. rubrum* has a complete complement of citric acid cycle enzymes led Gest *et al.* (1962) to the conclusion that the decomposition occurred via an anaerobic citric acid cycle. Such a cycle requires the regeneration of NAD$^+$ from NADH, and this apparently occurs by the evolution of H$_2$.

Some heterotrophic microorganisms and even some photosynthetic bacteria can evolve hydrogen in the dark at a slow rate via well-known mechanisms such as the ferredoxin-dependent phosphoroclastic reaction typical of the clostridia (Gest, 1951; Mortenson *et al.*, 1963; Bennett *et al.*, 1964).

The mechanism involved in the photoproduction of hydrogen by *R. rubrum* is not known, but let us consider some of the known observations. Bose and Gest (1963) found that the reaction was inhibited by inhibitors of cyclic electron transport (e.g., antimycin a), uncouplers of photophosphorylation (*m*-chlorocarbonylcyanide phenylhydrazone, pentachlorophenol, dicumarol) and reduced dyes, which presumably inhibit cyclic electron transport by "overreducing" the electron carriers. Although other explanations may be available, the most logical interpretation is that the photoevolution of hydrogen requires energy in the form of ATP or an intermediate in ATP formation. The electron donor for hydrogenase in clostridia is reduced ferredoxin, which in turn is reduced by NADPH via ferredoxin–NADP reductase (Jungermann *et al.*, 1971). This reaction does not require ATP and therefore the question is: Why does *R. rubrum* require energy to evolve hydrogen from the NADH generated during the oxidation of organic compounds via the anaerobic citric acid cycle?

Gest (1972) has recently considered the question in depth and the following is a short summary of his conclusions. The precursor of hydrogen is believed to be reduced pyridine nucleotide, NAD(P)H, generated by citric acid cycle oxidations. By analogy with other systems, ferredoxin may be the electron donor to nitrogenase although the function of the ferredoxins isolated from *R. rubrum* have not yet been elucidated (Shanmugam *et al.*, 1972). There is no theoretical reason why hydrogen cannot be produced from NAD(P)H without energy input even though the standard redox potential of hydrogen ($E'_0 = -0.42$ V) is lower than that of pyridine nucleotide ($E'_0 = -0.32$ V), provided a high ratio of reduced:oxidized pyridine nucleotide is maintained. *Rhodospirillum rubrum*, however, does not have a strongly reducing metabolism, and thus energy may be required to provide the high ratio of reduced:

oxidized pyridine nucleotide necessary. The maximal photoproduction of hydrogen is observed under conditions in which cells are probably generating ATP and NAD(P)H in excess of the cellular requirements for biosynthesis (e.g., resting cells in the light). Thus the energy-dependent production of molecular hydrogen can be viewed as a mechanism for the regulation of the reduced pyridine nucleotide level under various growth conditions. Gray and Gest (1965) have summarized this as:

$$\text{Light} \to \text{electron transport} \to {\sim}X \text{ or } ATP \to \begin{array}{c} NAD(P)H + H^+ \\ \downarrow \\ NAD(P^+) + H_2 \end{array}$$

Light is required to produce energy in the form of an energized intermediate (∼X) or ATP, which then is used to drive the reduction of NADPH from NADH via an energy-linked transhydrogenase (Keister and Yike, 1967b). In cell-free systems this enzyme has an equilibrium constant [(NADPH) (NAD⁺)/(NADH) (NADP⁺)] = ∼28. Thus this reaction could help to maintain the reducing conditions in the cell at levels necessary for hydrogen evolution. NAD(P)H is then oxidized by hydrogenase, regenerating NAD(P)⁺. Alternatively, energy could be required to promote the reduction of ferredoxin in an unknown energy-linked reaction that would serve a regulatory function. Ferredoxin is the probable electron carrier between NAD(P)H and hydrogenase, just as it has been found to be in clostridia (Thauer et al., 1969).

When molecular nitrogen is available, the electrons can be used to reduce the dinitrogen molecule. Thus, the photoproduction of hydrogen is a control system for regulating the redox potential and the energy charge (Atkinson, 1968; Gest, 1972) of the cell so that the intracellular environment is suitable for biosynthesis.

These studies and considerations provide a role for hydrogenase in the cellular metabolism of anaerobes, but have not provided any insight into the apparent correlation between hydrogenase and nitrogenase, except that an intermediate electron donor is apparently common to both reactions. In aerobic organisms like *Azotobacter*, which do not evolve hydrogen, no physiological role for hydrogenase is evident. Some speculations have been made, but this question awaits resolution.

3. Role of Nitrogen Fixation in Metabolism

Although ammonia was postulated to be a key intermediate in nitrogen fixation for a long time, only the discovery and use of $^{15}N_2$ made possible the confirmation of this postulate (Burris, 1942). This work indicated that, in *Azotobacter* and *Clostridium*, glutamic acid contained the highest level of label followed by ammonia after the introduction of $^{15}N_2$. This work was extended to include the photosynthetic bacteria *R. rubrum*, *Chlorobium* sp., and *Chromatium* sp. by Wall et al. (1952), and again it was found that glutamic acid had the highest amount of label followed by ammonia. Subsequently this work has been done with all major groups of nitrogen-fixing organisms and Wilson (1971) has summarized the pertinent observations: (1) The highest level of label is always found in glutamic acid, followed by ammonia and usually by aspartic acid. (2) The distribution of $^{15}N_2$ in various amino acids was similar whether $^{15}N_2$ or $^{15}NH_4^+$ was supplied. (3) Ammonia immediately supplants N_2 when supplied to a culture actively fixing N_2. (4) *Clostridium pasteurianum* and *A. vinelandii* excrete ammonia if the supply of dicarboxylic acids is low, and ammonia then has the highest ^{15}N concentration followed by glutamine and asparagine. The simplest conclusion that can be drawn is that N_2 is reduced to NH_3 and then converted to glutamic acid by the reductive amination of α-ketoglutaric acid by glutamic dehydrogenase. This has been considered to be the primary pathway for the incorporation of NH_3.

These conclusions must now be modified at least under some conditions. Tempest et al. (1970) recently discovered an alternate pathway of ammonia assimilation in *Aerobacter aerogenes* when the organism was grown under nitrogen-limited conditions, and it is now recognized that this pathway exists in a large number of microorganisms (Meers et al., 1970; Nagatani et al., 1971). One of the reactions of this pathway is catalyzed by a new enzyme (reaction 2), glutamate synthase [L-glutamate-NADP oxidoreductase (deaminating, glutamine forming)] (Miller and Stadtman, 1972), which when coupled with the reactions catalyzed by glutamine synthetase (reaction 1) and various transaminases provides the cell with

$$\text{L-Glutamate} + NH_4^+ + ATP \rightarrow \text{L-glutamine} + ADP + P_i \quad (1)$$

$$\text{L-Glutamine} + \alpha\text{-ketoglutarate} + NAD(P)H \rightarrow 2\text{L-glutamate} + NAD(P^+) \quad (2)$$

an ATP-dependent and therefore essentially irreversible pathway of ammonia assimilation. The K_m of glutamine synthetase for ammonia is much lower than glutamic dehydrogenase, suggesting that this pathway is valuable for ammonia assimilation in nitrogen-deficient environments.

Nagatani et al. (1971) recently presented evidence that this pathway is found in all major groups of nitrogen-fixing bacteria and was the prominent pathway in *K. pneumoniae* grown under nitrogen. The photosynthetic bacterium *Chromatium* contained larger amounts of glutamate synthetase activity than glutamic hydrogenase irrespective of the nitrogen source, suggesting that this pathway may be used exclusively for glutamate synthesis as has been found with *Bacillus megaterium* (Elmerich and Aubert, 1971).

Thus, the idea that glutamic dehydrogenase is the almost universal pathway of ammonia assimilation must be modified; to what extent remains to be determined.

Nitrogen fixation by photosynthetic bacteria is a light-dependent process and Pratt and Frenkel (1959) using mass spectrophotometric techniques found that N_2 uptake ceased abruptly upon turning off the light. More recently Schick (1971) found that after illumination, a delay of 20 minutes ensued before N_2 uptake ceased in the dark. These results are consistent with N_2 fixation being a dark reaction as in other microorganisms. Light is required to generate reducing power and ATP and N_2 fixation can proceed in the dark as long as an endogenous source of energy is available.

When ammonia is added to a cell culture that is actively fixing nitrogen, the uptake of N_2 gas is rapidly and for all practical purposes completely inhibited. The cells resume fixation of molecular nitrogen only after NH_3 is exhausted (Pratt and Fenkel, 1959; Schick, 1971). Thus it is obvious that the cell prefers to incorporate ammonia rather than carry out the energetically expensive reduction of N_2. It would be of interest to know whether a source of fixed nitrogen such as glutamate would also inhibit N_2 uptake since this amino acid does not cause the repression of nitrogenase synthesis, but this information is apparently not available in the literature.

4. Control of Nitrogenase

The nitrogenase system represents a considerable amount of the total cellular protein and consumes substantial amounts of energy (ATP) and reductant during nitrogen fixation. With these facts, one could deduce that the cell might have control mechanisms for the regulation of both the synthesis and function of the enzyme. Indeed, there are two types of control, inhibition of synthesis of the enzyme by repression, and a fine control in which the activity of the enzyme is controlled by chemical factors in the cell.

4.1 Control of Nitrogenase Synthesis

4.1.1 AMMONIA

Ammonia, which has long been known to suppress nitrogen fixation by most nitrogen-fixing systems (Burris and Wilson, 1946), was also found to inhibit nitrogen fixation and hydrogen evolution in photosynthetic bacteria (Gest et al., 1950; Pratt and Frenkel, 1959; Schick, 1971). Upon exhaustion of ammonia during growth of K. pneumoniae or A. vinelandii, a diauxic lag occurred before fixation of $^{15}N_2$ began and growth was resumed using molecular nitrogen (Pengra and Wilson, 1958; Strandberg and Wilson, 1967). Ormerod et al. (1961) observed that hydrogen evolution in R. rubrum was not evidenced until ammonia was exhausted. Since hydrogenase is correlated in some way with nitrogenase, it could be deduced that nitrogenase was also inhibited. Growth on certain amino acids did not inhibit hydrogen evolution, and in K. pneumoniae amino acids were found to dramatically shorten or eliminate the diauxic lag observed following ammonia exhaustion (Yoch and Pengra, 1966). It is probable that amino acids are not metabolized rapidly enough to give rise to an intracellular pool of ammonia. From these studies it was concluded that ammonia or a product of its metabolism caused repression of nitrogenase synthesis.

When cells were grown in a chemostat with low levels of ammonia, the nitrogenase complex was synthesized, in some cases in higher concentration than when the cells were grown on molecular nitrogen (Daesch and Mortenson, 1968; Munson and Burris, 1969; Dalton and Postgate, 1969). This unusual observation is thought to be due to the formation of a small internal pool of ammonia when the cells are grown on nitrogen which slightly represses enzyme synthesis; whereas, when the cells are grown on a low concentration of free ammonia, the ammonia is rapidly utilized and no internal pool of free ammonia is present to inhibit enzyme synthesis. Similar studies have indicated that molecular nitrogen is not necessary as an inducer (Parejko and Wilson, 1970; Daesch and Mortenson, 1972). Therefore, it is reasonable to conclude that nitrogenase synthesis is not controlled by an induction mechanism but by repression controlled by ammonia or a product of its metabolism and derepression in its absence.

4.1.2 OXYGEN

Nitrogen fixation is an anaerobic process, and all facultative anaerobes require anaerobic or semiaerobic conditions to fix nitrogen. Oxygen is a well-known inhibitor of nitrogen fixation *in vivo* and *in vitro*. *In vivo* studies have shown that oxygen is a competitive inhibitor (Parker and Scutt, 1960; Bergersen, 1962) of nitrogen reduction and *in vitro* studies have shown that the nitrogenase enzyme is inactivated by oxygen uncompetitively (Wong and Burris, 1972). It is not known whether oxygen may control the synthesis of the enzyme in addition to affecting the enzyme activity.

Oxygen or a product of oxidative metabolism is well known to have a controlling influence on many enzymes and enzyme systems (Hughes and Wimpenny, 1969), and in photosynthetic bacteria oxygen suppresses synthesis of the entire internal membrane system and the associated photosynthetic apparatus (Cohen-Bazire et al., 1957; Biedermann et al., 1967). Thus it is possible that oxygen may directly or indirectly cause repression of nitrogenase synthesis, and studies are needed to examine this point.

4.1.3 NITRATE

Nitrate has been shown to inhibit nitrogen fixation in whole cells of *Azotobacter* and blue-green algae and in root nodule symbiotic systems. Part of this inhibition is undoubtedly due to enyzme repression since nitrate is reduced by nitrate reductase and nitrite reductase to ammonia which is known to cause repression. To examine the effect of nitrate alone on nitrogenase, Sorger (1969) isolated mutants of *Azotobacter* which were deficient in nitrate reductase and were, therefore, incapable of reducing nitrate to nitrite and ammonia. With this system he was able to demonstrate that nitrate inhibited the activity of nitrogenase but did not induce repression.

4.2 Control of Nitrogenase Activity

4.2.1 OXYGEN

As discussed in Section 4.1.2, the oxygen lability of nitrogenase has been known for a long time and the inability for years to prepare active cell-free extracts was undoubtedly in large measure due to the oxygen lability of the enzyme. Recent work has demonstrated that in the Azotobacteriaceae, nitrogen reduction may be physiologically controlled by oxygen.

When cultures of *A. chroococcum* grown no a low pO_2 are exposed to higher pO_2, nitrogen and C_2H_2 reduction are "turned off" almost immediately. Lowering the pO_2 immediately restores most of the enzyme activity, provided the exposure to high pO_2 has not been prolonged, in which case irreversible inactivation may occur (Drozd and Postgate, 1970; Yates, 1970). In crude cell-free extracts, the particulate nitrogenase of *Azotobacter* can be insensitive to oxygen, but it becomes oxygen sensitive upon purification (Bulen and LeComte, 1966). These observations have led Dalton and Postgate (1969) to postulate that *Azotobacter* has two means of enzyme protection: "conformational protection," a condition in which the enzyme assumes a conformational state that is insensitive to oxygen; and "respiratory protection," a condition in which the enzyme is protected from oxygen by the respiratory system of the organism.

Oppenheim et al. (1970) have suggested a model of enzyme protection that does not require a "conformational protection." They found that if *Azotobacter* cells were ruptured by osmotic lysis, the nitrogenase was essentially soluble and free of membraneous proteins, whereas mechanical disruption of the cells in the French pressure cell led to a particulate enzyme associated with membranous proteins. The soluble enzyme was oxygen sensitive in contrast to the particulate enzyme. They postulated that the enzyme was protected from oxygen by encapsulation in the vesicles formed during cell rupture in the French pressure cell. Since the respiratory components of the cell are in these vesicular particles, their model requires only "respiratory protection."

This encapsulation model is questionable since disruption of the inner membranes probably leads to an "inside-out" orientation if the *Azotobacter* membranes vesiculate upon disruption in the same manner as other bacterial and mitochondrial membranes (Lee and Ernster, 1966; Mitchell, 1967; Crofts, 1969). The intracellular membranes of bacteria and the cristae membrane of mitochondria vesiculate after disruption so that the side of the membrane which was exposed to the inside of the cell or mitochondrion becomes the outside of the vesicle formed after comminution of the cellular membranes. Thus, it is not likely that a soluble intracellular nitrogenase would be encapsulated during cell disruption. This problem is not yet resolved and further discussion can be found in Postgate (1971).

4.2.2 ADP

In cell-free extracts of *Rhizobium lupini* bacteroids, ADP was found to inhibit acetylene reduction by 50% when the ADP:ATP ratio was 0.1 (Kennedy, 1970). This was considerably lower than the ratio necessary in other systems where a ratio of about 0.5 was required (Moustafa and Mortenson, 1967; Kelly, 1969). These results have led to the postulation that the *in vivo* activity of nitrogenase may be controlled by ADP or the ADP:ATP ratio.

4.2.3 Ammonia

In addition to repressing enzyme synthesis, ammonia has a direct effect on nitrogen fixation in *R. rubrum*. Pratt and Frenkel (1959) reported that N_2 uptake was inhibited within 20 minutes after the addition of ammonia. Schick (1971) reported an even shorter period (6 minutes) between the addition of ammonium chloride and inhibition of N_2 uptake. Since these time periods are much too short for repression to be noted, ammonia must be inhibiting N_2 uptake. Schick also found that the duration of the inhibition was strictly correlated to the amount of ammonia added. Small amounts of ammonia led to short periods of inhibition. He also reported that the total sum of nitrogen incorporated during a lengthy experiment in which either N_2 or ammonia or both were assimilated was constant. From these results it appears that ammonia acts as a competitor of N_2 uptake or as an allosteric inhibitor (Lyubimov et al., 1969) and that the cell prefers the energetically favorable incorporation of ammonia. Again, it would be of interest to know whether a source of fixed nitrogen such as glutamate would also inhibit nitrogen uptake since the amino acid does not repress nitrogenase synthesis.

4.2.4 FATE OF NITROGENASE DURING REPRESSION

Repression of enzyme synthesis occurs very rapidly after the addition of ammonia to a culture of *A. vinelandii* (Strandberg and Wilson, 1968). Shah et al. (1972) and Davis et al. (1972) recently studied in detail the fate of the enzyme during repression. They found that for 0.5 generation of growth, loss of enzyme activity, electron paramagnetic resonance signal characteristic of the iron–molybdenum protein (component I) and material immunochemically cross-reacting with component I exactly corresponds with the increase in cell mass; this suggests that the enzyme was simply diluted out. After 0.5 generation, the activity was lost more rapidly while the loss of cross-reacting material continued to be accounted for by simple dilution. Thus the cell appears to have a way of inactivating the enzyme, although it is not known whether this is an active or simply a passive mechanism. It may be that the enzyme is not protected against O_2 inactivation after the initial 0.5 generation of cell growth.

5. Enzymology of Nitrogen Fixation

5.1 General Properties of the Nitrogenase System

As early as 1934, attempts were being made to obtain nitrogen fixation with cell-free preparations from nitrogen-fixing microorganisms (Bach et al., 1934). But not until 1960 was a technique devised that could give high and reproducible rates of N_2 reduction. Carnhan et al. (1960) incubated soluble extracts of vacuum-dried *C. pasteurianum* with $^{15}N_2$ in the presence of pyruvate, and found substantial ^{15}N incorporation, primarily in the form of $^{15}NH_3$. At about the same time Schneider et al. (1960) reported that similar procedures could be used to demonstrate nitrogen fixation with soluble extracts of the photosynthetic bacterium *R. rubrum*.

During the past decade nitrogen fixation has been obtained with soluble extracts from a wide variety of microorganisms, ranging from blue-green algae (Schneider et al., 1960; Smith and Evans, 1970) to soybean root nodule bacteroids (Koch et al., 1967). In several cases, the proteins involved have been isolated and purified, and their mechanism of action studied in considerable detail.

A generalization that has emerged from these studies is that the nitrogenase systems from all the organisms examined are strikingly similar in both structure and function. The reduction of N_2 in every case has the following requirements: (1) two metalloproteins, one containing molybdenum and iron, the other containing only iron; (2) a low potential reductant; (3) adenosine triphosphate (ATP); (4) a divalent cation such as Mg^{2+}; and (4) anaerobic conditions. The first stable product of N_2 reduction is always NH_3. It would seem that the capacity for nitrogen fixation appeared early in the course of evolution, and that at least since the evolutionary divergence of the ancestors of photosynthetic bacteria and clostridia, organisms have been unable to improve the system substantially.

The reduction of N_2 coupled to the oxidation of organic substrates, such as glucose or H_2, is exergonic (Bayliss, 1956; Hardy et al., 1971). Yet hydrolysis of ATP is required for N_2 reduction by all known biological nitrogenase systems, presumably in order to overcome the rather high activation energy of the enzymatic reaction (14.6 kcal at temperatures greater than 21°C) (Burns, 1969). A more efficient system for the reduction of N_2 clearly would have conferred a selective advantage on an organism. The fact that none has yet evolved suggests that the design of a system capable of catalyzing N_2 reduction under physiological conditions is a task that is far from trivial. It also makes elucidation of the mechanism by which the existing nitrogenase functions an especially intriguing problem.

The nitrogenase components which have been isolated from photosynthetic bacteria thus far are similar in most respects to those of the more intensively studied microorganisms. In this section, we shall summarize properties which appear to be shared by all nitrogen fixing enzyme systems. The enzymology of nitrogen fixation has been discussed in more detail in recent reviews by Postgate (1970), Burris (1971), Hardy et al. (1971), Dalton and Mortenson (1972), and Benemann and Valentine (1972).

5.1.1 THE Mo-Fe PROTEIN AND THE Fe PROTEIN

Cell-free extracts of nitrogen-fixing microorganisms have been fractionated, always under anaerobic conditions, by a variety of procedures. Bulen and LeComte (1966, 1972), for example, employed differential precipitation by protamine sulfate and separation on DEAE-cellulose and gel columns.

Such separations yield two protein components which are indispensable for nitrogen fixation. One protein contains molybdenum and nonheme iron and is referred to by various authors as fraction or component 1, molybdoferredoxin, Mo-Fe protein, and azofermo. The other contains only nonheme iron, and has been variously designated fraction or component 2, azoferredoxin, Fe protein, and azofer. The reduction of N_2 requires both proteins acting in concert, but it has not been firmly established whether the two proteins form a physical complex. Silverstein and Bulen (1970) suggest that the two proteins are in dynamic equilibrium with a complex which has enzymatic activity. Optimal rates of N_2 reduction are found when the molar ratio of Mo-Fe protein to Fe protein is 1:2 (Vandecasteele and Burris, 1970; Jeng and Mortenson, 1969).

The similarity between the nitrogenase proteins of various organisms is underscored by the fact that active nitrogenase in some cases can be reconstituted by combining Mo-Fe and Fe proteins from different organisms (Detroy et al., 1968). For example, a system containing Mo-Fe protein from *R. rubrum* together with Fe protein from *K. pneumoniae* or *M. flavum* displayed nitrogenase activity comparable to that of a system in which both proteins were obtained from the same bacterium (Biggins et al., 1971). Complementary functioning of *R. rubrum* Mo-Fe protein with Fe protein from *B. polymyxa* and *A. chroococcum* was weaker, however, and no activity was found when the Fe protein was obtained from *C. pasteurianum*. Nitrogenase activity has also been obtained with a system containing Fe protein from the green photosynthetic bacterium *C. ethylicum* and Mo-Fe protein from the blue-green alga *Anabaena cylindrica* (Smith et al., 1971). The reciprocal complementation yielded negative results. In general, complementation is most successful when both proteins are from similar organisms; e.g., when both are facultative anaerobes.

The Mo-Fe proteins from various sources have most recently been reported to have molecular weights ranging from 182,000 (Bergerson and Turner, 1970) to 270,000 (Burns et al., 1970). *Clostridium pasteurianum* Mo-Fe protein, as an example is reported to have a molecular weight of 220,000 and to contain 2 molybdenums, 18 nonheme irons, 18 acid-labile sulfides, and 30 sulfhydryl groups per molecule (Dalton and Mortenson, 1972). The Mo-Fe proteins from other sources have similar, although not identical, properties. Analysis of partially purified Mo-Fe protein from the photosynthetic bacterium *Chromatium* indicates 1.5 Mo, 22 Fe, and 16 sulfides per molecule if a molecular weight (not yet measured) of 220,000 is assumed (M. C. W. Evans, personal communication, 1972). On the basis of disc gel electrophoresis experiments, it has been argued that the Mo-Fe proteins may be composed of from three to six subunits, some of which may be dissimilar (Nakos and Mortenson, 1971a). The Mo-Fe proteins are sensitive to oxygen inactivation. Although crystals of Mo-Fe protein from *A. vinelandii* have been reported by Burns *et al.* (1970), it has not yet been possible to prepare crystals of either nitrogenase protein which are suitable for X-ray crystallography, and indeed the spectrum of the crystals obtained by Burns *et al.* appeared to have a cytochrome component. Therefore, other physical techniques are being employed to determine the structure. Magnetic susceptibility measurements indicate that the Mo-Fe protein of *Azotobacter* contains predominantly high-spin ferric iron (Dalton and Mortenson, 1972). The Mo-Fe protein of *Chromatium* exhibits an electron paramagnetic resonance (EPR) absorption with g values of 4.3, 3.68, and 2.01 in the presence of excess hydrosulfite at temperatures below 20°K (M. C. W. Evans, personal communication, 1972). The signal is unaffected when ^{95}Mo is substituted in the protein, and it is tentatively attributed to an Fe-containing center with a net spin of 3/2. Upon oxidation with methylene blue, the signal disappears. Walker and Mortenson (Dalton and Mortenson, 1972) have found that analogous dye oxidation extracts two electrons from reduced *Azotobacter* Mo-Fe protein.

Upon air oxidation of hydrosulfite-reduced *Chromatium* Mo-Fe protein, EPR signals appear at g values of 4.4, 2.03, and 2.0 (M. C. W. Evans, personal communication, 1972). The $g = 4.4$ signal may be due to high spin ferric iron (Nakazawa *et al.*, 1969).

The Fe protein components of nitrogenase have molecular weights near 50,000 and have been estimated by various authors to contain 1 to 4 nonheme irons and 2 to 4 acid-labile sulfides per molecule. The Fe proteins, including the *Chromatium* Fe protein, exhibit ferredoxin-like EPR signals at $g = 1.94$ (M. C. W. Evans, personal communication, 1972). Dye oxidation is reported to extract a single electron from reduced *Azotobacter* Fe protein (Dalton and Mortenson, 1972). The *C. pasteurianum* protein apparently can be dissociated with sodium dodecyl sulfate (SDS) into two identical subunits with a molecular weight of 27,500 (Nakos and Mortenson, 1971b). Activity of the Fe proteins is lost much more rapidly at 0°C than at room temperature (Moustafa and Mortenson, 1969) even though the Fe proteins are quite stable at liquid nitrogen temperature (Kelly et al., 1967). The Fe protein of the photosynthetic bacterium *Chromatium* D may be an exception to the general cold lability of Fe proteins, for Winter and Arnon (1970) reported that crude *Chromatium* nitrogenase is not cold-labile. The Fe proteins are inactivated by O_2 more rapidly than are the Fe-Mo proteins (Moustafa, 1970).

5.1.2 FERREDOXIN AND FLAVODOXIN

In their successful nitrogen fixation experiments with cell-free extracts of *C. pasteurianum*, Carnahan et al. (1960) included pyruvate in the reaction mixtures. It was soon learned that one function of pyruvate was to provide electrons for the reduction of a nonheme iron-containing protein found in the *C. pasteurianum* extract (Mortenson, 1964a) which was required for the reduction of N_2 by the nitrogenase. The term "ferredoxin" was coined to designate those nonheme iron proteins which function in mediating electron transfer. Ferredoxins subsequently have been isolated from most varieties of nitrogen-fixing microorganisms. They appear to be the immediate donors of electrons to nitrogenase under physiological conditions in nitrogen-fixing cells in which the electron donor has been defined.

Bacterial ferredoxins have molecular weights ranging from 6,000 to 15,000, contain 2–8 nonheme irons per molecule, depending on the source, and an equal amount of acid-labile sulfide, and are characterized by exceptionally low redox potentials. For example, a ferredoxin isolated from *Chromatium* has 8 irons and has a midpoint redox potential of -490 mV (Sasaki and Matsubara, 1967). The characteristics of ferredoxins isolated from several photosynthetic bacteria are summarized in Table I.

Ferredoxins participate in several biochemical pathways other than nitrogen fixation. Their properties and functions have been the subject of several recent reviews (Yoch and Valentine, 1972; Buchanan and Arnon, 1970; Hall and Evans, 1969).

The biosynthesis of some ferredoxins, like that of nitrogenase, may in some cases be regulated by the physiological state of the bacteria. A membrane-bound 6-iron ferredoxin is found in photosynthetically grown *R. rubrum* cells but not in cells grown heterotrophically in the dark, whereas a 2-iron ferredoxin is present under both growth conditions (Shanmugam et al., 1972). *C. ethylicum* and *Chromatium* also recently have been shown to develop 2-iron ferredoxins when growing on N_2 (Evans et al., 1971; M. C. W. Evans, personal communication, 1972).

A type of protein capable of carrying out some of the metabolic functions of ferredoxin but using flavin mononucleotide (FMN) as a prosthetic group instead of iron has been isolated from several micro-

TABLE I
Properties of Ferredoxins from Photosynthetic Bacteria

Organism	Molecular weight	Iron	References
Rhodospirillum rubrum	8,700	6	Shanmugam et al. (1972)
R. rubrum	7,500	2	Shanmugam et al. (1972)
Chromatium	5,600	8	Bachofen and Arnon (1966)
	10,000		Sasaki and Matsubara (1967)
Chromatium		2	M. C. W. Evans (personal communication, 1972)
Chloropseudomonas ethylicum			Rao et al. (1969)
		2	Evans et al. (1971)
Chlorobium thiosulfatophilum	6,000		Hall and Evans (1969)
			Rao et al. (1969)

organisms (Yoch and Valentine, 1972). Designated "flavodoxin," its cellular concentration is often enhanced in bacteria grown on iron-deficient media, a condition which inhibits ferredoxin biosynthesis. The redox potential of the FMN semiquinone–FMN hydroquinone couple is comparable to the ferredoxin redox potential (Mayhew et al., 1969). Such a flavodoxin, of molecular weight 23,000, has been isolated from R. rubrum by Cusanovich and Edmondson (1971), but it has not been shown whether it can replace ferredoxin in mediating electron transfer to nitrogenase.

5.1.3 Electron Acceptors

In the presence of ATP, Mg^{2+}, and a suitable strong reducing agent, nitrogenase can reduce a number of substrates in addition to N_2 and protons. These substrates contain NN, NC, or CC triple bonds and include acetylene, cyanide, azide, N_2O, and alkyl isocyanides. The product invariably is formed by the addition of an even number of electrons to the substrate. The reduction of acetylene to ethylene has become the basis of the most convenient assay for nitrogenase activity since ethylene can be measured very sensitively by gas chromatography (Koch and Evans, 1966). Nitrogenases from all sources thus far examined reduce acetylene in a reaction exhibiting the same requirements (ATP, hydrosulfite or reduced ferredoxin, divalent cation) as N_2 reduction, and at about 4 times the rate of N_2 reduction.

5.1.4 The ATP Requirement

In the cell-free nitrogen fixing system from Chromatium devised by Carnahan and his associates (1960), pyruvate provided electrons for the reduction of ferredoxin by the phosphoroclastic reaction:

$$\text{Pyruvate} + P_i + \text{ferredoxin (oxidized)} \xrightarrow[\text{pyruvate dehydrogenase}]{\text{coenzyme A}} \text{acetyl phosphate}$$
$$+ CO_2 + \text{ferredoxin (reduced)} \quad (1)$$

$$\text{Acetyl phosphate} + \text{ADP} \xrightarrow{\text{acetyl phosphate kinase}} \text{acetate} + \text{ATP} \quad (2)$$

There were indications that N_2 reduction also required a metabolic energy source (McNary and Burris, 1962), and it was suggested that the phosphoroclastic oxidation of pyruvate might furnish the energy (reaction 2) as well as supplying electrons for the reduction of ferredoxin. This was confirmed by the demonstration that the pyruvate requirement in cell-free reduction of N_2 could be eliminated if ATP was added to the reaction mixture (Hardy and D'Eustachio, 1964; Mortenson, 1964b; Bulen et al., 1964) and if ferredoxin was reduced by H_2 in a hydrogenase-mediated reaction (D'Eustachio and Hardy, 1964). ADP has been found to inhibit nitrogenase, so ATP is ordinarily furnished by an ATP-generating system, such as creatine phosphate plus creatine kinase.

This elucidation of the energy requirement made it possible to search for alternative electron donors. Bulen and his co-workers (1965a) found that hydrosulfite could donate electrons directly to nitrogenase, thus permitting a significant reduction in the complexity of cell-free nitrogen fixing systems.

Although the requirement for ATP is well documented, the function of ATP is still a matter of controversy. For discussions of some of the possibilities that have been suggested, we refer the reader to papers by Bulen et al. (1965b), Hardy et al. (1971), and Benemann and Valentine (1972).

Nitrogenase catalyzes a slow hydrolysis of ATP to ADP and inorganic phosphate even when electron transfer is not occurring, i.e., in the absence of a reductant. But in the presence of a reductant such as hydrosulfite or reduced ferredoxin, the rate of ATP hydrolysis increases as much as 7-fold (Hadfield and Bulen, 1969). Simultaneously electrons are transferred from the reductant to an electron acceptor such as N_2 or acetylene or to protons. Both nitrogenase proteins are required for ATP hydrolysis and the associated electron transfer reactions (Bulen and LeComte, 1966). The rate of ATP hydrolysis and the total rate of electron flow are independent of the nature or the concentration of the electron acceptor. If the acceptor concentration is decreased or if a poorer electron acceptor is substituted, a larger fraction of the electron flow is diverted to H_2 prodution, but the total rate of electron flow remains constant. Therefore, the rate-limiting step in the ATP-coupled electron flow is not the transfer of electrons from nitrogenase to the electron acceptor.

The situation may be somewhat different in intact cells, where H_2 evolution attributable to nitrogenase is not ordinarily observed. Postgate (1969) has suggested that the hydrogen-evolving site on the nitrogenase protein may be formally anhydrous in living cells.

Hadfield and Bulen (1969) have suggested that nitrogenase can catalyze the hydrolysis of ATP by two mechanisms. One is coupled to the transfer of electrons from a reductant to an electron acceptor while the other is not. If it is assumed that the rate of the ATP hydrolysis reaction which is not coupled to electron transfer remains the same in the presence or in the absence of electron flow, the ratio of ATP molecules hydrolyzed in the coupled reaction to electrons transferred is 1:1. But it has not been shown unequivocally that the rate of noncoupled ATP hydrolysis is the same in the presence and absence of a reductant.

Uncouplers of oxidative and photosynthetic phosphorylation have no effect on the coupled electron flow and ATP hydrolysis that is catalyzed by nitrogenase. In the presence of D_2, ATP, N_2, and a suitable reductant, nitrogenase catalyzes the formation of HD (Jackson et al., 1968).

The detailed mechanism of nitrogenase activity remains an enigma. There are conflicting reports about which of the nitrogenase proteins interacts with ATP; the sequence of electron flow through the Fe protein and the Mo-Fe protein is uncertain; no intermediates at levels of reduction between N_2 and NH_3 have been isolated or even determined. There is much speculation, but little direct evidence, on the binding of N_2 and other electron acceptors to nitrogenase and the role of Mo and Fe in the binding. Still, preliminary reports of research currently underway seem encouraging. M. C. W. Evans (personal communication, 1972) found that the EPR signal at $g = 4.3$ and 3.68 in the Mo-Fe protein of *Chromatium* disappears when ATP is added to a mixture of Mo-Fe protein and Fe protein in the presence of hydrosulfite. Since the signal reappears when hydrosulfite or ATP is exhausted, he suggested that the disappearance of the signal may represent a further, ATP-driven, reduction of the Mo-Fe protein. Orme Johnson et al. (1972) reported similar results with *C. pasteurianum* and *A. vinelandii* nitrogenase. They speculated that a complex of reduced Fe protein and Mg·ATP reduces the Mo-Fe protein, and that in the presence of Mg^{2+} and ATP, the reduced Fe and Mo-Fe proteins are oxidized by the substrate.

5.2 Nitrogen Fixation in Specific Cell-Free Extracts

5.2.1 *Rhodospirillum rubrum*

Schneider and his associates first reported nitrogen fixation in cell-free extracts of *R. rubrum* in 1960. Their extracts were supplemented with pyruvate, as were those of Carnahan et al. in the *Clostridium* experiments reported in the same year. It is now known that *R. rubrum* does not possess the phosphoroclastic reaction, so the source of energy and electrons in these experiments is not clear.

Subsequent work with extracts of photosynthetic bacteria has paralleled the studies with other organisms. Thus Bulen et al. (1965a) in a single publication reported that hydrosulfite could serve as an electron donor for N_2 reduction in *A. vinelandii* and *R. rubrum* extracts, and that both extracts displayed ATP-dependent H_2 evolution.

Burns and Bulen (1966) found that the stability of *R. rubrum* extracts could be improved if particulate matter was removed by ultracentrifugation and the supernatant then passed through Sephadex. They found that the extracts were cold-labile, in contrast to the cold stability of *Chromatium* extracts reported by Winter and Arnon (1970).

Munson and Burris (1969) employed continuous culture of *R. rubrum* cells to obtain reproducible nitrogenase activity in cell-free extracts. Cells were broken by sonication in an H_2 atmosphere. The broken cell suspensions were centrifuged and the supernatants were passed anaerobically through a Sephadex G-25 column. Hydrosulfite supplied electrons and the creatine phosphate–creatine kinase system with a catalytic amount of ATP (5 mM) served to generate ATP. Specific activities of the order of 3 nmoles of N_2 or 10 nmoles of acetylene reduced per milligram of protein per minute were typical. The usual range of substrates was found to be reducible. Activity was found only in extracts of cells in which nitrogenase had been induced by growth in medium containing limiting amounts of fixed nitrogen. The rate of N_2 reduction was measured as a function of N_2 partial pressure. A Lineweaver-Burk plot of the data indicated the enzyme system to be half saturated at a pN_2 of 0.071 atm, which is close to that reported for other systems.

Nitrogenases from photosynthetic bacteria seem to be similar to those from other microorganisms in almost every respect. Thus the photosynthetic bacteria are unique and interesting not because of the nitrogenase protein, but because the reducing power and ATP required for N_2 fixation may be generated photochemically in reactions unique to these organisms.

Whether photochemical reactions in fact provide ATP or electrons directly for nitrogen fixation in *R. rubrum* has not been established. *Rhodospirillum rubrum* readily forms ATP in a reaction coupled to cyclic photosynthetic electron transport (Frenkel, 1954), and it seems quite probable that this photophosphorylation furnishes at least part of the ATP required for N_2 reduction in intact cells. It is less likely that the electrons for N_2 reduction are supplied photochemically. Photochemical reduction of ferredoxin by *R. rubrum* extracts has not been demonstrated, and since the midpoint redox potential of the primary photochemical electron acceptor—presumably the strongest reductant that is generated photochemically—is -145 mV (Cramer, 1969), photochemical reduction of ferredoxin seems improbable in this bacterium. It is more probable that ferredoxin is reduced by reduced pyridine nucleotide or by substrate-linked pathways, as in nonphotosynthetic bacteria. Pyridine nucleotide is reduced in *R. rubrum* by a dark, energy-driven reversal of electron transport. Succinate is the probable source of electrons and energy can be supplied by light, ATP, or inorganic pyrophosphate (Keister and Yike, 1967a). As discussed in Section 3, Schick (1971) found that N_2 uptake continued for up to 20 minutes in the dark at a constantly decreasing rate after illumination. Therefore, it is apparent that both ATP and electrons can be supplied in dark reactions provided endogenous substrate is available.

5.2.2 *Chromatium*

The first attempts to demonstrate nitrogen fixation with cell-free extracts of *Chromatium* were reported in 1961 by Arnon and his co-workers, and Winter and Arnon (1970) have continued these studies in more defined and sophisticated systems. In most respects their procedures and results paralleled the *R. rubrum* experiments of Munson and Burris (1969). Winter and Arnon were able to use H_2 as a source of electrons if catalytic amounts of viologen dyes were present, since hydrogenase was present in the soluble extract. In addition, these workers demonstrated an ATP-dependent H_2 evolution by the *Chromatium* extracts. Unlike classical hydrogenase, this H_2 evolution was insensitive to CO but was inhibited by N_2. The total rate of electron flow during N_2 reduction, i.e., the sum of the rates of electron flow to N_2 and H^+, was approximately equal to the rate of electron flow to protons alone under argon in the absence of N_2.

In an experiment which has primarily heuristic value, Yoch and Arnon (1970) were able to drive N_2 or acetylene reduction in cell-free *Chromatium* extracts with ferredoxin photochemically reduced with spinach chloroplasts, and ATP formed by photophosphorylation with *R. rubrum* chromatophores. As in the case of *R. rubrum*, and by a similar argument, it seems reasonable to suppose that some of the ATP required for N_2 reduction in *Chromatium* is supplied by the photosynthetic apparatus, but the phosphoroclastic reaction would appear to be a more probable electron source. Bennett et al. (1964) have shown that pyruvate can be used to drive N_2 reduction in *Chromatium* extracts.

5.2.3 *Chloropseudomonas ethylicum*

Evans and Smith (1971) have recently demonstrated nitrogen fixation in cell-free extracts of this green photosynthetic bacterium. In most respects the system behaved similarly to the cell-free *R. rubrum* and *Chromatium* extracts. Evans and Smith observed that in the *C. ethylicum* system, N_2 reduction could be driven by pyruvate if ferredoxin was present. However, if exogenous ATP was supplied, electrons could be furnished by ferredoxin that had been photoreduced by *C. ethylicum* particles. Particles from the green photosynthetic bacterium *Chlorobium thiosulfatophilum* can also reduce ferredoxin photochemically at a low rate (Evans and Buchanan, 1965). Evans and Smith speculated that in the green photosynthetic bacteria ferredoxin may be reduced photochemically when cells are grown photoautotrophically or on 2-carbon substrates, but by pyruvate when cells are grown heterotrophically on other substrates.

The generation of reducing power in bacterial photosynthesis is the subject of a recent article by Gest (1972), and the reader is referred to this source for a thoughtful discussion of the subject.

6. Ecological Significance of Nitrogen Fixation in Photosynthetic Bacteria

Photosynthetic bacteria are widely distributed in nature and under appropriate conditions can fix nitrogen and CO_2. The study of the ecology of this group of organisms is scanty, and we can only guess at their contribution in the natural fixation of molecular nitrogen. There are, however, a few studies, mainly from Kobayashi and his associates (Okuda *et al.*, 1959, 1969; Kobayashi *et al.*, 1967; Kobayashi and Hague, 1971), which indicate that their contribution may be significant.

Photosynthetic bacteria are particularly widely distributed in soils of southeast Asia. Kobayashi *et al.* (1967) found that these microorganisms are more abundant than nitrogen-fixing blue-green algae and that collectively the photosynthetic nitrogen-fixing forms are more abundant than the heterotrophic forms. It is difficult to determine which organisms are responsible for gains in total nitrogen which occur, and the assumption could be made that since the bacteria require anaerobic conditions for nitrogen fixation, the blue-green algae are primarily responsible. However, Okuda *et al.* (1960) and Kobayashi and Hague (1971) have found that *Rhodopseudomonas capsulata*, when grown in mixed culture with *Propionibacterium*, *A. vinelandii*, or *Bacillus megaterium*, fixed nitrogen even under what appeared to be aerobic conditions. Further, they found that a correlation exists between the growth of photosynthetic bacteria and reproduction of the rice plant. Such studies have led this group to conclude that photosynthetic bacteria contribute significantly to soil fertility and improvement of plant growth. Obviously our current knowledge in this area is infantile.

7. Concluding Remarks

It has been stated that nitrogen fixation is one of the least exploited of the important biological discoveries of modern times (Hedén, 1964). Recent progress in unveiling the chemical mechanism of nitrogen fixation leads us to believe that an understanding of this important process may come in the near future, and this understanding may enable the chemist to design an efficient synthetic catalyst effective in catalyzing dinitrogen reduction. The benefits for mankind associated with this are manifold.

For the photobiologist, photosynthetic bacteria, with their remarkable adaptability, and blue-green algae would seem to offer considerable potential for studying the physiological control mechanisms involved in nitrogen fixation and the interactions between the nitrogenase and the energy-generating systems.

For the molecular biologist, the recent reports on the successful transfer of the genetic information for nitrogen fixation from nitrogen-fixing bacteria (*K. pneumoniae*) to nonfixing bacteria (*E. coli* or nonfixing strains of *K. pneumoniae*) (Dixon and Postgate, 1971, 1972; Streicher *et al.*, 1971) are an important milestone, for they open the door for further studies concerning the genetic control of nitrogen fixation. An understanding of the molecular biology of this system may lead eventually to the development of super effective strains of rhizobia and to the development of new nitrogen-fixing strains of bacteria with potential agronomic value.

Acknowledgments

The authors express their appreciation to Drs. William Bulen, Berger Mayne, Dan Reed, and Gerald Peters for helpful criticism of the manuscript.

References

Allen, M. B. (1972). In "Photophysiology" (A. C. Giese, ed.), Vol. 7, pp. 73–84. Academic Press, New York.
Arnon, D. I., Lasada, M., Nozaki, M., and Tagawa, K. (1961). *Nature (London)* **190**, 601.
Atkinson, D. E. (1968). *Biochemistry* **7**, 4030.
Bach, A. N., Jermolieva, Z. V, and Stepanian, M. P. (1934). *Dokl. Akad. Nauk SSSR* **1**, 22.
Bachofen, R., and Arnon, D. I. (1966). *Biochim. Biophys. Acta* **120**, 259.
Baltscheffsky, H., Baltscheffsky, M., and Thore, A. (1971). *Curr. Top. Bioenerg.* **4**, 273–325.
Bayliss, N. S. (1956). *Aust. J. Biol. Sci.* **9**, 364.
Benemann, J. R., and Valentine, R. C. (1971). *Advan. Microbiol. Physiol.* **5**, 135.
Benemann, J. R., and Valentine, R. C. (1972). *Advan. Microbiol. Physiol.* **8**, 59.
Bennett, R., Rigopoulos, N., and Fuller, R. C. (1964). *Proc. Nat. Acad. Sci. U. S.* **52**, 762.
Bergersen, F. J. (1962). *J. Gen. Microbiol.* **29**, 113.
Bergersen, F. J. (1971). *Annu. Rev. Plant Physiol.* **22**, 121.

Bergerson, F. J., and Turner, G. L. (1970). *Biochim. Biophys. Acta* **214**, 28.
Biedermann, M., Drews, G., Marx, R., and Schröder, J. (1967). *Arch. Mikrobiol.* **56**, 133.
Biggins, D. R., Kelly, M., and Postgate, J. R. (1971). *Eur. J. Biochem.* **20**, 140.
Bose, S. K., and Gest, H. (1963). In "Energy-Linked Functions of Mitochondria" (B. Chance, ed.), p. 207. Academic Press, New York.
Buchanan, B. B., and Arnon, D. I. (1970). *Advan. Enzymol.* **33**, 119.
Bulen, W. A., and LeComte, J. R. (1966). *Proc. Nat. Acad. Sci. U. S.* **56**, 979.
Bulen, W. A., and LeComte, J. R. (1972). In "Methods in Enzymology" (A. San Pietro, ed.), Vol. 24, p. 456. Academic Press, New York.
Bulen, W. A., Burns, R. C., and LeComte, J. R. (1964). *Biochem. Biophys. Res. Commun.* **17**, 265.
Bulen, W. A., Burns, R. C., and LeComte, J. R. (1965a). *Proc. Nat. Acad. Sci. U. S.* **53**, 532.
Bulen, W. A., LeComte, J. R., Burns, R. C., and Hinkson, J. (1965b). In "Non-Heme Iron Proteins" (A. San Pietro, ed.), pp. 261–274. Antioch Press, Yellow Springs, Ohio.
Burns, R. C. (1969). *Biochim. Biophys. Acta* **171**, 253.
Burns, R. C., and Bulen, W. A. (1966). *Arch. Biochem. Biophys.* **113**, 461.
Burns, R. C., Holsten, R. D., and Hardy, R. W. F. (1970). *Biochem. Biophys. Res. Commun.* **39**, 90.
Burris, R. H. (1942). *J. Biol. Chem.* **143**, 509.
Burris, R. H. (1971). In "The Chemistry and Biochemistry of Nitrogen Fixation" (J. R. Postgate, ed.), pp. 105–160. Plenum, New York.
Burris, R. H., and Miller, E. C. (1941). *Science* **93**, 114.
Burris, R. H., and Wilson, P. W. (1946). *J. Bacteriol.* **52**, 505.
Carnahan, J. E., Mortenson, L. E., Mower, H. F., and Castle, J. E. (1960). *Biochim. Biophys. Acta* **38**, 188.
Cohen-Bazire, G., Sistrom, W. R., and Stanier, R. Y. (1957). *J. Cell. Comp. Physiol.* **49**, 25.
Collins, V. G. (1958). *Proc. Int. Congr. Microbiol., 7th, 1958* Abstracts, p. 71.
Cramer, W. A. (1969). *Biochim. Biophys. Acta* **189**, 54.
Crofts, A. R. (1969). In "Electron Transport and Energy Conservation" (J. M. Tager et al., eds.), pp. 221–228. Adriatica Editrice, Bari, Italy.
Cusanovich, M. A., and Edmondson, D. E. (1971). *Biochem. Biophys. Res. Commun.* **45**, 327.
Daesch, G., and Mortenson, L. E. (1968). *J. Bacteriol.* **96**, 346.
Daesch, G., and Mortenson, L. E. (1972). *J. Bacteriol.* **110**, 103.
Dalton, H., and Mortenson, L. E. (1972). *Bacteriol. Rev.* **36**, 231.
Dalton, H., and Postgate, J. R. (1969). *J. Gen. Microbiol.* **54**, 463.
Davis, L. C., Shah, V. K., Brill, W. J., and Orme-Johnson, W. H. (1972). *Biochim. Biophys. Acta* **256**, 512.
Detroy, R. W., Witz, D. F., Parejko, R. A., and Wilson, P. W. (1968). *Proc. Nat. Acad. Sci. U. S.* **61**, 537.
D'Eustachio, A. J., and Hardy, R. W. F. (1964). *Biochem. Biophys. Res. Commun.* **15**, 319.
Dixon, R. A., and Postgate, J. R. (1971). *Nature (London)* **234**, 47.
Dixon, R. A., and Postgate, J. R. (1972). *Nature (London)* **237**, 102.
Dixon, R. O. D. (1968). *Arch. Microbiol.* **62**, 272.
Drozd, J., and Postgate, J. R. (1970). *J. Gen. Microbiol.* **63**, 63.
Eisenberg, M. A. (1953). *J. Biol. Chem.* **203**, 815.

Elmerich, C., and Aubert, J.-P. (1971). *Biochem. Biophys. Res. Commun.* **42**, 371.
Evans, M. C. W., and Buchanan, B. B. (1965). *Proc. Nat. Acad. Sci. U. S.* **53**, 1420.
Evans, M. C. W., and Smith, R. V. (1971). *J. Gen. Microbiol.* **65**, 95.
Evans, M. C. W., Smith, R. V., Telfer, A., and Cammack, R. (1971). *Proc. Eur. Biophys. Congr., 1971* Vol. IV, p. 115.
Fedorov, M. V., and Kalininskaya, T. A. (1961). *Mikrobiologiya* **30**, 9.
Frenkel, A. W. (1954). *J. Amer. Chem. Soc.* **76**, 5568.
Frenkel, A. W. (1970). *Biol. Rev. Cambridge Phil. Soc.* **45**, 569.
Gest, H. (1951). *Bacteriol. Rev.* **15**, 183.
Gest, H. (1972). *Advan. Microbiol. Physiol.* **7**, 243.
Gest, H., and Kamen, M. D. (1949). *Science* **109**, 558.
Gest, H., Kamen, M. D., and Bregoff, H. M. (1950). *J. Biol. Chem.* **182**, 153.
Gest, H., Ormerod, J. G., and Ormerod, K. S. (1962). *Arch. Biochem. Biophys.* **97**, 21.
Grau, F. H., and Wilson, P. W. (1962). *J. Bacteriol.* **83**, 490.
Gray, C. T., and Gest, H. (1965). *Science* **148**, 186.
Hadfield, K. L., and Bulen, W. A. (1969). *J. Amer. Chem. Soc.* **8**, 5103.
Hall, D. O., and Evans, M. C. W. (1969). *Nature (London)* **223**, 1342.
Hamilton, P. B., and Wilson, P. W. (1955). *Ann. Acad. Sci. Fenn., Ser. A2* p. 139.
Hardy, R. W. F., and D'Eustachio, A. J. (1964). *Biochem. Biophys. Res. Commun.* **15**, 314.
Hardy, R. W. F., Burns, R. C., and Parshall, G. W. (1971). *Advan. Chem. Ser.* **100**, 219.
Hedén, C.-G. (1964). *In* "The Population Crisis and the Use of World Resources" (S. Mudd, ed.), p. 478. Junk, The Hague.
Hoch, G. E., Little, H. N., and Burris, R. H. (1957). *Nature (London)* **179**, 430.
Hughes, D. E., and Wimpenny, J. W. T. (1969). *Advan. Microbiol. Physiol.* **3**, 197.
Jackson, E. K., Parshall, G. W., and Hardy, R. W. F. (1968). *J. Biol. Chem.* **19**, 4952.
Jeng, D., and Mortenson, L. E. (1969). *Abstr. 158th Nat. Meet., Amer. Chem. Soc.* p. 277.
Jones, B. K. (1956). *Sewage Ind. Wastes* **28**, 883.
Jungermann, K., Rupprecht, E., Ohrloff, C., Thauer, R. K., and Decker, K. (1971). *J. Biol. Chem.* **246**, 960.
Kamen, M. D., and Gest, H. (1949). *Science* **109**, 560.
Keister, D. L., and Yike, N. J. (1967a). *Arch. Biochem. Biophys.* **121**, 415.
Keister, D. L., and Yike, N. J. (1967b). *Biochemistry* **6**, 3847.
Kelly, M. (1969). *Biochim. Biophys. Acta* **171**, 9.
Kelly, M., Postgate, J. R., and Richards, R. L. (1967). *Biochem. J.* **102**, 1C.
Kennedy, I. R. (1970). *Biochim. Biophys. Acta* **222**, 135.
Kobayashi, M., and Haque, M. Z. (1971). *In* "Biological Nitrogen Fixation in Natural and Agricultural Habitats" (T. A. Lie and E. G. Mulder, eds.), *Plant and Soil*, Special Volume, pp. 443–456. Nijhoff, The Hague.
Kobayashi, M., Takahashi, E., and Kawaguchi, K. (1967). *Soil Sci.* **104**, 113.
Koch, B., and Evans, H. J. (1966). *Plant Physiol.* **41**, 1748.
Koch, B., Evans, M. C. W., and Russell, S. (1967). *Proc. Nat. Acad. Sci. U. S.* **58**, 1343.
Kondrat'eva, E. N. (1965). "Photosynthetic Bacteria." Israel Program Scientific Translations, Jerusalem.
Lee, C.-P., and Ernster, L. (1966). *In* "Regulation of Metabolic Processes in Mitochondria" (J. M. Tager *et al.*, eds.), pp. 218–234. Amer. Elsevier, New York.

Lee, S. B., and Wilson, P. W. (1943). *J. Biol. Chem.* **151**, 377.
Lie, T. A., and Mulder, E. G., eds. (1971). "Biological Nitrogen Fixation in Natural and Agricultural Habitats," *Plant and Soil*, Special Volume. Nijhoff, The Hague.
Lindstrom, E. S., Burris, R. H., and Wilson, P. W. (1949). *J. Bacteriol.* **58**, 313.
Lindstrom, E. S., Tove, S. R., and Wilson, P. W. (1950). *Science* **112**, 197.
Lindstrom, E. S., Lewis, S. M., and Pinsky, M. J. (1951). *J. Bacteriol.* **61**, 481.
Lyubimor, V. I., L'vov, N. P., Kirshteine, B. E., and Kretovich, V. L. (1969). *Izv. Akad. Nauk SSSR, Ser. Biol.* No. 4, p. 505.
McNary, J. E., and Burris, R. H. (1962). *J. Bacteriol.* **84**, 589.
Mayhew, S. G., Foust, G. P., and Massey, V. (1969). *J. Biol. Chem.* **244**, 803.
Meers, J. L., Tempest, D. W., and Brown, C. M. (1970). *J. Gen. Microbiol.* **64**, 187.
Miller, R. E., and Stadtman, E. R. (1972). *J. Biol. Chem.* **247**, 7407.
Mitchell, P. (1967). *Fed. Proc., Fed. Amer. Soc. Exp. Biol.* **26**, 1370.
Mortenson, L. E. (1964a). *Biochim. Biophys. Acta* **81**, 473.
Mortenson, L. E. (1964b). *Proc. Nat. Acad. Sci. U. S.* **52**, 272.
Mortenson, L. E., Valentine, R. C., and Carnahan, C. (1963). *J. Biol. Chem.* **238**, 794.
Moustafa, E. (1970). *Biochim. Biophys. Acta* **206**, 178.
Moustafa, E., and Mortenson, L. E. (1967). *Nature (London)* **216**, 1241.
Moustafa, E., and Mortenson, L. E. (1969). *Biochim. Biophys. Acta* **172**, 106.
Munson, T. O., and Burris, R. H. (1969). *J. Bacteriol.* **97**, 1093.
Nagatani, H., Shimizu, M., and Valentine, R. C. (1971). *Arch. Mikrobiol.* **79**, 164.
Nakazawa, T., Nozaki, M., Yamano, T., and Hayaishi, O. (1969). *J. Biol. Chem.* **244**, 119.
Nakos, G., and Mortenson, L. (1971a). *Biochim. Biophys. Acta* **229**, 431.
Nakos, G., and Mortenson, L. (1971b). *Biochemistry* **10**, 455.
Oelze, J., and Drews, G. (1972). *Biochim. Biophys. Acta* **265**, 209.
Okuda, A., Yamaguchi, M., and Kobayashi, M. (1959). *Plant Soil Food* **5**, 73.
Okuda, A., Yamaguchi, M., and Kobayashi, M. (1960). *Plant Soil Food* **6**, 35.
Oppenheim, J., Fisher, R. J., Wilson, P. W., and Marcus, L. (1970). *J. Bacteriol.* **101**, 292.
Orme-Johnson, W. H., Hamilton, W. D., Jones, T. L., Tso, M.-Y., Burris, R. H., Shah, V. K., and Brill, W. J. (1972). *Proc. Nat. Acad. Sci. U. S.* **69**, 3142.
Ormerod, J. G., Ormerod, K. S., and Gest, H. (1961). *Arch. Biochem. Biophys.* **94**, 449.
Parejko, R. A., and Wilson, P. W. (1970). *Can. J. Microbiol.* **16**, 681.
Parker, C. A., and Scutt, P. B. (1960). *Biochim. Biophys. Acta* **38**, 230.
Pengra, R. M., and Wilson, P. W. (1958). *J. Bacteriol.* **75**, 21.
Pfennig, N. (1967). *Annu. Rev. Microbiol.* **21**, 285.
Pine, M. J., and Barker, H. A. (1954). *J. Bacteriol.* **68**, 589.
Postgate, J. R. (1969). *Proc. Roy. Soc., Ser. B* **172**, 355.
Postgate, J. R. (1970). *Nature (London)* **226**, 25.
Postgate, J. R. (1971). *In* "The Chemistry and Biochemistry of Nitrogen Fixation" (J. R. Postgate, ed.), pp. 161–190. Plenum, New York.
Pratt, D. C., and Frenkel, A. W. (1959). *Plant Physiol.* **34**, 333.
Rao, K. K., Matsubara, H., Buchanan, B. B., and Evans, M. W. C. (1969). *J. Bacteriol.* **100**, 1411.
Sasaki, R. M., and Matsubara, H. (1967). *Biochem. Biophys. Res. Commun.* **28**, 467.
Schick, H.-J. (1971). *Arch. Mikrobiol.* **75**, 89.
Schneider, K. C., Bradbeer, C., Singh, R. N., Wang, L. C., Wilson, P. W., and Burris, R. H. (1960). *Proc. Nat. Acad. Sci U. S.* **46**, 726.

Shah, V. K., Davis, L. C., and Brill, W. J. (1972). *Biochim. Biophys. Acta* **256**, 498.
Shanmugam, K. T., and Arnon, D. I. (1972). *Biochim. Biophys Acta* **256**, 487.
Shanmugam, K. T., Buchanan, B. B., and Arnon, D. I. (1972). *Biochim. Biophys. Acta* **256**, 477.
Silverstein, R., and Bulen, W. A. (1970). *Biochemistry* **9**, 3809.
Sisler, F. D., and ZoBell, C. E. (1951). *Science* **113**, 511.
Smith, R. V., and Evans, M. C. W. (1970). *Nature (London)* **225**, 1253.
Smith, R. V., Telfer, A., and Evans, M. C. W. (1971). *J. Bacteriol.* **107**, 574.
Sorger, G. J. (1969). *J. Bacteriol.* **98**, 56.
Stanier, R. Y. (1961). *Bacteriol. Rev.* **25**, 1.
Stewart, W. D. P. (1966). "Nitrogen Fixation in Plants." Oxford Univ. Press (Athlone), London and New York.
Strandberg, G. W., and Wilson, P. W. (1967). *Can. J. Microbiol.* **14**, 25.
Streicher, S., Gurney, E., and Valentine, R. C. (1971). *Proc. Nat. Acad. Sci. U. S.* **68**, 1174.
Tempest, D. W., Meers, J. L., and Brown, C. M. (1970). *Biochem. J.* **117**, 405.
Thauer, R. K., Jungermann, K., Rupprecht, E., and Decker, K. (1969). *FEBS Lett.* **4**, 108.
Vandecasteele, J.-P., and Burris, R. H. (1970). *J. Bacteriol.* **101**, 794.
Vernon, L. P. (1968). *Bacteriol. Rev.* **32**, 243.
Wall, J. S., Wagenknecht, A. C., Newton, J. W., and Burris, R. H. (1952). *J. Bacteriol.* **63**, 563.
Wilson, P. W. (1940). "The Biochemistry of Symbiotic Nitrogen Fixation." Univ. of Wisconsin Press, Madison.
Wilson, P. W. (1971). *In* "The Chemistry and Biochemistry of Nitrogen Fixation" (J. R. Postgate, ed.), pp. 1–18. Plenum, New York.
Wilson, P. W., and Umbreit, W. W. (1937). *Arch. Mikrobiol.* **8**, 440.
Winter, H. C., and Arnon, D. I. (1970). *Biochim. Biophys. Acta* **197**, 170.
Wong, P. P., and Burris, R. H. (1972). *Proc. Nat. Acad. Sci. U. S.* **69**, 672.
Yates, M. G. (1970). *J. Gen. Microbiol.* **60**, 393.
Yoch, D. C., and Arnon, D. I. (1970). *Biochim. Biophys. Acta* **197**, 180.
Yoch, D. C., and Pengra, M. (1966). *J. Bacteriol.* **92**, 618.
Yoch, D. C., and Valentine, R. C. (1972). *Annu. Rev. Microbiol.* **26**, 139.

Chapter 6

PARALLEL AND SEQUENTIAL PROCESSING OF VISUAL INFORMATION IN MAN: INVESTIGATION BY EVOKED POTENTIAL RECORDING

D. Regan

Department of Communication, University of Keele, Keele, Staffordshire, England

1. Introduction 185
2. Methods of Recording Evoked Potentials in Man 188
3. Examples of Evoked Potential Studies of Visual Information Processing . 192
   3.1 Correlations between EP Features and Visual Perception 192
   3.2 Parallel Functional Channels; Differentiation by EP Recording . . 196
   3.3 Peripheral and Central Mechanisms in the Visual Pathway . . . 201
4. Summary 207
   References 207

1. Introduction

Up to the last few years, access to human visual function has been almost entirely by means of subjective reports and psychophysical measurements. Objective measures, both of the gross organization of neural pathways and of the electrical activities of individual nerve cells, have been developed, but at the present time these methods are not appropriate for studies on man. Microelectrode recordings of the firing of one or two nerve cells in the visual pathway has already provided revealing clues to visual function, even though this technique is still at an early stage of development. Objective demonstrations of the existence of a number of specialized neural organizations within the visual pathways of cat and monkey have been made possible by microelectrode recording. Among neural organizations so far found are those that seem to be specialized for handling visual information of spatial pattern, or color, or retinal disparity.

Although the existence of specialized organizations within the human visual pathway had been suspected on psychophysical grounds long before the current flood of electrophysiological studies, microelectrode recording has greatly extended our knowledge of these specialized mechanisms. However, almost all microelectrode studies have been on animals. Whether the microorganization of the human visual pathway and that of the cat or monkey differ greatly, or only in detail, is not known. However, it is known that the gross neuroanatomy of visual cortex differs markedly between man and even rhesus monkey.

All this indicates that extrapolation across the interspecies gap from monkey or cat to man will be uncertain until there is available some means of bridging the gap by practical experiment. The subject of this chapter, evoked potential recording, is an experimental method that promises to offer an objective means of bridging this interspecies gap.

Before describing the techniques of recording evoked potentials (EP's), it is well to outline both the types of scientific question that are accessible to EP methods and the types of question that can be attacked by EP methods more advantageously than by either the psychophysical or the single-neuron approach.

The EP's discussed here are the minute electrical signals elicited by sensory stimulation that can be recorded from electrodes on the human scalp. These signals offer objective measures of the electrical activities of the normally functioning human brain. They are most probably generated by cortical neurons. However, it does not follow that their properties are entirely those of cortical cells. Although EP's may be generated by cortical cells, their properties may be determined by more peripheral structures. For example, scalp EP's may exhibit the Stiles-Crawford effect (Section 3.3.2) characteristic of retinal cones, or may even be used to measure the chromatic aberration of the eye (see Fig. 5). In contrast, binocular depth EP's may be determined by neural processing central to the stage at which the signals from the two eyes interact with each other. Thus, by manipulating the stimulus and the method of signal analysis, EP recording from scalp electrodes can be used as a tool to study visual information processing at a number of restricted sites between the cornea and the cortical cells that generate EP's.

A second way in which EP's may be used is to allow objective studies of the parallel functional channels that exist within the organization of the visual pathway. The various features of EP's that are specific to the processing of spatial contrast (spatial patterning), brightness, color, and retinal disparity can be separately identified and used as tools to study the processing of neural signals that carry information about these stimulus parameters, even though to some extent the different signals are processed in parallel.

One possibility offered by EP recording is that of investigating the processing of visual information not available to perception. A second possibility is that of separately studying the successive transformations that visual information undergoes as it travels centrally. These two possibilities form one side of the coin, whose reverse is the accumulation of findings that in many experimental situations there is a poor correlation between EP's and perception. In these areas EP, methods offer a complementary approach to that of psychophysics.

The most obvious differences between single neuron recording and EP recording are that: (1) single neuron records are often in the form of pulses whereas EP's take the form of graded slow waves; (2) EP's may conveniently be recorded from man and single neuron records may not. A more fundamental distinction than these is that single neuron records give a precise account of the activity of one or a very few neurons over a period of a few hours. There are, however, some ten thousand million cells in the human brain. In contrast, EP's offer information about the behavior of large populations of cells. Although EP methods are developing the capability of studying small regions of the brain, these regions must always be much larger than a single cell: one could hardly hope to study a cortical cell population comprising less than some tens of thousands of cells. It is not yet known how serious are the restrictions imposed on single neuron data by the difficulty of investigating relationships between the simultaneous activities of large numbers of neurons. Neither is it known what restrictions are placed upon the value of EP data by the comparatively gross nature of the recordings. If some aspects of visual information find physical representation in the state of large populations of neurons, then EP methods may prove to have the advantage over single-neuron recording in those studies. On the other hand, if in certain experimental situations neighboring cells represent quite different visual information, then a sample of the simultaneous activities of many cells would look like a meaningless jumble, and single neuron recording would have the advantage over EP methods in those experimental situations. To the question, "which sensory information is represented by the activities of a few cells, and which sensory information is represented by the states of large populations of cells?" no general and (probably) no specific answers are available, so that the single neuron and EP methods can best be regarded as complementary until proved otherwise.

This chapter discusses the use of EP's as a tool for investigating the neural processing of visual information. The emphasis is on steady-state EP's. The use of EP's for a number of quite different purposes has been reviewed elsewhere (Regan, 1972b), including EP studies of psychophysical variables such as attention, slow (dc) potentials that reflect the subject's conditional expectancy, brain activities that accompany movements of parts of the body, and changes in EP's that are characteristic of neurological diseases.

2. Methods of Recording Evoked Potentials in Man

A variety of electrodes have been used in EP recording. One common type is a hollowed silver disk some 1 cm in diameter. The disk is coated with silver chloride and held onto the scalp by glue or by a flexible cap. Contact resistance is reduced either by injecting saline jelly underneath the electrode or by surrounding the metal with a saline-soaked pad.

Fluctuations of the potential difference between two such electrodes can be recorded by connecting them to the input terminals of a high-impedance differential amplifier. This electrical activity is called the electroencephalogram (EEG). It has a bandwidth of dc to some 100 Hz,* but because of the technical difficulty of dc recording, ac-coupled amplifiers are commonly used. In many subjects the most prominent feature of the EEG is an oscillation of frequency roughly 10 Hz (alpha activity) whose amplitude fluctuates from moment to moment. Most subjects' EEG has a peak-to-peak amplitude which does not exceed roughly 20 μV to 100 μV.

The EEG gives a diffuse picture of the gross electrical activity of large areas of the brain. The activities of more restricted regions of the brain can be recorded by extracting from the EEG, evoked potentials elicited by sensory stimulation. The visual pathway can be studied by recording EP's elicited by stimuli such as a flash of light or a change of color; the somatosensory pathway can be studied by, for example, recording EP's elicited by tactile stimulation of a finger; the auditory pathway can be studied by recording EP's elicited by such stimuli as a click or an amplitude-modulated tone.

The amplitudes of EP's are usually no more than a few microvolts so that they are completely buried in the 20–100 μV of general EEG activity. In order to study EP's, it is first necessary to extract them from the much larger EEG. In general, this is done by utilizing prior knowledge of the EP in order to "recognize" it in the EEG. "Recognition" can be carried out by hand calculations on recorded samples of EEG, but it is more usual to use some electronic device which, when

---

* Hertz (Hz) was adopted by international agreement for cycles per second (cps) and is in use in optical journals.

fed with the EEG, extracts the EP and displays it on an oscilloscope or pen recorder. "Recognition" of the EP is essentially a process of extracting a signal (i.e., the EP) from noise (i.e., everything except the EP). In effect it is a process of "throwing away" information contained in the EEG until the amount of information left is sufficiently small to be handled. Clearly, the decision about what to keep and what to throw away is vital. The design of the device which enhances the signal-to-noise effectively *defines* what is signal and is therefore to be kept (i.e., the EP) so that the experimenter is blinded to all forms of electrical activity which his device defines as noise. The importance of this point is that there is at present no general agreement as to what the human brain regards as signal and what the human brain regards as noise.

Two distinct types of EP are often called "transient EP's" and "steady-state EP's." This classification stems from physics or electrical engineering and corresponds to the distinction between time-domain and frequency-domain methods of systems analysis (Fig. 1). Transient EP's are elicited by an abrupt change in some stimulus parameter (e.g., an increase in brightness) (Fig. 2). The sample of EEG following the instant of stimulation is stored in an instrument's memory. Typically, the length of the EEG sample might be 0.5 second. The stimulus is repeated, and the sample of EEG immediately following the second stimulus is added to the EEG sample already in store. The process is repeated until (say) 100 stimuli have been applied. If an EP occurs at exactly the same interval after each stimulus, then the sum of

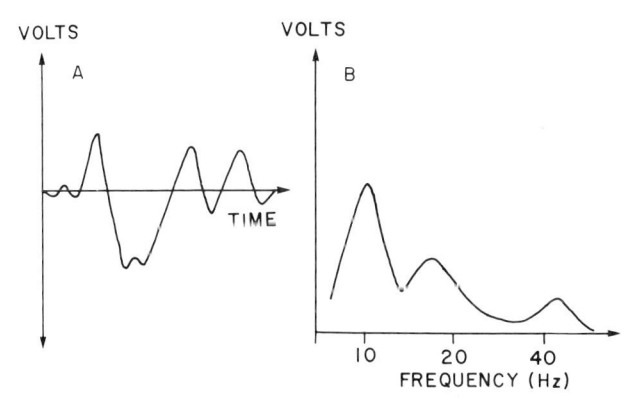

FIG. 1. Transient and steady-state analyses. (A) The response of a system to slowly repeated transient stimulus presented as a plot of volts vs time. (B) The responses of the system to steady-state stimulation presented as a plot of volts vs the repetition frequency of the stimulus. Here it is assumed that a sine wave output results from a sine wave input. In such a linear system *either* the transient or the steady-state description is adequate. (Modified from Regan, 1969.)

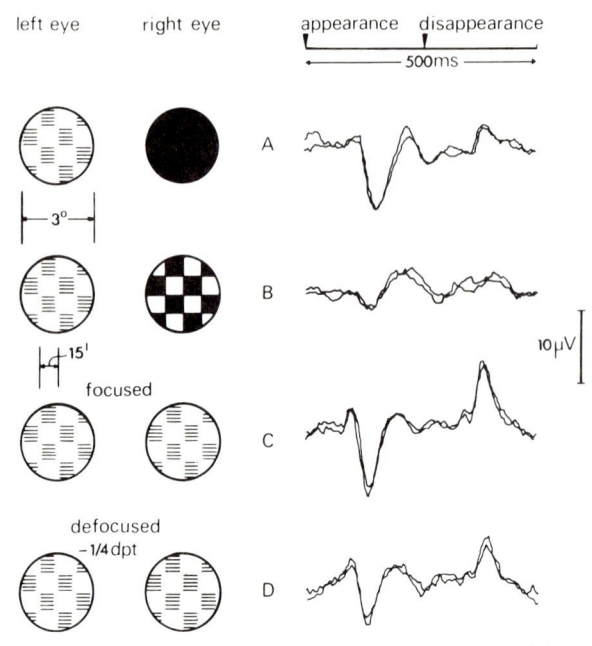

FIG. 2. Monopolar (inion-ear) recordings of evoked potentials (EP's) to the appearance and the disappearance of a checkerboard pattern; to indicate variability superimposed averages are shown of 200 recordings each. Figure A shows the clearly distinguishable responses to the appearance and disappearance of a checkerboard pattern presented to the left eye. Figure B demonstrates the suppression of this response by simultaneous presentation of a steady high-contrast identical pattern to the right eye. Figure C shows that particularly the response to the disappearance of the pattern is enhanced by simultaneous stimulation of both eyes. The response is already diminished (Fig. D) when the pattern is not sharply focused—by only 0.25 diopter misadjustment—on the retina. The defocusing was achieved by focusing at a point immediately in front of the checkerboard-patterned mirror. For all conditions reported in this paper, the contrast was 20% and the luminance of the checkerboard field, after passing through a yellow Wratten 15 filter, was approximately 200 asb. Contrast is defined as $(I_{max} - I_{min})/(I_{max} + I_{min})$. where I is the stimulus luminance. In all records an upward deflection corresponds to a negative voltage of the scalp electrode. (From Spekreijse et al., 1972.)

100 EP's will be 100 times larger than a single response. On the other hand, background EEG activity (i.e., noise) has no fixed relation with the instants of stimulation so that these noise voltages are accumulated in the store as positive voltages just as often as negative voltages. Although noise voltages do not sum to zero, as one might expect at first, the summed noise contribution does accumulate much more slowly than the summed EP. Roughly speaking, the amplitude of the summed noise increases proportionally to N (where N is the number of EEG

samples.) For example, an EP of amplitude 10 μV will be invisible in an EEG of 50 μV since the signal-to-noise ratio is $10/50 = 0.2$. After summing the responses to 100 stimuli, the signal-to-noise ratio will (ideally) be enhanced by a factor of $N/\sqrt{N} = 100/10 = 10$, so that the signal-to-noise ratio will be $0.2 \times 10 = 2$ after summation. The EP will now be evident, although strongly contaminated by noise.

For transient recording, the intervals between successive stimuli should be sufficiently long to ensure that the effect of one stimulus has died away before the next one occurs. If the intervals between stimuli are progressively reduced, successive responses overlap to an increasing extent. When repetitive stimuli follow each other sufficiently frequently, no response cycle can be associated with a particular stimulus cycle; the EP is a wavetrain of identical cycles. The transient regime has been succeeded by what can be called a dynamic steady-state regime (Regan, 1966, 1972b). Such steady-state EP's need not be of sinusoidal waveform and indeed generally show strong harmonic distortion (Regan, 1968a; Spekreijse, 1966; Van Der Tweel and Verduyn Lunel, 1965).

Transient EP's are often displayed as plots of voltage vs time (Fig. 1). Steady-state EP's may be displayed as plots of voltage (and phase) vs stimulus repetition frequency (Fig. 1). Transient EP's can be analyzed into a number of waves (i.e., components) of different latencies (Fig. 1). In adopting this approach many experimenters hope that different temporal components will prove to reflect different physiological processes (Regan, 1972b). A steady-state EP waveform can be analyzed into a number of frequency components.* At first, the reasons for analyzing into different frequency components was not the notion that different frequency components might reflect different physiological processes, but that this method of analysis allowed powerful mathematical techniques to be used in extracting the EP from the background EEG (Regan, 1966). However, as a bonus, it has turned out that nonlinear processes in the visual pathway can be studied separately from linear processes. This was done by measuring frequency components of the EP other than that at the stimulus frequency, since these higher harmonics are by definition entirely due to nonlinear processing (see below) (Spekreijse, 1966). A further bonus was the evidence that the properties of a frequency component of the EP are determined by its absolute frequency (Regan, 1968a, 1970a,b).

There are a number of alternative methods of displaying, analyzing and computing EP's. The choice can be determined by the experi-

---

* In this chapter "EP component" is used to mean both (a) one fixed-latency wave of a transient EP, and (b) one frequency component of a steady-state EP.

menter's opinion of which features of the EP have physiological significance. For example, rather than analyzing a transient EP into waves of different latencies, it may be analyzed into a small number of elementary waveforms, each of which is characteristic of electrical activity in some localized region of the brain.

Transient (time domain) and steady-state (frequency domain) analyses of the visual system would be quite equivalent if the visual system were linear. It is clear, however, that the visual system is far from linear, so that the two methods of analysis can give complementary information.

3. Examples of Evoked Potential Studies of Visual Information Processing

3.1 Correlations between EP Features and Visual Perception

It is clear that there is no simple answer to the apparently simple question: What is the correlation between the EP and visual perception? In the first place there is no such thing as "the" EP. Evoked potentials are made up of a number of components that have quite different properties. In the second place, even if the question specified a particular EP component, the answer generally would depend not only on the particular visual stimulus chosen, but also on whether the stimulus strength were near sensory threshold or well above threshold. These various points are illustrated below.

An example of good correlations between EP amplitude and perception is given by EP's elicited by changes in the spatial luminance contrast of patterned stimuli. A plot of the amplitudes of the EP's elicited by grating patterns versus the log of stimulus contrast is a straight line which, when extrapolated to zero EP amplitude, cuts the contrast axis near the point at which the subject just fails to see the grating pattern (Campbell and Maffei, 1970). This relation also holds for check patterns and for bar and check patterns of chromatic contrast (Regan, 1972a,b, 1973). The straight line relation holds only for low contrast levels; at contrast levels exceeding 30–50% EP amplitude may fail to increase further (saturation) or may even grow smaller (oversaturate). Campbell and Maffei (1970) reported that EP's elicited by changing the contrast of grating patterns reflected a specificity to the orientation of grating pattern stimuli. Now orientation specificity is a well-established feature of psychophysical responses to grating patterns in man. They pointed out that single neuron studies in cat and monkey have demonstrated the presence of orientation-sensitive cells and that these cells are not found more peripherally than cortex. Hence they suggest that the orientation-specific properties of grating EP's may reflect the properties of cortical cells. In the quite different experimental situations of retinal rivalry and sustained interocular suppression, close correlations between the amplitudes of EP's elicited by pattern stimuli and perception have been reported (Cobb et al., 1967; Mackay, 1968; Spekreijse et al., 1972; Van Der Tweel et al., 1969). One example of such a correlation is shown in Fig. 2.

There are experimental situations in which psychophysical measures do not correlate with EP's elicited by patterned stimuli. One of these is when prisms are used to alter the degree of convergence of the eyes. This produces marked changes in the apparent size of objects even though their retinal images do not change in size. Under these circumstances EP's elicited by check patterns depend on retinal image size, not on the apparent size of the checks (Regan and Richards, 1971). A second situation arises when the check size of a checkerboard stimulus pattern is altered. The apparent brightness of the checks has a maximum at a particular check size; this check size is considered to be determined by the dimensions of the inhibitory, and excitatory regions of retinal receptive fields. EP amplitude also has a maximum for a particular check size. EP amplitude and apparent brightness do not seem to be determined by the same lateral inhibitory-excitatory physiological mechanism, since different values of check size give EP and brightness maxima, respectively, and the two check sizes are altered differently by blurring the image and by converging the eyes (Regan and Richards, 1973).

A further example of EP components which do not correlate with perception are those whose absolute frequency fall in the region 45–60 Hz (high-frequency components). Figure 3A shows summed EP waveforms elicited by sinusoidally modulating (i.e., flickering) the brightness of a patch of light. Similar near-sinusoidal EP waveforms were elicited by flickering the light either at roughly twice that frequency. Figure 3B illustrates tuning or resonancelike curves; the largest EP was produced either by 24 Hz flicker or by 48 Hz flicker, and the EP was at 48 Hz in both cases. The 24-Hz flicker therefore produced a frequency-doubled EP. A dissociation between perception and EP amplitude is clear from Fig. 3B. EP amplitude is *rising* as the flicker frequency is raised from 30 to 40 Hz, although subjective sensitivity to flicker *falls* as flicker frequency is raised through the same range (Regan, 1968a; Spekreijse, 1966; Van Der Tweel and Verduyn Lunel, 1965). In this frequency region EP's can be recorded for depths of flicker which are far below subjective threshold (e.g., 20-fold below); such EP components do not

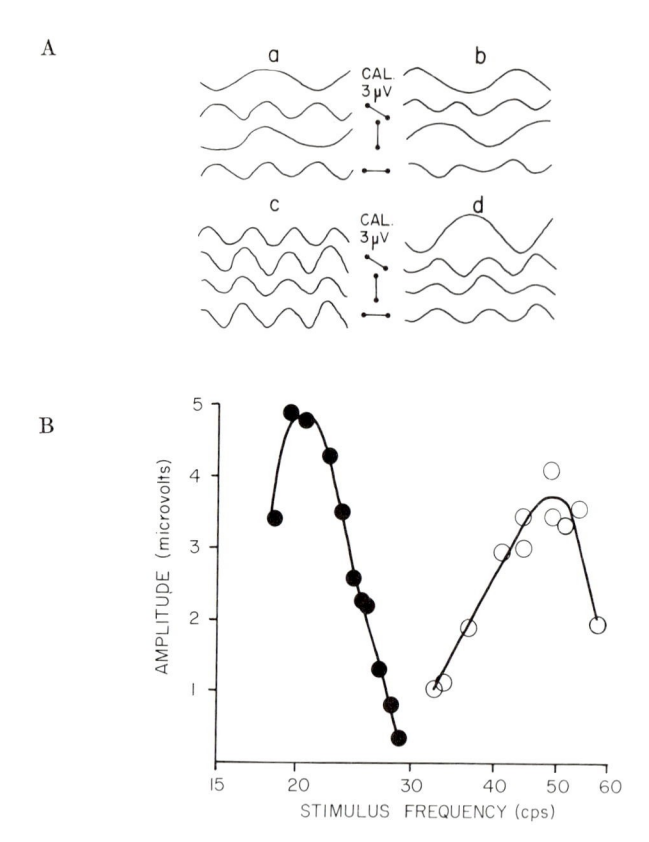

Fig. 3. (A) Averaged potentials evoked by a red stimulus field of 60° subtense sinusoidally modulated to a depth of 33%. The retinal illumination was about 9300 trolands for a, b, and c and about 930 trolands for d. The averaging time was 1 minute, the sweep time was 62.5 msec, and the number of sweeps was 960 in all cases. The top channel is a calibration of 3 µV peak to peak derived from the stimulus modulation wave. The lines indicate that the second trace was recorded between electrodes 2 and 4, the third trace between 2 and 1, and the fourth trace between 2 and 3 (see below for electrode positions). (a) Stimulus modulation frequency 24.4 Hz. Subject J. S. Note second harmonic high-frequency response in traces 2 and 4. (b) Same stimulus as for (a) but different subject, F.M.S. (c) Subject J. S. Stimulus frequency 55.5 Hz. Note clear responses at fundamental frequency. (d) Subject J. S. Same stimulus as for (a) except that it is 10 times less intense. (From Regan, 1968a.)

(B) Amplitude (in microvolts peak to peak) vs stimulus modulation frequency in Hz (cps) for the fundamental (○—○) and second harmonic (●—●) components of the steady-state high-frequency evoked potentials. The cross-correlator was used to analyze bipolar recordings made between electrodes 2 and 4 for subject J. S. Same stimulus as for Fig. 3A (a). Note the logarithmic frequency scale. (From Regan, 1968a). Electrode 1 was 1 cm above the inion, 2 was 7 cm in front of the inion, 3 was 5 cm from 2 perpendicular to midline, and 4 was 7 cm from 1 and 2.

seem to have any threshold at all. On the other hand, although they do not correlate with flicker perception, they can be used as a tool to measure the photopic (bright light or cone) spectral sensitivity of the eye with high precision (e.g., ±0.05 log unit) (Regan, 1970a). These findings raise the question: What (if any) is the function of the high-frequency flicker information that is not available to perception and yet can be detected at cortical level by EP methods? One speculation is that differences between high-frequency EP components and components in lower-frequency ranges reflect differences between the processing of visual signals in the classical geniculocalcarine visual pathway and in other nonclassical visual pathways.

Apart from the possibility that some EP components may reflect information that is not available to perception, the finding that EP's correlate with perception in some situations and do not correlate in other situations can be understood in the following way. Although scalp EP's may be generated by cortical cells, the properties of these EP's are determined (constrained) not by one, but by several, mechanisms (constraints). Similarly, perceptual phenomena are constrained by several different mechanisms. Now some of these mechanisms exert similar effects upon EP's and visual perception. One example of a common constraint is that which underlies the Stiles-Crawford effect (DeVoe et al., 1968). On the other hand, there are mechanisms which exert different effects upon EP's and perception. For example, EP amplitude may fall off at high stimulus strengths even though estimates of perceived sensory magnitude continue to increase (Regan, 1972b).* Good correlations between EP's and perception may be shown when the experimental design ensures that the only effective constraints are those which are common to EP and perception. If the experimental design is loosened to allow an additional constraint to be effective, and if this additional constraint is one that is not common to EP and perception, then the correlation between EP and perception will break down. Constraints not common to EP's and perception include bipolar versus monopolar EP recording (Regan, 1972a), the site of active and reference electrodes (Regan, 1972b), EP saturation, foveal vs extrafoveal and upper field vs lower field location of the visual stimulus (Halliday and Michael, 1970; Jeffreys, 1971; Regan, 1972b).

---

*Other examples of constraints common to EP's and perception are those that underlie orientation specificity (Campbell and Maffei, 1970), threshold spatial acuity (Campbell and Maffei, 1970; Regan, 1972a), interocular suppression (Cobb et al., 1967; Mackay, 1968; Spekreijse et al., 1972; Van Der Tweel et al., 1969).

3.2 Parallel Functional Channels; Differentiation by EP Recording

If a subject views an unpatterned patch of light which abruptly becomes patterned, an averaged transient EP is elicited by the change. If now, after allowing time for the visual system to settle down, the patterned stimulus abruptly becomes unpatterned, a second averaged transient EP can be recorded. Figure 2 shows that quite different transient EP's are elicited by the appearance of pattern and by the disappearance of pattern. These two types of EP differ in their properties as well as in their waveform (Spekreijse et al., 1972, 1973). For example, the pattern-disappearance response seems to show more interocular summation than the pattern-appearance response (Fig. 2); again, the rate of change of spatial contrast within the stimulus pattern affects the appearance-response more than the disappearance-response (Van Der Tweel et al., 1969; Regan, 1972b, p. 54). These and further EP findings suggest that the human visual pathway has different ways of handling information that contrast has increased and information that contrast has decreased (Spekreijse et al., 1973); furthermore, that these different items of information are processed in parallel channels. These different properties of the visual pathway may be accessible to EP methods of investigation.

Quite apart from this parallel processing of pattern-appearance and pattern-disappearance information, is the segregation of pattern information into parallel color channels. EP evidence supports the notion that if the luminance contrast within a patterned field changes, then signals are generated by a neural organization which is, in effect, viewing the stimulus through a red filter. Side by side with this "red" contrast channel is a similar channel which is, in effect, viewing the stimulus through a green filter. There may also be a third parallel channel, this time most sensitive to blue light. Some of the EP evidence for this is as follows (Regan and Sperling, 1971; Regan, 1973).

Figure 4 shows the steady-state EP's elicited by a two-colored pattern of checks. Alternate checks were red and green. The color of adjacent checks exchanged six times per second. The net luminance of every point of the pattern did not change at any time. The stimulus can therefore be regarded as a repetitive reversal of chromatic contrast across the sharply accommodated boundaries within the stimulus pattern. Figure 4 shows that this stimulus elicited clear steady-state EP's, even though the net (red plus green) luminance was constant at every point so that luminance contrast was always zero (at 0.6 on the abscissa).

It is not easy to show unequivocally that the EP's of Fig. 4 were entirely determined by contrast changes. Evidence for contrast-specificity

6. EP STUDIES OF VISUAL INFORMATION PROCESSING

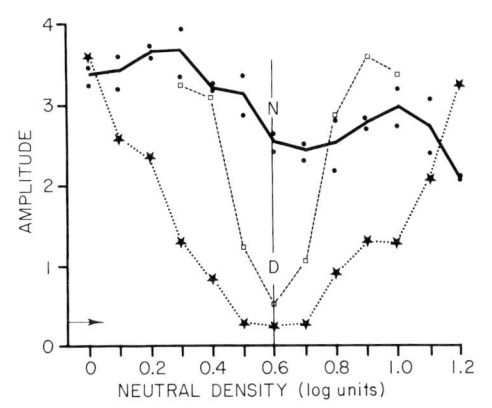

FIG. 4. The full line is a plot of the amplitudes of the 6 Hz component of the EP elicited by a pattern of red-green checks. The red and green checks exchanged places six times per second, i.e., chromatic contrast reversed six times per second. The brightness of the green checks was altered in steps of 0.1 log unit (abscissa). In this situation the net luminance at every point on the pattern was constant with time. Even so, clear EP's were recorded. The dotted line shows EP's recorded in a control experiment. The only difference between the two experiments was that in the control the red filters were replaced by green filters. Therefore at 0.6 log unit on the abscissa there were no reversals either in chromatic contrast or in luminance contrast. The difference between the full curve and the dotted curve at the equal-luminance point (0.6) is entirely due to changes in chromatic contrast. The chain line plots EP's recorded from a deuteranopic subject. The stimulus was similar to that used with a normal subject (N) to obtain the full line. The deuteranopic subject (D) could not distinguish red from green lights. The figure shows that for the color-blind subject EP's elicited by reversals of chromatic contrast in a patterned field fell to zero when red and green luminances were made equal. The stimulus field subtended 3° and was composed of 11 minute checks.

is that EP amplitude was attenuated by blurring the image (Fig. 5), though this evidence is not conclusive. This behavior is quite different from that of EP's elicited by extensive unpatterned stimuli, which are little affected by changes in ocular accommodation. Further items of evidence that the chromatic contrast EP's of Fig. 4 were indeed contrast-specific are first that only small EP's were elicited by flickering a blank field of a similar size to the pattern field (horizontal pairs of lines, Fig. 5A), and second, that a plot of the amplitudes of chromatic contrast EP's vs check size is an inverted U-shaped function which peaks at about 12–18 minutes of arc.

When red and green checks exchange places in the stimulus of Fig. 4, there are equal and opposite changes in the luminance contrast of the red and green components of the light. There would be no EP if these equal and opposite changes in luminance contrast were processed by a single mechanism. Therefore, a contrast-sensitive mechanism sensitive to green light must function in parallel to a second contrast-sensitive mechanism sensitive to red light; not until the two contrast signals have been generated can the "red" and "green" responses converge (Regan, 1972a, 1973; Regan and Spekreijse, 1973).

The chromatic aberration of the eye has not yet been considered. It could be argued that as a result of ocular chromatic aberration the red-green stimulus of Fig. 4 could elicit contrast-specific EP's, even though red and green contrast signals were not segregated into parallel color channels. This could occur if, for example, the green component of the stimulus light were focused to a sharp image on the retina while the red component of the light was blurred owing to the chromatic aberration of the eye. The more sharply focused green components of the stimulus would then be expected to be physiologically more effective than the blurred red component, so that EP's would be generated which were due to changes in the luminance contrast of the green light. In order to guard against this possibility, ocular chromatic aberration was measured by EP means (Fig. 5A) and a method was developed to cancel chromatic aberration so that the red and green lights were both sharply in focus on the retina (Fig. 6).* The well-known method of viewing the stimulus through a lens whose chromatic aberration is equal and opposite to that of the eye has the disadvantage that longitudinal and transverse chromatic aberrations are not corrected simultaneously. The method outlined in Fig. 6 does not have this drawback. The red checks were placed farther from the eye than the green checks so that longitudinal chromatic aberration was canceled. Since the red checks were farther away than the green checks, both the red checks and field stop had to be proportionately larger to ensure that the red and green retinal images were identical. Figure 5B shows how, even when ocular chromatic aberration was canceled, clear contrast-specific EP's were elicited by changes in chromatic contrast. Hence, the argument above that there are parallel red and green channels for processing spatial contrast information still holds.

An alternative way of explaining the results of Figs. 4 and 5 is to postulate a neural organization that is specifically sensitive to changes in chromatic contrast. However, the approach outlined here seems more parsimonious.

Evoked potentials elicited by spatially unpatterned visual stimuli provide evidence of further parallel channels for processing visual in-

* An essentially similar method of canceling chromatic aberration has recently been used by F. W. Campbell (personal communication).

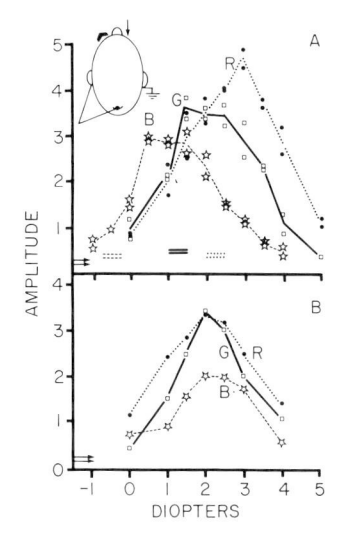

FIG. 5. (A) Effect of blurring on the amplitudes of 6-Hz component of steady-state EP's. The EP's were elicited by six reversals per second of the luminance contrast in a pattern of monochromatic checks. The subject's accommodation was paralyzed by 1% Mydriacyl. Ordinates are amplitudes in microvolts. The abscissa are the powers (in diopters) of lenses placed before the subject's eye. A 3-mm exit pupil was used. The figure shows that, in order to obtain EP's of maximum amplitudes, different lens powers were required for red, green, and blue checks. The image appeared subjectively sharpest for the lens power which gave maximum EP amplitude. The EP curves for red (R), green (G), and blue (B) checks, therefore, indicated the value of the eye's chromatic aberration.

The two arrows are noise values, recorded with the stimulus occluded. The pairs of dotted lines, full lines, and chain lines give the EP amplitudes elicited by flickering a blank stimulus field of the same size.

(B) Both the transverse chromatic aberration (chromatic difference in magnification) and the longitudinal chromatic aberration (chromatic difference focus) of the eye were simultaneously canceled by the method outlined in Fig. 6. The R (red checks) and G (green checks) curves show that both red and green lights were simultaneously in focus on the retina. The retinal image sizes of red and green checks were also identical. The red and green checks were then superposed to give red-green chromatic contrast reversal as in Fig. 4, except that in this case there was no chromatic aberration (this red-green curve is the one marked D in Fig. 4B). This figure shows that clear EP's were elicited by reversals of chromatic contrast, even when chromatic aberration was canceled. Furthermore, the R − G EP's were largest when the retinal image was sharpest.

formation. High-frequency flicker (45–60 Hz) evokes potentials that can be used in a kind of heterochromatic flicker photometry. An EP is generated by viewing a patch of light whose wavelength alternates 45 times per second between say, red (630 nm) and yellow (590 nm). By setting the relative intensities of the red and green lights to some

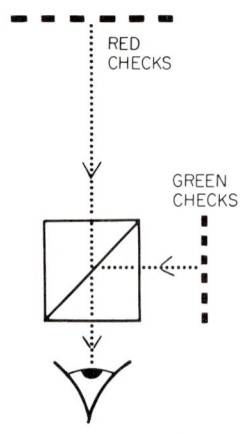

FIG. 6. Principle of method for simultaneously canceling longitudinal and transverse chromatic aberrations of the eye. The red pattern is placed farther from the eye than the green pattern; the relative distances are such that both patterns are sharply focused onto the retina at the same time. The difference (in diopters) of the eye's lens power for the two colors can be obtained either from subjective measurements, or from EP curves such as those of Fig. 5A. The physical red pattern must be of larger physical dimensions than the green pattern to ensure that the red and green retinal images are identical in size. In this way the relative sharpness of focus and the relative sizes of the red and green retinal images can be controlled independently. In order to cancel chromatic aberration for a range of stimulus wavelengths, it is more convenient in practice to produce the red-green check patterns of this figure by variable-power optical projection.

fixed ratio, the amplitude of the EP can be reduced to zero. This ratio is a measure of the relative sensitivity of the high-frequency EP to red and yellow lights. By using a range of different wavelengths, the relative spectral sensitivity of the EP can be measured to a precision of 0.1 and even to within 0.05 log unit. This relative spectral sensitivity curve agrees closely with the photopic spectral sensitivity curve of the eye measured by conventional subjective flicker photometry (Fig. 7B). It is a curiosity that the EP curve can be measured at least 10 times below subjective threshold so that the subject does not see flicker at any time (Regan, 1970a).

EP's elicited by unpatterned stimuli provide evidence for more than one channel of information processing. Brightness flicker can elicit EP components whose frequencies fall in a range 13–30 Hz. The properties of these EP components are quite different from the properties of components in the 45–60 Hz range. For example, the two types of components do not seem to be generated in the same regions of the brain. One particularly clear difference is that the 13–30 Hz components do not have the photopic spectral sensitivity of the 45–60 Hz components; in fact, an indefinite number of spectral sensitivities can be defined so that the notion loses its point (Regan, 1970b). These 13–30 Hz components show phase shifts between the responses to different colors which depend strongly on stimulus wavelengths, but may be remarkably independent of flicker frequency. The phase shifts can be attributed to antagonistic processing of color information. More speculatively, 13–30 Hz EP components reflect opponent-color processing, while 45–60 Hz reflect luminance processing.

Summarizing, then, different types of EP component provide evidence that different types of visual information are processed in parallel channels. These channels are separately sensitive to (1) appearance of pattern (contrast increase); (2) disappearance of pattern (contrast decrease); (3) spatial contrast for long wavelengths; (4) spatial contrast for medium wavelengths; (5) (probably) spatial contrast for short wavelengths; (6) photopic brightness changes; (7) antagonistic processing of color information.

3.3 Peripheral and Central Mechanisms in the Visual Pathway

This section will discuss how EP's can be used to investigate mechanisms at different levels of the visual pathway. Starting with the most peripheral level the examples given below are ocular chromatic aberration, photoreceptor properties, processing between photoreceptors and spike initiation, the processing of retinal disparity, and the gross neuroanatomy of the visual cortex.

3.3.1 Chromatic Aberration

Figure 5A shows how EP's can be used to measure ocular chromatic aberration. If checks of less than some 20 minutes subtense are used, the amplitude of the EP gives an index of the sharpness of the retinal image (Regan, 1973; Regan and Richards, 1973). The chromatic aberration for the green (530 nm) and red (630 nm) stimuli is roughly 0.7 diopter. This is in accord with psychophysical measurements. The aberration between blue (460 nm) and red (630 nm) is some 1.5 diopters (Fig. 5A).

3.3.2 Stiles-Crawford Effect

Moving more centrally, a well-known property of retinal cones is that they are less sensitive to light that strikes the retina obliquely than to light that strikes the retina normally (Stiles-Crawford effect). Transient EP's have been reported to show this effect.

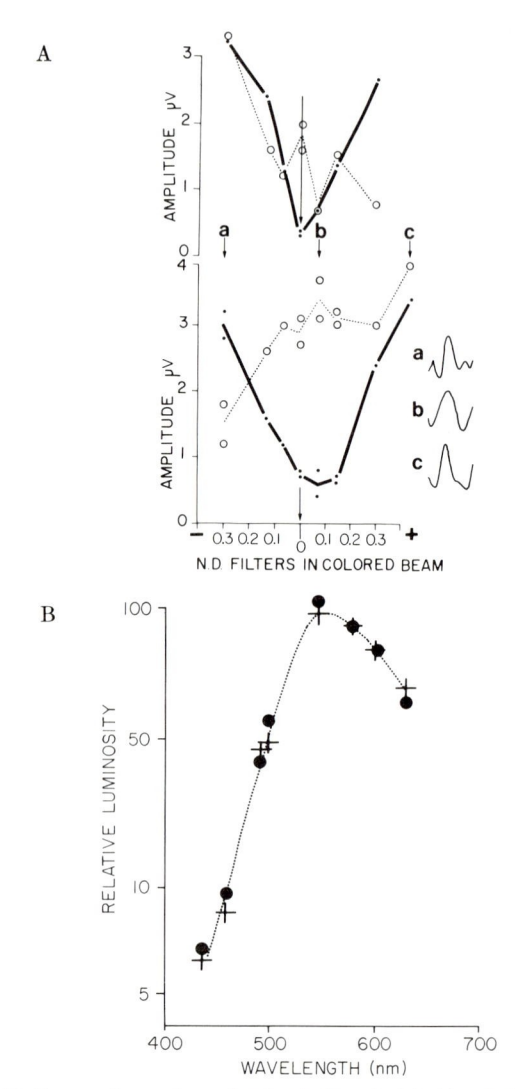

Fig. 7. (A) Comparison of psychophysical and objective heterochromatic flicker-photometry data. A standard white light was alternated at 24 Hz in antiphase with a colored light whose luminance was adjusted by the subject until he saw minimum flicker (zero on the abscissa). The luminance of the colored light was then altered in increments of 0.07 log unit. The EEG was processed by an on-line Fourier analyzer, and the amplitudes of the fundamental and second-harmonic components of the evoked potential were plotted vs the luminance of the colored beam for 580 nm (upper half of the figure) and 460 nm (lower half). The position of the minimum of the second-harmonic component i.e., 46-Hz component (filled circles, full lines) agreed with the point of minimum subjective flicker, to within the smallest increment of luminance used (0.07 log unit, 18%). In contrast, the funda-

3.3.3 SPIKE GENERATION

Moving a further step from the periphery, EP's have been found which seem to reflect the processes which underlie spike generation in the retina. The properties of spike generation which impress themselves upon subsequently generated scalp EP's are properties which follow from the fact that positive rates of spike firing are possible but negative rates are not possible. This fact means that successive half-cycles of a flickering stimulus may be handled in different ways by the retina. In the extreme case, when in the absence of stimulation there is no spontaneous firing, alternate half-cycles may produce firing, while the remaining half-cycles produce no firing at all. This phenomenon, which is akin to rectification, has the result that a sine-wave modulation of stimulus intensity (flicker) can produce a response which contains strong harmonic components of twice the flicker frequency. These frequency-doubled components can appear in the scalp EP's. Spekreijse, Tweel, and collaborators used these frequency-doubled components as objective indications of peripheral rectification (Spekreijse, 1966; Spekreijse and Oosting, 1970; Van Der Tweel and Verduyn Lunel, 1965; Van Der Tweel and Spekreijse, 1969). In this way it was possible to study processes which were peripheral to the site of the spike generation. The method they adopted was to add a second auxiliary signal to the stimulus sine wave so as to reduce the amount of frequency doubling. This is similar to the familiar device of adding a high-frequency wave (auxiliary signal) to the music or voice signal so as to improve the

---

mental component of the evoked potential did not generally have a minimum at all (open circles, dotted lines). The peak-to-peak amplitude of the evoked-potential waveform obtained with an averaging computer gave no reliable indication of the point of minimum subjective flicker (averaged waveforms shown for points a, b, c in lower half of the figure). The stimulus field size was 15°, with a 30° white surround whose luminance was equal to the luminance of the standard white component of the central stimulus field. The retinal illuminance produced by the standard white component was 100 trolands. Both beams were sine-wave modulated to depths of 100%. Subject J. S. Bipolar recordings between an electrode 1 cm above the inion and a second electrode on the midline 7 cm ahead of the first electrode. Half-widths of interference filters were roughly 10 nm; amplifier bandwidth 1.0–75 Hz (−3 dB). Fourier analysis carried out during 1 minute of stimulation, starting roughly 20 seconds after onset of stimulation. An averaging computer was run simultaneously with the Fourier analyzer; 800 sweeps, 62-msec sweep time, 100 points.

(B) Crosses: relative luminosities measured by psychophysical heterochromatic flicker photometry. Full circles: Spectral sensitivity determined by EP method. The psychophysical and objective determinations agreed to within the smallest luminance increment used (0.07 log unit, 18%). The dotted curve was drawn smoothly through the psychophysical data points. Subject J. S. (From Regan, 1970a.)

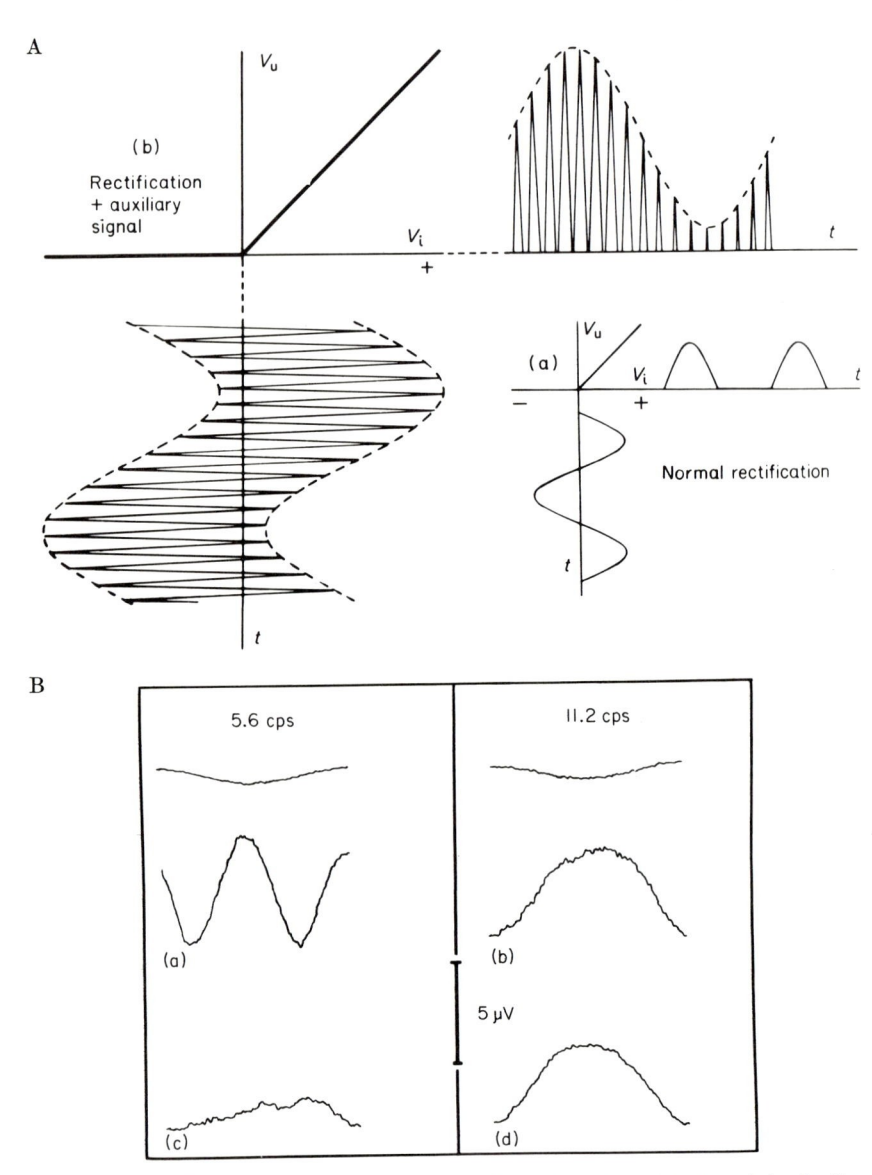

Fig. 8. (A) The principle of linearizing. (From Van Der Tweel and Spekreijse, 1969).

(B) Occipital responses recorded with electrodes on the midline for 5.6 Hz and 11.2 Hz (subject D). Diffuse stimulus field; modulation depth, 10%; natural pupil; brightness, 200 asb; signal-to-noise ratio, $a^2/2\sigma^2 = 0.04$. The upper curves represent an averaged photoelectric signal caused by the light. The middle curves show the responses to sinusoidally modulated light plus gaussian noise (C and D). (From Van Der Tweel and Spekreijse, 1969.)

quality of reproduction in a domestic tape recorder. The auxiliary signal could be, for example, a second sinewave, a triangular wave or a random noise waveform.

Figure 8A illustrates this procedure for reducing the frequency-doubling (nonlinear) distortion caused by certain peripheral stages of visual-information processing within the visual pathway. The smaller sketch marked "normal rectification" shows how the sinewave visual signal (y axis) passes through an information-processing stage that can be represented as half-wave rectification (e.g., spike generation), giving an output (x axis) that is a strongly distorted representation of the input sine wave. Therefore, although the input sine wave contains only one frequency, the output waveform contains several frequency components, and in particular one whose frequency is twice the frequency of the input sine wave. The sketch (Fig. 8A) marked "rectification plus auxiliary signal" shows how the waveform distortion produced by the rectifier stage can be removed by adding a high-frequency auxiliary signal to the input sine wave. This procedure works by shifting the mean position of the net input signal away from the discontinuity in the rectifier's characteristic (at $x = 0$, $y = 0$ in Fig. 8A) so as to maintain the net input signal on the linear branch of the rectifier's characteristic. Signals to the left of the origin give no output as shown in Fig. 8A. Hence the effect of adding the high-frequency auxiliary signal is to make the input sinewave and the output sinewave more similar (i.e., to reduce distortion), which includes in this case the abolition of frequency-doubled components in the output. In practice the auxiliary signal in the output (Fig. 8A) is removed either by filtering or by the averaging process. Figure 8B (waveform a) shows how in human subjects a light whose intensity changes sinusoidally at 5.6 Hz can generate an output waveform (evoked potential) that is so strongly distorted as to resemble a sine wave of doubled frequency (11.2 Hz). Figure 8B (waveform c) shows how the frequency-doubled distortion can be removed by adding an auxiliary light signal to the sine wave visual stimulus; all that is left is a sine wave whose frequency is the same as that of the sine wave light stimulus (5.6 Hz). Figure 8B (right) illustrates that the effect of the auxiliary signal is specific to distortion, since adding an auxiliary signal has no effect on the evoked potential when no distortion is evident.

Spekreijse (1969) has reported that the phenomena shown in Fig. 8B occur not only in human EP's, but also in the firing of retinal ganglion cells in goldfish and monkey. This, and further evidence, associates the rectifier stage of Fig. 8A with the process of spike generation at and peripheral to ganglion cell level in man.

The effectiveness of the auxiliary light signal in reducing waveform distortion is a function of the amplitude of the auxiliary neural signals at their point of arrival at the rectifier. Spekreijse took advantage of this point to investigate visual information processing at sites peripheral to the rectifier stage in man, i.e., at sites peripheral to spike generation. In this way he showed that even before the rectifier stage there is a functional separation between the signals which cause EP components of 45–60 Hz and of 9–12 Hz (Spekreijse, 1966).

This EP approach also offers a possibility for tests on human subjects of the notion that a "wired in" organization of nerve cells need not necessarily have a single fixed mode of operation. For example, the brain itself may use auxiliary signals (e.g., neural noise) to systematically alter the way in which neural mechanisms process sensory information (for example, color information).

3.3.4 Processing after Convergence of the Signals from the Left and Right Eyes

Moving again more centrally, EP's can be used as a tool to study interactions between signals from the left and right eyes. Regan and Spekreijse (1970) approached this by recording EP's which were specific to changes in retinal disparity and correlated with binocular depth perception.

Their stimulus was based on a pair of patterns made up of randomly arranged squares. In one of the patterns a central area was bodily displaced to one side by a fixed number of squares. When these Julesz patterns were viewed in binocular fusion, the central area appeared to stand out in depth. An optical arrangement gave rise to the illusion that the central area of the stimulus moved back and forth in depth. Evoked potentials could be recorded which depended on changes in retinal disparity and not on movements of the retinal image.

3.3.5 Visual Cortex

Finally, moving still further centrally, EP's offer a tool for studying the gross neuroanatomy of the visual cortex in normal man. By using knowledge of the retinocortical projection in man Halliday and Michael (1970) were able to show that the major component of the transient EP elicited by their patterned stimuli was generated outside the region of the calcarine fissure in the human visual cortex. Jeffreys (1971) reported evidence that, among the components of his transient EP's to pattern appearance, the earliest (80 msec) originated in and was immediately adjacent to the calcarine fissure, whereas a later (100–120 msec) component was generated outside the calcarine fissure.

EP evidence from brain-injured patients suggests that the occipital pole is the site of generation of steady-state EP components elicited both by patterned stimuli and by unpatterned stimuli flickering at rates of 45–60 Hz (Milner et al., 1972). On the other hand, EP components generated by unpatterned stimuli flickering at 9–12 Hz are generated in more extensive and anterior regions of visual cortex (Milner et al., 1972).

There is a substantial body of work in which evoked potentials have been used to study the evaluation and processing of visual information at even higher levels in the human brain. These studies differ in both aim and method from the experiments described in this chapter. They have been reviewed elsewhere (Regan, 1972b; Sutton, 1969).

4. Summary

The brain continually generates electrical signals that can be recorded from electrodes attached to the human scalp. Buried in this background activity are minute electrical responses to sensory stimuli. These responses are called evoked potentials (EP's). EP's provide an objective means for investigating the different ways in which the human eye and brain process different forms of visual information. For example, EP's have been used to study the processing of spatial contrast (pattern), color, brightness, retinal disparity (stereoscopic depth), and motion information.

One EP feature seems to reflect the processing of brightness information, another feature seems to reflect chromatically opponent processes, and the properties of a third EP feature suggest that the eye handles spatial contrast (pattern) information in parallel color channels.

Even EP's recorded from the scalp can be used to study information processing at a variety of central and peripheral sites within the visual pathway. At the most peripheral level, ocular chromatic aberration can be measured by EP recording; more centrally, information processing at sites peripheral to retinal ganglion cells can be studied by EP methods; at a more central level still, EP's can elucidate the gross neuroanatomy of the visual areas of the functioning human brain.

Acknowledgments

I thank the Medical Research Council for their support. Robert F. Cartwright provided invaluable technical assistance. Equipment used in some of the experiments described here was constructed by Mr. H. Wardell, Mr. R. Morrall, and Mr. H. Slynn.

References

Campbell, F. W., and Maffei, L. (1970). *J. Physiol. (London)* **207**, 635–652.
Cobb, W. A., Morton, H. B., and Ettlinger, G. (1967). *Nature (London)* **216**, 1123–1125.

DeVoe, R. G., Ripps, H., and Vaughan, H. G., Jr. (1968). *Vision Res.* **8**, 135–147.
Halliday, A. M., and Michael, W. F. (1970). *J. Physiol. (London)* **208**, 499–513.
Jeffreys, D. A. (1971). *Nature (London)* **229**, 502–504.
Mackay, D. M. (1968). *Nature (London)* **217**, 81–83.
Milner, B. A., Regan, D., and Heron, J. R. (1972). *Advan. Exp. Med. Biol.* **24**, 171–187.
Regan, D. (1966). *Electroencephalogr. Clin. Neurophysiol.* **20**, 238–248.
Regan, D. (1968a). *Electroencephalogr. Clin. Neurophysiol.* **25**, 231–237.
Regan, D. (1968b). *Percept. Psychophys.* **4**, 347–350.
Regan, D. (1969). *Proc. I.S.C.E.R.G. Symp., Istanbul, 1969* pp. 37–50.
Regan, D. (1970a). *J. Opt. Soc. Amer.* **60**, 856–859.
Regan, D. (1970b). *Vision Res.* **10**, 163–178.
Regan, D. (1972a). *Advan. Exp. Med. Biol.* **24**, 171–187.
Regan, D. (1972b). "Evoked Potentials in Psychology, Sensory Physiology and Clinical Medicine." Wiley, New York.
Regan, D. (1973). *Vision Res.* (in press).
Regan, D., and Richards, W. (1971). *Vision Res.* **11**, 679–684.
Regan, D., and Richards, W. (1973). *J. Opt. Soc. Amer.* (in press).
Regan, D., and Spekreijse, H. (1970). *Nature (London)* **255**, 92–94.
Regan, D., and Spekreijse, H. (1973). In preparation.
Regan, D., and Sperling, H. G. (1971). *Vision Res.* **11**, 173–176.
Spekreijse, H. (1966). "Analysis of EEG Responses in Man." Junk, The Hague.
Spekreijse, H. (1969). *Vision Res.* **9**, 1461–1472.
Spekreijse, H., and Oosting, H. (1970). *Kybernetik* **7**, 22–31.
Spekreijse, H., Van Der Tweel, L. H., and Regan, D. (1972). *Vision Res.* **12**, 521–526.
Spekreijse, H., Van Der Tweel, L. H., and Zuidema, Th. (1973). *Vision Res.* (in press).
Sutton, S. (1969). *NASA Spec. Publ.* **NASA SP-191**.
Van Der Tweel, L. H., and Spekreijse, H. (1969). *Ann. N. Y. Acad. Sci.* **156**, 678–695.
Van Der Tweel, L. H., and Verduyn Lunel, H. F. E. (1965). *Electroencephalogr. Clin. Neurophysiol.* **18**, 587–598.
Van Der Tweel, L. H., Regan, D., and Spekreijse, H. (1969). *Proc. I.S.C.E.R.G. Symp., Istanbul, 1969* pp. 1–12.

Chapter 7

INHIBITION OF GROWTH AND RESPIRATION BY VISIBLE AND NEAR-VISIBLE LIGHT

B. L. Epel

Department of Botany, Tel Aviv University, Tel Aviv, Israel

1. Introduction	209
2. Light and Growth: General Studies	210
2.1 Prokaryotes	210
2.2 Eukaryotic Microorganisms	211
2.3 Animal Tissues	214
2.4 Higher Plant Tissues	215
3. Mechanisms of Photoinhibition	215
3.1 Studies with the Colorless Alga *Prototheca zopfii*	215
3.2 Studies with the Yeast *Saccharomyces cerevisiae*	222
3.3 Studies with Mammalian Mitochondria	226
3.4 Studies with Purified Cytochrome Oxidase	227
4. Concluding Remarks	228
References	228

1. Introduction

Visible light is one of the dominant factors in determining the growth, development, and reproductive patterns in most organisms. During the course of evolution one of the most consistent environmental characteristics has been the repetitive diurnal variation in light intensity. The potential information content of the light environment was high, allowing for the development of various photoreceptor systems which permitted the organism to sense impending changes in the environment and to respond accordingly.

The light environment, however, is not entirely beneficial. Reports have long been accumulating in the literature that light, particularly in the blue and neighboring near ultraviolet, is inhibitory to growth and reproduction in certain plant and animal tissues and in numerous microorganisms, both prokaryotic and eukaryotic. In this chapter are discussed recent studies on the nature of the inhibitory actions of blue light and near-ultraviolet radiation on growth and respiration in various prokaryotic and eukaryotic organisms.

2. Light and Growth: General Studies

2.1 Prokaryotes

Among the prokaryotes, visible plus near-ultraviolet radiation, primarily in the range 330–490 nm, has been reported to kill or inhibit the growth of a wide variety of bacteria, including such heterotrophic bacteria as *Streptococcus salivarius* (Buchbinder et al., 1941), *Escherichia coli* (Hollaender, 1943; Kashket and Brodie, 1962; Jagger et al., 1964), *Bacterium prodigiosum* (Swart-Fuchtbauer and Rippel-Baldes, 1951), *Pseudomonas aeruginosa* (Kashket and Brodie, 1962), *Mycobacterium phlei* (Kurup and Brodie, 1966), and *Hemophilus influenzae* (Jagger and Stafford, 1962) and certain carotenoidless mutants of bacteria, such as *Sarcina lutea* (Mathews and Sistrom, 1959), *Mycobacterium* sp. (Wright and Rilling, 1963), *Halobacterium salinarium* (Dudas and Larsen, 1962), and *Myxococcus xanthus* (Burchard and Dworkin, 1966; Burchard et al., 1966), as well as the chemoautotrophic bacteria *Nitrosomonas europaea* (Schon and Engel, 1962; Block, 1965) and *Nitrobacter winogradskyi* (Block, 1965; Muller-Neugluck and Engel, 1961) and the denitrifying bacterium *Micrococcus denitrificans* (Mutze, 1963; Harm and Engel, 1965).

Buchbinder et al. (1941) found *Streptococcus* to be killed by natural daylight, direct sunlight, and even 2 hours of low intensity fluorescent light from a 15-W "daylight" fluorescent lamp. The portion of the emitted spectrum responsible for the killing was not determined. *Bacterium prodigiosum* was likewise found to be killed by sunlight (Swart-Fuchtbauer and Rippel-Baldes, 1951). In experiments using filters and sunlight, Swart-Fuchtbauer and Rippel-Baldes (1951) demonstrated that this bacterium was most sensitive to the sun's radiation in the spectral range 365–405 nm.

Mathews and Sistrom (1959) reported that the carotenoidless mutant of the nonphotosynthetic bacterium *Sarcina lutea* is also killed in an oxygen-dependent reaction by sunlight, although the carotenoid-containing wild type is not. Wright and Rilling (1963) reported similar results for a carotenoidless species of *Mycobacterium;* the active portion of the spectrum was 360–590 nm. Dudas and Larsen (1962) found that a carotenoidless mutant of the salt-water bacterium *Halobacterium salinarium* is not killed by exposure to very high intensity white light, but its growth is strongly inhibited. The portion of the spectrum responsible for the inhibition, however, was not determined.

Hollaender (1943) in early studies with starved cultures of *Escherichia coli* reported that this bacterium is killed by radiation in the near-ultraviolet and visible regions of the spectrum. In subsequent studies with actively growing cultures, Kashet and Brodie (1962) presented evidence which suggested that it is the near-ultraviolet radiation which is specifically inhibitory and that it acts on oxidative metabolism. A detailed discussion of this and subsequent works on the specific effects of near-ultraviolet radiation on *E. coli* and other bacterial systems whose oxidative metabolism has also been shown to be sensitive only to near-ultraviolet radiation, but not visible light, has been recently reviewed elsewhere (Jagger, 1972) and will not be presented here.

The chemoautotrophic bacteria *Nitrosomonas europaea* and *Nitrobacter winogradskyi* were found to be killed by high intensities of filtered sunlight (Schon, and Engel, 1962; Muller-Neugluck and Engel, 1961). Block (1965) found that very high intensities of light from an unfiltered high-pressure mercury lamp significantly inhibit NH_4^+ oxidation in *N. europaea*. He reported similar results for nitrite oxidation in *N. winogradskyi*. In both bacteria, after 24 hours of irradiation both cytochromes c and a_2 are destroyed. Mutze (1963) reported that high-intensity visible plus near-ultraviolet radiations is lethal to the denitrifying bacterium *Micrococcus denitrificans*. Harm and Engel (1965) in later studies showed that both mixotrophic and heterotrophic growth were sensitive either under aerobic or anaerobic conditions. As in the case of the autotrophic bacteria, the cytochromes were reported to be destroyed. The cytochromes were found to be more sensitive to photodestruction under anaerobic than under aerobic conditions, suggesting that the photodestruction was not due to a photooxidative-type mechanism.

Burchard and Dworkin (1966) found that dark-grown carotenoidless cells of the soil bacterium *Myxococcus xanthus* lysed when exposed to visible light while in stationary phase. The photoinduced lysis appears to be due to photooxidation, sensitized by an endogenous pigment which, apparently, accumulates only during the stationary state (Burchard and Dworkin, 1966). Burchard et al. (1966) report that the spectrum resembles the spectrum of a porphyrinlike pigment, the main active region of the spectrum being between 395 and 430 nm, with a less active secondary peak around 525 nm.

2.2 Eukaryotic Microorganisms

Among eukaryotic microorganisms light has been reported to be inhibitory to the green algae *Chlorella pyrenoidosa* (Sorokin and Krauss, 1959) and *Euglena gracilis* (Padilla and Cook, 1964; Cook, 1968), to the colorless algae *Prototheca zopfii* (Epel and Krauss, 1966), *Prototheca portoricensis* (Kowallik, 1965), and *Astasia longa* (Cook, 1968), to certain colorless mutants of *Chlamydomonas* (Gross and Dugger, 1969), to the yeast *Saccharomyces cerevisiae* (Matile and Frey-Wyssling, 1962; Matile, 1962a; Epel and Krauss, 1966; Ehrenberg, 1966a,b, 1968; Sulkowski *et al.*, 1964; Guerin and Sulkowski, 1966), and to the plasmodia of the myxomycete *Physarum polycephalum* (Daniel, 1966) and to be lethal to the fungi *Dacryopirax spathularia* (Goldstrohm and Lilly, 1965) and *Sporidiobolus johnsonii* (Goldstrohm, 1964) and to the protozoan *Tetrahymena pyriformis* (Epel and Krauss, 1966).

2.2.1 Green Algae

Sorokin and Krauss (1959) showed that the onset of division of light-dark synchronized cultures of *Chlorella pyrenoidosa* is inhibited if the cells are left in the light at the end of the light period when they normally would go into the dark. From this observation, plus the fact that the light saturation level for growth of synchronized cultures is greater than that of exponential steady-state cultures, they postulated that light inhibited some process essential to division but not to photosynthetic growth. Padilla and Cook (1964) and Cook (1968) have also claimed that the division of exponentially growing cultures of *Euglena gracilis* is inhibited by light. Cook (1968) has shown that if *Euglena gracilis* cells are grown in the dark on organic medium and then exposed to an irradiance of 1000 fc of incandescent light, no inhibition is detected; if however, the cells are exposed to 300 fc of fluorescent light, an immediate inhibition of division is detected, followed by a recovery that is never complete. Cook (1968) has interpreted these results as indicating a blue light-induced inhibition of some growth or division process.

2.2.2 Colorless Algae

Cook (1968) has also reported that the division of both the naturally colorless euglenoid *Astasia longa* and a colorless mutant of *Euglena gracilis*, *E. gracilis* var. *bacillaris* SM-LI, are inhibited when transferred from the dark to 300 fc of fluorescent light. The degree of inhibition observed was greater for cells grown on salt media with acetate than for cells grown on 1% proteose peptone.

Gross and Dugger (1969) have shown that visible light inhibits the growth of a yellow and a colorless mutant of the alga *Chlamydomonas reinhardi*. They reported that room light or moderate intensities of white light severely inhibit the growth of the white form while only moderately inhibiting the growth of the yellow form. Blue light was found to stop motility and growth in the white form and to induce clumping, whereas in the yellow form it only reduced the growth rate. In the white mutant but not the yellow, red light was also found to be inhibitory to growth but to a lesser extent than with blue light.

Epel and Krauss (1966) found the growth of the colorless alga *Prototheca zopfii* to be inhibited by white light from fluorescent lamps; the authors suggested that the inhibition was primarily related to a light-induced inhibition of cell division. Later studies by Epel and Butler (1970b) showed that the inhibition of cell division was not a primary effect (see discussion in Section 3.1).

2.2.3 FUNGI OTHER THAN YEAST

Goldstrohm and Lilly (1965) reported that unpigmented cells of the dark-grown fungi *Dacryopirax spathularia* and *Sporidiobolus johnsonii* are killed in an O_2-dependent reaction by sunlight whereas cells which had been induced to form pigment by exposure to low light intensities are not. The spectral region of the radiation responsible was not determined.

Daniel (1966) has shown that light depresses the growth and respiration in the plasmodia of the myxomycete *Physarum polycephalum*. The respiratory response is rapid and reversible with changes in the rate of oxygen uptake being detectable within 15 seconds. The inhibition of respiration has been demonstrated with isolated mitochondria as well as with the intact plasmodia.

2.2.4 YEAST

Sulkowski *et al.* (1964) have shown that the respiratory adaptation, protein synthesis, and growth of the colorless yeast *Saccharomyces cerevisiae* is inhibited by low-intensity light when anaerobically grown cells are placed in air. Fermentation, in contrast, is unaffected (Guerin and Sulkowski, 1966). The cells were sensitive only for a short period after the admission of oxygen (1 hour), and fully adapted cells were not inhibited at all (Guerin and Sulkowski, 1966). In earlier studies by Matile and Frey-Wyssling (1962) and Matile (1962a), it was demonstrated that the growth and respiration of fully adapted cells is inhibitable if they are exposed to much higher intensities of light. Under high-light conditions, using starved fully respiratory adapted cells, Matile (1962a) showed that O_2 uptake is inhibited, but CO_2 output remains constant. In addition an initial stimulation of endogenous respiration, followed by inhibition, is observed. Ehrenberg (1966a) in studies with growing cultures, showed that growth, protein synthesis, respiration, and cell propagation are all inhibited by strong visible light, the blue portion of the visible spectrum being the most inhibitory. She reported that respiration is most inhibited during growth on ethanol, less so during growth on glucose and least during stationary phase (Ehrenberg, 1968). Epel and Butler (1969) and Ninnemann et al. (1970a) have studied the mechanism of these light-induced inhibitions in yeast, the results of which are presented in a later section.

2.3 Animal Tissues

Wells and Giese (1950) have shown that sea urchin sperm become immobilized when irradiated with the 435 nm mercury line. Norman and Goldberg (1959) and Norman et al. (1962) have reported similar results for bovine, human, and cock sperm. In a crude action spectrum for the light-induced inactivation of sperm, Norman et al. (1962) demonstrated that blue and yellow light were the most effective in inactivating the sperm.

The developing embryos of brook and rainbow trout from pale eggs have been shown by Perlmutter (1961) to be killed by very low-intensity light from a 40-W fluorescent lamp, the most active wavelengths being in the blue-violet region of the spectrum. Isolated rat thymocytes have also been reported to be killed upon exposure to high intensities of visible light (Myers and Dewolfe-Slade, 1963).

Klein and Edsall (1967) have claimed that the growth of cultured HeLa cells is inhibited upon continuous exposure to either near ultraviolet or green radiation for prolonged periods; it is claimed that red light reverses the detrimental effects of green, but not those of near ultraviolet radiation.

Santamaria and Prino (1964) have shown that Yoshida hepatoma ascites cells when irradiated under oxygen but not nitrogen with wavelengths greater than 320 nm show a marked decrease in both respiration and glycolysis. They found that isolated calf retina, in contrast, showed a marked decrease in respiratory activity only. The portion of the emitted spectrum responsible for the reported inhibitions was not determined.

High-intensity blue, green, or red light has been shown by Rounds and Olson (1967) to inhibit the respiration of isolated rat cerebellar cells. Although it was not experimentally demonstrated, these workers suggested that the light inhibition was the direct result of photodestruction of one or more of the respiratory cytochromes. They based this suggestion on the observation that the pulsing of a suspension of cells with 50–200 kW of power from a red laser (=6096 + 6013 Å) resulted in an inhibition in the reduction of cytochrome oxidase (Rounds and Olson, 1965). However, the observation made with the laser may not be relevant to the other situation since the extremely high intensity of the laser pulses may have led to local heating effects; these could have resulted in the observed inhibition of the cytochrome oxidase.

2.4 Higher Plant Tissues

The respiration and growth of higher plant tissues have also been shown to be inhibited by light. Ninnemann and Epel (1968) have shown that blue-violet light is inhibitory to mitosis, but not meiosis, in synchronized tissue cultures of *Trillium* and lily microsporocytes. Near ultraviolet or green radiation is reported to slow the growth of *Parthenocissus* plant tissue culture and of *Ginkgo* pollen, and to interfere with mitosis in onion root tip meristem (Klein and Edsall, 1967; Wolff et al., 1967; Klein, 1964).

Bjorn et al. (1963) have shown that in excised wheat roots visible light induces multiple responses. From action spectra studies, Bjorn and co-workers found that blue and red light inhibited cell elongation, with peaks in the action spectrum at 430 and 650 nm, while blue-violet light inhibited cell division, with the main action spectrum peak around 400 nm.

In studies made with isolated mitochondria from cauliflower, Matile (1962b) has reported that the oxidation of Krebs cycle intermediates, as measured by oxygen uptake, is slowed by intense visible light. Cytochrome c oxidase activity was reported to have been inhibited, the enzyme becoming completely reduced in time. Bjorn et al. (1963) interpreted this as being due to a direct light-induced reduction of the cytochromes of the mitochondria and concluded that the primary lesion caused by light is in the Krebs cycle. In similar studies Mosolova and Sisakyan (1964) showed that although in mitochondria isolated from etiolated pea plants the oxidation of succinate is inhibited by light, in mitochondria isolated from green pea plants oxidation proceeded normally. In neither the studies of Bjorn nor those of Mosolova and Sisakyan was it determined whether intense visible light produces similar effects *in vivo*.

3. Mechanisms of Photoinhibition

3.1 Studies with the Colorless Alga *Prototheca zopfii*

3.1.1 Growth Inhibition

In 1966 Epel and Krauss demonstrated that when growing cultures of the "colorless" alga *Prototheca zopfii* were irradiated with "white light" from "cool-white" fluorescent lamps, growth was inhibited. With

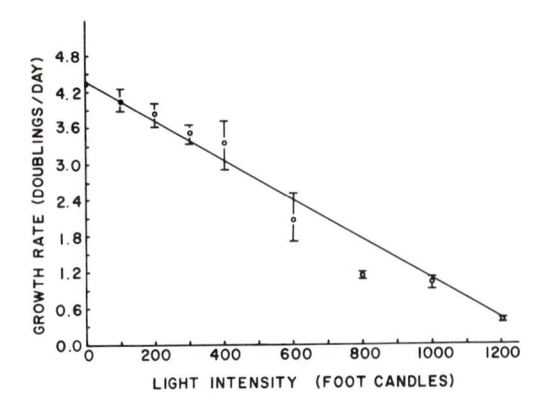

FIG. 1. Growth inhibition of *Prototheca zopfii* induced by light from "cool-white" fluorescent lamps as a function of light intensity reported in foot candles. Growth rates are in doublings per day. (After Epel and Krauss, 1966.)

increasing light intensity, the growth rate of the alga decreased linearly from about 4.5 doublings per day in the dark to less than half a doubling per day at 1200 fc (Fig. 1). The authors suggested that the inhibition of growth was primarily related to a light-induced inhibition of cell division.

Subsequent studies by Epel and Butler (1970b) showed, however, that the inhibition of cell division was a secondary effect. Measurements of the action of white light on protein synthesis, nucleic acid synthesis, respiration, and cell division performed with growing cultures of *P. zopfii* revealed no differential inhibition on any of these parameters either with respect to the onset or kinetics of the inhibition. All four physiological parameters measured showed the same degree of inhibition over the 12-hour irradiation period (Fig. 2). These experiments suggested that the primary locus of photoinhibition lay elsewhere than in cell division. From the observation that the major physiological growth parameters were equally inhibited by light, it seemed probable that the primary physiological system affected must be tightly coupled to the major synthetic and growth functions. The authors reasoned that the most likely target would prove to be the respiratory apparatus since it contains enzymes (the cytochrome and flavoproteins) that absorb light in the visible region of the spectrum. Furthermore, the action spectrum studies of Epel and Krauss (1966) for the inhibition of growth implicated a porphyrinlike substance as the light receptor.

3.1.2 Light Inhibition of Respiration

This conjecture, which was eventually shown to be correct, was reinforced from the results of studies by Epel and Butler (1970b) on

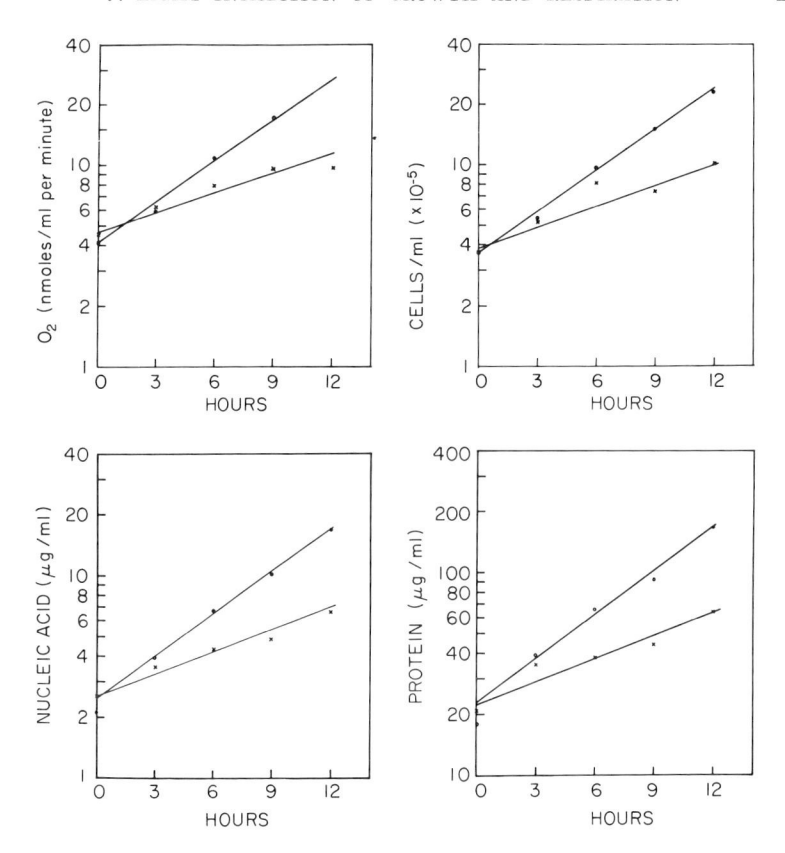

FIG. 2. Inhibition of respiration, cell division, nucleic acid synthesis, and protein synthesis in growing cultures of *Prototheca zopfii* induced by light from "cool-white" fluorescent lamps (1–2.9×10^4 ergs cm^{-2} sec^{-1}). Upper left: Rate of oxygen uptake per milliliter of culture versus hours of growth in light ($\times$) and dark ($\bigcirc$); upper right: number of cells per milliliter of culture vs hours of growth in light ($\times$) and dark ($\bigcirc$); lower left: concentration of nucleic acid in micrograms per milliliter of culture vs hours of growth in light ($\times$) and dark ($\bigcirc$); lower right: concentration of protein in micrograms per milliliter of culture versus hours of growth in light ($\times$) and dark ($\bigcirc$). K_D and K_L are the growth rate constants for the various parameters for dark grown and light grown cultures, respectively (After Epel and Butler, 1970b.)

the effect of light on the respiratory capacity of a starved culture of the alga. Cultures were starved initially in the dark for 24 hours, and at various periods thereafter aliquots were removed and irradiated with moderate intensities of white light from fluorescent lamps. The respiratory capacity of the cells was then determined by measuring the rate of oxygen uptake upon the addition of substrate to a test aliquot. Figure 3 shows the results of such an experiment. Cells left in the dark for as long as 7 days showed no loss in their capacity to respire on added

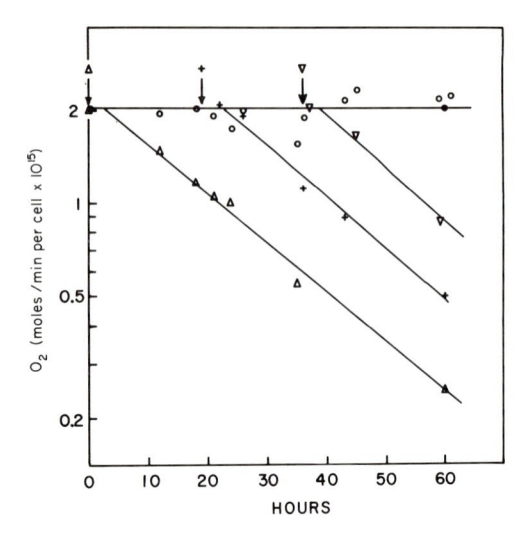

Fig. 3. Inhibition of capacity of starved cells of *Protothec a zopfii* (3 × 10⁶ cells/ml) to respire on added substrate (0.25% ethanol) induced by white light from "cool-white" fluorescent lamps (1–2.9 × 10 ergs cm² sec⁻¹). Rate of oxygen uptake per cell after the addition of substrate vs hours starved. Nonirradiated control (○); cells irradiated continuously from start of starvation (□); cells irradiated after 19 hours of dark starvation (×): cells irradiated after 36 hours of dark starvation (△). Arrows indicate start of irradiation periods. (After Epel and Butler, 1970b.)

substrates. Continuous irradiation with white light from fluorescent lamps, however, was shown to result in an exponential inhibition of respiratory capacity. The degree of inhibition proved to be independent of the time of previous dark starvation. Although it could have been argued that the decrease in respiratory capacity resulted from a killing of cells, viability determinations, performed by comparing the colony-forming ability of irradiated cells to those of control cells, showed almost no loss in viability of the cells due to irradiation.

3.1.3 THE CYTOCHROME OF *P. zopfii*

The physiological and action spectra data of Epel and Butler (1970b) of Epel and Krauss (1966) strongly suggested that the target site for the photoinhibition lay in the respiratory electron transport chain. The use of special spectroscopic techniques and instrumentation in studies (Epel and Butler, 1970a) made with both intact cells and isolated mitochondria from *P. zopfii* established that the respiratory electron transport chain of the alga contained at least 7 cytochromes: two c-type cytochromes, a soluble form, c-549, and a membrane-bound form, c-551; three b-type cytochromes, b-556, b-559, and b-564; and two a-type cytochromes, a and a_3 (cytochrome oxidase).

3.1.4 Photosensitivity of Cytochrome a_3

Of the seven cytochromes associated within the respiratory electron transport chain, three—cyt c-551, b-559, and a_3—were shown to be partially photolabile; of these, a_3 was shown to be by far the most labile. The destruction of cytochrome a_3 could be shown by a number of different spectral assays. In Fig. 4 the difference spectrum of intact cells irradiated for 2 hours with high-intensity blue light is compared with that of a dark control. Obvious qualitative differences are seen in the absorption spectra between the two samples in the spectral region 430–450 nm, the region of the Soret band of cytochrome oxidase. In addition to the large loss in absorption observed in the Soret region, there was a shift in the cytochrome oxidase α-band from 598 to 601.

Epel and Butler (1970b) were able to establish that these spectral changes were due to the destruction of the a_3 component of cytochrome oxidase (a + a_3) through a direct spectral assay for cytochrome a_3.

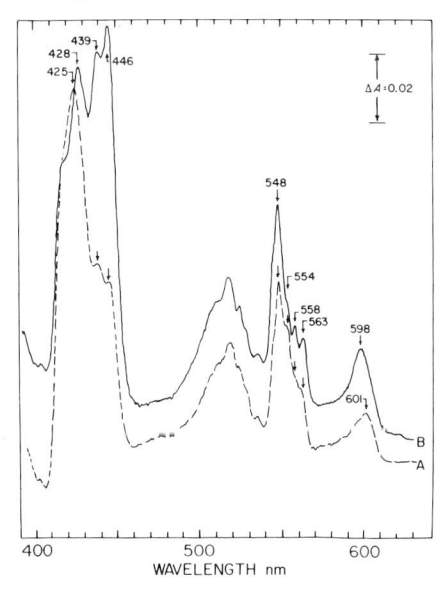

Fig. 4. Low-temperature difference spectra (dithionite versus substrate-respiring aerated cells) of cells of *Prototheca zopfii* (8.7×10^6 cells/ml) before irradiation (curve B) and after 2 hours of irradiation (curve A) (AH-6 super-pressure mercury lamp, Corning filters 5433 + 3850, 1–2.5 × 10^5 ergs cm^{-2} sec^{-1}) with added light-scattering agent (0.33 gm of $CaCO_3$/ml). [After B. Epel and W. L. Butler, *Science* **166**, 621 (1969). Copyright 1969 by the American Association for the Advancement of Science.]

CO, a classical inhibitor of respiration, is known to bind to the a_3 component of cytochrome oxidase and to cause a spectral shift in the absorption spectrum of the cytochrome. Furthermore, this binding is photoreversible. Chance et al. (1964) showed that at liquid nitrogen temperature, irradiation for a few second results in an irreversible photodissociation of the cytochrome a_3–CO complex thus allowing for the measurement of the photodissociation difference spectrum. That blue irradiation causes a destruction of the a_3 component of cytochrome oxidase was shown using this assay (Fig. 5). While the nonirradiated control cells exhibited a normal difference spectrum for the CO–cytochrome a_3 complex, the irradiated cells exhibited essentially no indication of the complex (Epel and Butler, 1970b).

The effects of high-intensity blue light on cytochrome oxidase was demonstrated even more directly through the use of a new spectral assay which directly distinguishes between cytochrome a and a_3. Epel and Butler (1970a) found that the α-bands of cytochrome a and a_3

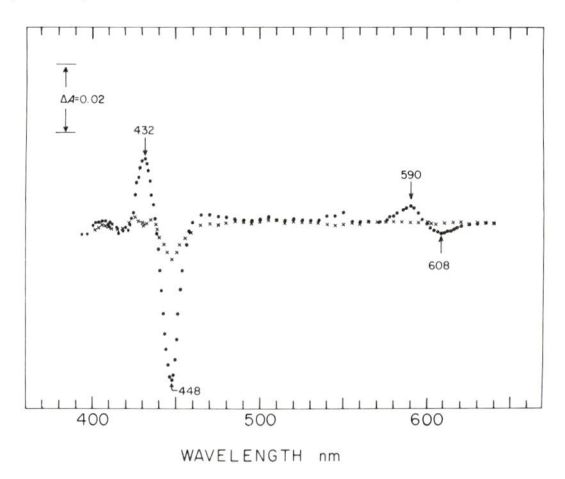

FIG. 5. Low-temperature carbon monoxide photodissociation difference spectra of nonirradiated (●) and 2-hour irradiated (AH-6 lamp, Corning filters 5433 + 3850, 1–2.5 × 10⁵ ergs cm⁻² sec⁻¹) cells of *Prototheca zopfii* (×). Starved cells (8.7 × 10⁷ cell/ml) suspended in starvation medium, pH 6.9, with added scattering agent (0.33 gm of CaCO₃ per milliliter) were reduced with dithionite, and humidified CO was blown across the surface of the reduced suspension. Cells were frozen in both sample and reference cuvettes, and a base line was recorded. The reference cuvette was irradiated for 10 seconds with white light from a Unitron LKR lamp to dissociate the carbon monoxide–cytochrome a_3 complex, and the spectrum was recorded. The spectra presented in the figure were computed by subtracting the base line from the recorded spectra. [After B. Epel and W. L. Butler, *Science* **166**, 621 (1969). Copyright 1969 by the American Association for the Advancement of Science.]

could be resolved if the spectrum was measured in the presence of methanol, cyanide, and dithionite at low temperatures. In the absence of methanol, the absorption band of reduced cytochrome oxidase appears at low temperature ($-196°C$) as a single band with a maxima at 598 nm. If methanol is included with cyanide and the strong reductant dithionite, the oxidase band is split into two district bands with maxima at 595 nm and 602 nm (Fig. 6). The methanol causes a red shift of the α-band of cytochrome a to 602 nm and permits dithionite to fully reduce the cyanide–cytochrome a_3 complex. It is the reduced cytochrome a_3–cyanide complex that gives rise to the 595 nm band (Epel and Butler, 1970a). In the absence of methanol, the cyanide–cytochrome a_3 complex remains largely oxidized even in the presence of dithionite (Yonetani, 1960). This spectral assay provided a convenient assay for distinguishing effects of light on cytochromes a and a_3. With this assay, Epel and Butler (1969, 1970a) were able to show that while the a_3 component of the oxidase was nearly completely destroyed after 1 hour of irradiation, the a component was unaffected

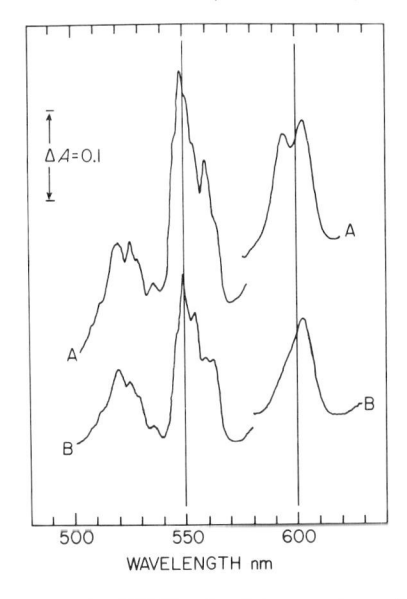

Fig. 6. Low-temperature ($-196°C$) absolute spectra of cells of *Prototheca zopfii* (3×10^7 cell/ml) suspended in starvation medium, pH 6.9, with added scattering agent (0.5 gm of Al_2O_3 per milliliter); cells reduced with dithionite in the presence of MeOH (2.5%) and KCN ($2.5 \times 10^{-4} M$). Curve A, nonirradiated cells; curve B, cells irradiated for 1 hour (AH-6 lamp, Corning filters 5433 + 3850. $1-2.5 \times 10^6$ ergs cm^{-2} sec^{-1}). (After B. Epel and W. L. Butler.) *Science* **166**, 621 (1969). Copyright 1969 by the American Association for the Advancement of Science.]

(Fig. 6). Cytochrome c-551 and b-559 were also shown to be photolabile as can be seen on inspection of Fig. 6, but to a much lesser degree than cytochrome a_3 (Epel and Butler, 1970b).

In most of their studies Epel and Butler irradiated the starved cells with extremely high intensities of blue light in order to shorten the irradiation period. However, irradiation with lower intensity of light for correspondingly longer irradiation periods gave identical results (Epel and Butler, 1970b).

3.1.5 Oxygen Requirement for Photodestruction

The respiratory apparatus was protected if irradiation was carried out under anaerobic conditions. The capacity to respire added substrates under aerobic conditions after an irradiation treatment under anerobic conditions was found to be essentially unaffected (7% inhibition), while after a similar irradiation treatment under aerobic conditions the exogenous respiration was almost completely inhibited (88%). A comparison of the absorption spectra of cells irradiated under aerobic and anaerobic conditions confirmed that only in the presence of oxygen were the three cytochromes photolabile (Epel and Butler, 1970b).

Cyanide, a classical inhibitor of cytochrome oxidase, is known to compete with oxygen for the active site of cytochrome a_3 (Yonetani, 1960). Epel and Butler (1970b) demonstrated that $2.5 \times 10^{-4} M$ cyanide protected cytochrome a_3 from photodestruction under aerobic conditions. These data are highly suggestive of a photooxidative type mechanism and furthermore suggest that the photooxidation probably occurs at the O_2 binding site of cytochrome a_3.

3.2 Studies with the Yeast *Saccharomyces cerevisiae*

3.2.1 Inhibition of Growth and Respiration

Although during the latter half of the 19th century and the early part of the present century numerous but often contradictory reports appeared concerning the effects of visible radiation on growth division and respiration in yeast and other fungi (Ninnemann et al., 1970a), more recent studies (see Section 2.2.4) have clearly established that visible radiation exerts a marked inhibitory effect on growth, cell division, respiration and protein synthesis in yeast. Figures 7A and B shows an example of the nature of visible light effect on both exogenous and endogenous respiration on previously starved yeast cells. As was first shown by Matile and Frey-Wyssling (1962) and later confirmed by Ninnemann et al. (1970a), irradiation of starved yeast cells with white light inhibits the capacity of the cells to respire on added substrates. Under the conditions employed in the experiments of Ninnemann et al. (1970a) depicted in Fig. 7A, the inhibition caused by irradiation with moderate intensities of white light was found to be approximately first order after an initial lag period. Dark control cultures in comparison showed no loss of respiratory capacity over a period of 100 hours. Under much higher light intensities, where the rate of inhibition was more rapid ($t_{1/2} = 8$ minutes), Ninnemann et al. (1970a) reported that no lag phase was observable (Fig. 7B). As in the case of *P. zopfii*, photoinhibition requires the presence of oxygen (Ninnemann et al., 1970a). Irradiation under anaerobic conditions had no effect on the capacity of the yeast cells to respire substrates when the cells were subsequently returned to aerobic conditions (Fig. 7B).

Matile and Frey-Wyssling (1962) showed that the response of the endogenous respiration of the yeast cells to irradiation is more complex and, as pointed by Ninnemann et al. (1970a), is a reflection of a different mechanism. During the initial part of the irradiation period, either

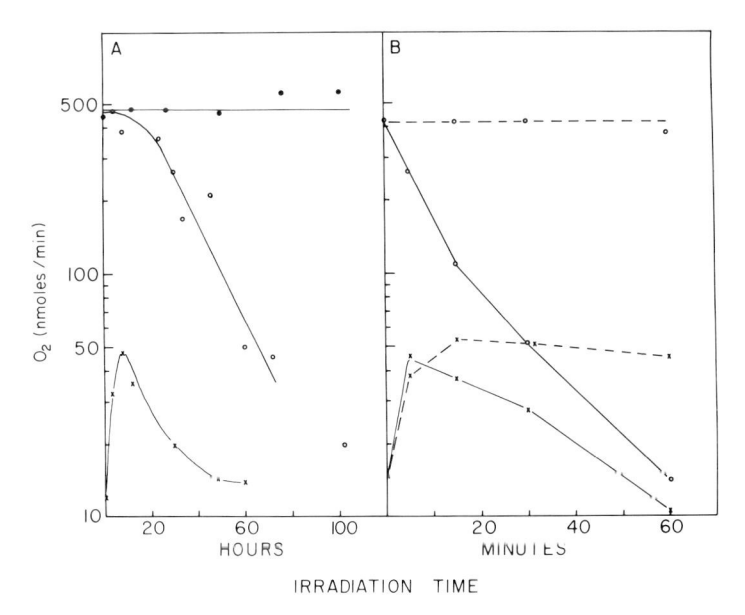

Fig. 7. Effect of light on respiration of yeast (1×10^8 cells/ml) irradiated (A) with moderate-intensity (3×10^4 ergs cm^{-2} sec^{-1}) white light from fluorescent lamps and (B) with high-intensity (1×10^8 ergs cm^{-2} sec^{-1}) blue light from high-pressure mercury lamp and Corning 5562 filter. Rate of exogenous (0.076% ethanol) respiration (○—○) and endogenous respiration (×—×) vs time of irradiation under aerobic conditions. Rate of exogenous respiration (○- - -○) and endogenous respiration (×- - - -×) vs time of irradiation under anaerobic conditions. Exogenous respiration of nonirradiated cells vs time in dark (●). (After Ninnemann et al., 1970a.)

under high- or moderate-intensity irradiation, endogenous respiration is stimulated (Matile and Frey-Wyssling, 1962; Ninnemann et al., 1970a) be it under aerobic or anaerobic conditions. Only under aerobic conditions is a subsequent inhibition noted (Ninnemann et al., 1970a). Examples of such complex responses are shown in Figs. 7A and 7B. The light stimulation of respiration thus appears to be a separate phenomenon unrelated to the inhibitory phenomenon. Further studies of these very interesting phenomena have not yet been pursued.

3.2.2 Photosensitivity of Cytochromes a and a_3

The nature of the light inhibition in yeast appears to be very similar to that for *P. zopfii*. Epel and Butler (1969), and Ninnemann et al. (1970a) in spectroscopic studies modeled after those made with *Prototheca*, established that irradiation with blue light results in the destruction of both components of cytochrome oxidase, i.e., cytochromes a and a_3, and part of cytochrome b. An example of this can be seen in Fig. 8A. In this experiment, yeast cells were irradiated with high-intensity blue light for various periods, and the cytochrome content was measured spectroscopically at low temperature ($-196°C$). The bands at 548, 554, and 559 nm are the reduced α-bands of cytochrome

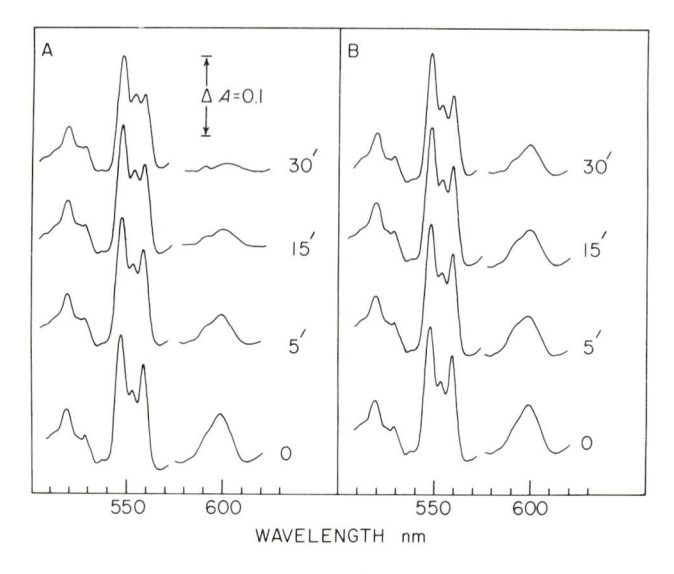

Fig. 8. Absorption spectra of yeast at 77°K after various periods (minutes) of irradiation (pH 5.5) with blue light (1.5×10^6 ergs cm^{-2} sec^{-1}). (A) In the absence of azide. (B) In the presence of 1 mM NaN$_3$. All spectra were measured in the presence of 1 mM NaN$_3$, 1 mM KCN, and dithionite with Al$_2$O$_3$ added as a light-scattering agent. (After Ninnemann et al., 1970a.)

c, c₁, and b, respectively, while those at 590 and 603 are the reduced α-bands of the cytochrome a₃–CN complex and of cytochrome a, respectively. As can be seen in the example depicted in Fig. 8A, increasing periods of irradiation resulted in a progressive diminution in the α-bands of cytochromes a, a₃, and b. Although cytochromes a and a₃ were nearly completely destroyed within 30 minutes as determined by the disappearance of their α-bands, cytochrome b was still partially present, indicating that it is less photolabile than either cytochromes a or a₃. Ninnemann *et al.* showed that the kinetics of the inactivation of cytochrome oxidase follow very closely the inactivation of respiration, as can be seen on comparing Figs. 9A and 9B with Figs. 7A and 7B.

3.2.3 OXYGEN REQUIREMENT FOR PHOTOSENSITIVITY

Epel and Butler (1969) and Ninnemann *et al.* (1970a) found that, as in the case of *P. zopfii*, oxygen is required for the photoinactivation of the yeast cytochromes. Under anaerobic conditions the cytochromes were found to be fully protected (Ninnemann *et al.*, 1970a). Furthermore

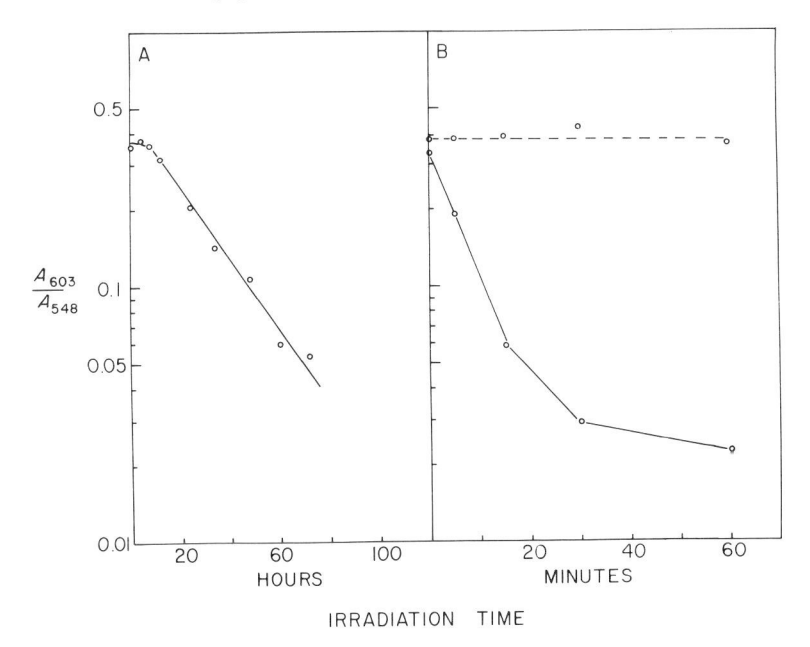

FIG. 9. Effect of light on cytochrome oxidase of yeast as measured by the ratio of the absorbances at 603 and 548 nm vs time of irradiation. Starved cultures of yeast irradiated (A) with moderate-intensity (3×10^4 ergs cm^{-2} sec^{-1}) white light from fluorescent lamps and (B) with high intensity (1×10^6 ergs cm^{-2} sec^{-1}) blue light from a high pressure mercury lamp and Corning 5562 filter. ○—○, Aerobic conditions; ○---○, anaerobic conditions. (After Ninnemann *et al.*, 1970a.)

cyanide and azide, ligands which compete with oxygen for the active site of cytochrome oxidase, were shown to protect respiration from photoinactivation. The protective action of azide can be seen in the example depicted in Fig. 8B.

3.3 Studies with Mammalian Mitochondria

As noted in Section 2.3, there have appeared in the literature a number of reports concerning the inhibitory nature of light on respiration in a number of mammalian tissues.

Epel and Butler (1969) and Ninnemann et al. (1970b) in light of their findings on the inhibitory nature of visible radiation in *Prototheca* and yeast reinvestigated the question of the nature of the photoinhibition in a mammalian system. Ninnemann et al. (1970b) found that O_2 uptake by isolated beef-heart mitochondria with succinate as substrate was inhibited by irradiation with blue light. Spectroscopic studies similar to those made with *Prototheca* and yeast established that the irradiation treatment destroyed the a_3 component of cytochrome oxidase (Fig. 10). The mammalian oxidase appeared to be of the *Prototheca* type with respect to its light sensitivity in that cytochrome a was not

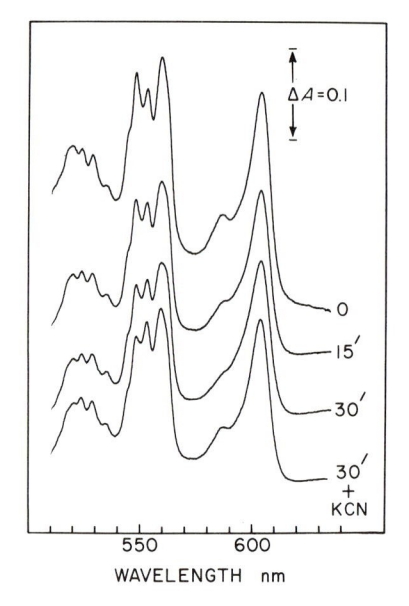

FIG. 10. Absorption spectra of beef heart mitochondria after various periods (minutes) of irradiation in the absence or presence of 1 mM KCN duing the irradiation as indicated. All spectra were measured in the presence of dithionite, 2.5 mM KCN, and 2.5% methanol at 77°K. (After Ninnemann et al., 1970b.)

affected. As in the case of *Protothera*, cytochrome a₃ was protected if the irradiation was carried out in the presence of cyanide (Fig. 10).

3.4 Studies with Purified Cytochrome Oxidase

Proof that the light acted directly on the oxidase came from studies of Ninnemann et al. (1970b). The effect of irradiation on the activity of purified cytochrome oxidase from beef heart measured as the rate of O_2 uptake with reduced cytochrome c as substrate is shown in Fig. 11. Irradiation of the oxidase with strong blue light under aerobic condition was shown to result in an inhibition of the oxidase activity. Irradiation under anaerobic conditions was almost without effect. Spectroscopic studies confirmed that the a₃ component was inactivated under these conditions. Although the lack of photoinhibition under anaerobic conditions is highly suggestive of a photooxidative-type mechanism, it could be argued that since the photoreceptor is a redox type pigment only the oxidized form of the pigment is photolabile and that oxygen itself is not directly required for the photoinhibition. Such a possibility was

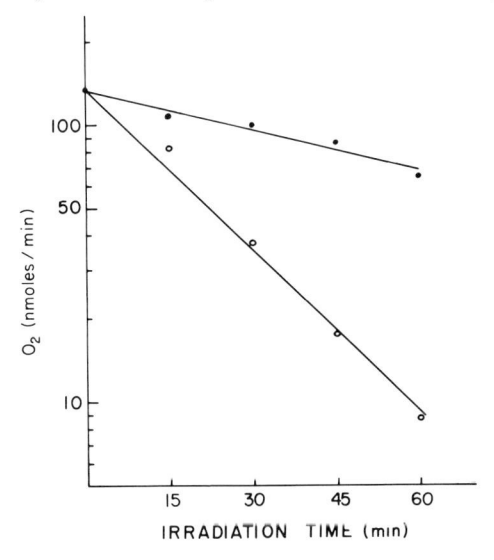

Fig. 11. Influence of oxygen on cytochrome oxidase activity as a function of the time of irradiation with light from a high-pressure mercury lamp and Corning filter 5562 (1–2 × 10⁶ ergs cm⁻² sec⁻¹). Irradiation was performed at 0°C. Oxidase activity was measured as rate of oxygen uptake at 27°C with reduced cytochrome c as substrate [20 µg of oxidase protein per milliliter with 30 µM cytochrome c and 1.5 mM ascorbate; the reaction mixture contained 0.05 M Tris·HCl (pH 8.0), 1 mM histidine and 0.66 M sucrose]. ●—●, Anaerobic; ○—○, aerobic. (After Ninnemann et al., 1970b.)

ruled out, however, by the observation that the purified oxidase remains in its oxidized state when made anaerobic by bubbling with argon even for as long as 2 hours. Since no inactivation of the oxidase was found under anaerobic conditions with the oxidase in its oxidized form, it was concluded that photoinactivation is oxygen dependent.

4. Concluding Remarks

Although the extent to which the photodestruction of cytochrome oxidase, and the resultant inhibition of respiration by visible light occurs in the biological world is difficult to estimate, since definitive work in intact organism has been done only with the alga *P. zopfii* and the yeast *C. cerevisiae*, the observation that the oxidase of isolated beef heart mitochondria is also sensitive argues strongly in favor of the premise that this phenomenon is probably very widespread. If it is conceded that this phenomenon is universal, it is then intuitively obvious that most organisms must process a protective or active repair mechanism since sunlight appears to be directly detrimental to few organisms. How such protective and repair mechanisms function provides an interesting area for future investigations.

Finally, it should be pointed out that inactivation of respiration by near-ultraviolet light probably reflects a different mechanism from that described here. Unpublished studies by the author comparing the action spectrum for the inactivation of respiration with the inactivation of cytochrome oxidase in intact cells show that blue light inactivated both cytochrome oxidase and respiration, while near-ultraviolet irradiation inactivated only respiration. For a recent review of studies into the nature of such near-ultraviolet induced inhibition of growth and respiration, the reader is referred to a recent review by Jagger (1972).

References

Bjorn, L. O., Suzuki, Y., and Nilsson, J. (1963). *Physiol. Plant.* **16**, 132.
Block, E. (1965). *Arch. Mikrobiol.* **51**, 18.
Buchbinder, L., Soloway, M., and Phelps, E. B. (1941). *J. Bacteriol.* **42**, 353.
Burchard, R. P., and Dworkin, M. (1966). *J. Bacteriol.* **91**, 535.
Burchard, R. P., Solon, A. G., and Dworkin, M. (1966). *J. Bacteriol.* **91**, 896.
Chance, B., Schoener, B., and Yonetani, T. (1964). *In* "Oxidases and Related Redox Systems" (T. E. King, H. S. Mason, and M. Morrison, eds.), pp. 609–614. Wiley, New York.
Cook, J. R. (1968). *J. Cell. Physiol.* **71**, 177.
Daniel, J. W. (1966). *In* "Cell Synchrony" (I. L. Cameron and G. M. Padilla, eds.), pp. 117–152. Academic Press, New York.
Dudas, I., and Larsen, H. (1962). *Arch. Mikrobiol.* **44**, 233.
Ehrenberg, M. (1966a). *Arch. Mikrobiol.* **54**, 358.
Ehrenberg, M. (1966b). *Arch. Mikrobiol.* **55**, 26.

Ehrenberg, M. (1968). *Arch. Mikrobiol.* **61**, 20.
Epel, B., and Krauss, R. W. (1966). *Biochim. Biophys. Acta* **120**, 73.
Epel, B. L., and Butler, W. L. (1969). *Science* **166**, 621.
Epel, B. L., and Butler, W. L. (1970a). *Plant Physiol.* **45**, 723.
Epel, B. L., and Butler, W. L. (1970b). *Plant Physiol.* **45**, 728.
Goldstrohm, D. D. (1964). *Proc. West Va. Acad. Sci.* **36**, 17.
Goldstrohm, D. D., and Lilly, V. G. (1965). *Mycologia* **57**, 612.
Gross, R. G., and Dugger, W. M. (1969). *Photochem. Photobiol.* **10**, 243.
Guerin, B., and Sulkowski, E. (1966). *Biochim. Biophys. Acta* **129**, 193.
Harm, H., and Engel, H. (1965). *Arch. Mikrobiol.* **52**, 224.
Hollaender, A. (1943). *J. Bacteriol.* **46**, 531.
Jagger, J. (1972). In "Research Progress in Organic-Biological and Medical Chemistry" (U. Gallo and L. Santamiria, eds.), Vol. III, Part I, pp. 383–401. North-Holland Publ., Amsterdam.
Jagger, J., and Stafford, R. S. (1962). *Photochem. Photobiol.* **1**, 245.
Jagger, J., Wise, W. C., and Stafford, R. S. (1964). *Photochem. Photobiol.* **3**, 11.
Kashket, E. R., and Brodie, A. F. (1962). *J. Bacteriol.* **83**, 1094.
Klein, R. M. (1964). *Plant Physiol.* **39**, 536.
Klein, R. M., and Edsall, P. C. (1967). *Photochem. Photobiol.* **6**, 841.
Kowallik, W. (1965). *Flora (Jena), Abt. A* **156**, 231.
Kurup, C. K., and Brodie, A. F. (1966). *J. Biol. Chem.* **241**, 4016.
Mathews, M. M., and Sistrom, W. R. (1959). *Nature (London)* **184**, 1892.
Matile, P. (1962a). *Ber. Schweiz. Bot. Ges.* **72**, 236.
Matile, P. (1962b). *Experientia* **13**, 649.
Matile, P., and Frey-Wyssling, A. (1962). *Planta* **58**, 154.
Mosolova, I. M., and Sisakyan, N. M. (1964). *Dokl. Akad. Nauk SSSR* **156**, 702.
Muller-Neugluck, M., and Engel, H. (1961). *Arch. Mikrobiol.* **39**, 130.
Mutze, B. (1963). *Arch. Mikrobiol.* **46**, 402.
Myers, D. K., and Dewolfe-Slade, D. (1963). *Can. J. Biochem.* **24**, 529.
Ninnemann, H., and Epel, B. (1968). *Abstr. Int. Congr. Photobiol., 5th, 1968* p. 12.
Ninnemann, H., Butler, W. L., and Epel, B. L. (1970a). *Biochim. Biophys. Acta* **205**, 499.
Ninnemann, H., Butler, W. L., and Epel, B. L. (1970b). *Biochim. Biophys. Acta* **205**, 507.
Norman, C., and Goldberg, E. (1959). *Science* **130**, 624.
Norman, C., Goldberg, E., and Porterfield, I. D. (1962). *Exp. Cell Res.* **28**, 69.
Padilla, G. M., and Cook, J. R. (1964). In "Synchrony in Cell Division and Growth" (E. Zeuthen, ed.), p. 521. Wiley, New York.
Perlmutter, A. (1961). *Science* **133**, 1081.
Rounds, D. E., and Olson, R. S. (1965). *NEREM Rec.* **7**, 106.
Rounds, D. E., and Olson, R. S. (1967). *Life Sci.* **6**, 359.
Santamaria, L., and Prino, B. (1964). *Res. Progr. Org.-Biol. Med. Chem.* **1**, 259.
Schon, G. H., and Engel, H. (1962). *Arch. Mikrobiol.* **42**, 415.
Sorokin, C., and Krauss, R. W. (1959). *Proc. Nat. Acad. Sci. U. S.* **45**, 1740.
Sulkowski, E., Guerin, B., De Faya, J., and Slonimski, P. P. (1964). *Nature (London)* **202**, 36.
Swart-Fuchtbauer, H., and Rippel-Baldes, A. (1951). *Arch. Mikrobiol.* **16**, 358.
Wells, P. H., and Giese, A. C. (1950). *Biol. Bull.* **99**, 163.
Wolff, E. G., Fiver, D. W., and Klein, R. M. (1967). *Bull. Torrey Bot. Club* **94**, 411.
Wright, L J., and Rilling, H. C. (1963). *Photochem. Photobiol.* **2**, 339.
Yonetani, T. (1960). *J. Biol. Chem.* **235**, 845.

Chapter 8

DENATURATION IN ULTRAVIOLET-IRRADIATED DNA*

Ronald O. Rahn

Biology Division, Oak Ridge National Laboratory, Oak Ridge, Tennessee

1. Introduction 231
2. Types of Photodamage to DNA 232
3. Qualitative Detection of Defects 233
   3.1 Physical Methods 234
   3.2 Chemical Methods 240
   3.3 Biochemical Methods 242
4. Quantitative Detection of Defects 244
   4.1 Kinetic Formaldehyde Method (KFM) 244
   4.2 Thermal Melting Analysis 247
   4.3 Comparison of the Kinetic Formaldehyde Method with
       the Thermal Melting Method 248
5. Nature of Defect Region 249
   5.1 Size of Defects 249
   5.2 Distribution of Defects 250
6. Summary and Future Experiments 254
   References 254

1. Introduction

   Ultraviolet (UV) irradiation of DNA results in the formation of various kinds of photoproducts that may have a disruptive influence on the local integrity of the DNA structure. In this chapter we devote our attention to these localized altered regions or defects and will examine a variety of techniques that have emerged in recent years for studying them. Such studies have several benefits: (1) It is valuable from a physical point of view to have methods capable of examining photodamage in intact DNA, since the hydrolysis procedures normally used in the isolation of photoproducts destroy certain kinds of photoproducts. (2) Photochemical alterations of the structure of DNA may prove useful in DNA sequence studies as determined by electron

* Research sponsored by the U. S. Atomic Energy Commission under contract with the Union Carbide Corporation.

microscopy. (3) A study of the distribution of photodamage may give information on the mechanism of formation of damage which might include long-range energy transfer processes.

From a biological point of view, the study of the nature of defects is directed toward the question of whether repair enzymes recognize the photoproduct or the structural modification caused by the photoproduct. Hence a study of the physical nature of defects should provide useful information for an understanding of the repair process.

2. Types of Photodamage to DNA

A wide variety of photoproducts are formed in UV-irradiated DNA. [For recent reviews see Varghese (1972) and Rahn (1972).] The relative yields of these photoproducts are given in Table I for native and denatured DNA. Values for acetophenone sensitization are included for comparison. In native DNA, the major photoproduct is $\widehat{TT}_1$, the cis-syn form of the cyclobutane dimer of thymine, and for UV irradiation it

TABLE I
Relative Efficiency of Formation of Various Photoproducts in *Escherichia coli* DNA[a]

Class of defect	Type of lesion	Denatured DNA	Native DNA	Acetophenone sensitization (313 nm) Native DNA
		Ultraviolet irradiation (254 nm)		
I	Double-strand chain break	—	<0.0001	<0.001
II	Single-strand chain break	—	0.01–0.001	0.01[c]
III	Cross-link	0–(0.05)[b]	0.01–0.001	—
IV	a. Pyrimidine dimer: $\widehat{TT}_1$	1.3	1	1
	$\widehat{TT}_2$	0.2	<0.02	—
	$\widehat{CT}$	0.3	0.3	0.03
	$\widehat{CC}$	—	0.2	<0.0025
	b. Cytosine hydrate	.1–.3	0.01–0.03	<0.003
	c. Pyrimidine adduct: 6-4'-(pyrimidin-2'-one)-thymine	—	0.12	—

[a] Adapted from data given in tables by Rahn (1972), Rahn and Landry (1971), and Rahn *et al.* (1973).

[b] There is ~5-fold enhancement in cross-linking when DNA is ~20% denatured (either with heat or acid) in the presence of high salt (0.2 *M* NaCl).

[c] There is a 4- to 10-fold increase in the rate of single-strand chain breaking when denatured DNA is used.

amounts to about 60% of the observed products. It is assigned a relative yield of unity, although the quantum yield is actually 0.01. Acetophenone sensitization is a more highly selective method of introducing thymine dimers into DNA; 313 nm radiation is preferentially absorbed by acetophenone, which has a triplet state above thymine but below the other three bases in DNA. Hence, triplet transfer occurs only to thymine, and the photodamage is nearly exclusively $\widehat{TT}_1$.

Of particular interest is the change in the efficiency of forming certain photoproducts when DNA is irradiated in the denatured state. Cytosine hydration, for example, is 10 times more efficient in denatured than in native DNA; i.e., the photoaddition of water to cytosine is enhanced when the bases are more accessible to the solvent as in the denatured form. Another photoproduct, $\widehat{TT}_2$, is also favored 10-fold more by the denatured form of DNA. In this case, the loss of structure in the denatured form allows two adjacent thymines to achieve the proper geometry for the trans-syn form to occur. When DNA is partially denatured ($\sim$20%) by either heat or pH, the yield of cross-links goes up 5-fold, provided the salt concentration is high enough. Presumably, when the DNA structure is weakened, bases on opposite strands are more likely to interact photochemically and the yield of cross-links increases.

These results indicate that if there is some denaturation of DNA during irradiation, there will be an enhancement of forming certain photoproducts in these denatured regions. It is reasonable that localized denatured regions should occur in irradiated DNA, since many of the photoproducts will alter internucleotide spacings, disrupt normal hydrogen bonding, and cause a loss of base stacking. Hence, as the radiation dose increases, the likelihood of forming photoproducts favored by single-stranded DNA increases. Furthermore, since the yield of thymine dimers cannot exceed the photo-steady state concentration, nondimer photoproducts will accumulate at high doses. The ratio of thymine dimers to other photoproducts will also depend strongly on the wavelength of irradiation as well as on the $(A + T)/(G + C)$ content of the DNA. Hence, it is very difficult to estimate which photoproducts will contribute most to the denaturation of DNA when the dose is high. In the following section, we will examine the various methods that have been used to detect denatured regions or defects in ultraviolet-irradiated DNA.

3. Qualitative Detection of Defects

A wide range of methods are available for detecting defects; these methods for the most part depend upon some kind of specificity for

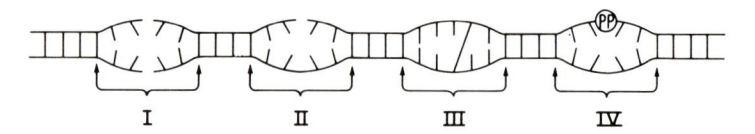

FIG. 1. Pictorial model of double-stranded DNA containing defects due to (I) double-strand chain break; (II) single-strand chain break; (III) cross-link; (IV) photoproduct, such as a pyrimidine dimer or a cytosine hydrate. Arrows point to unwinding points.

denatured DNA. We have arranged the methods according to physical, chemical, and biochemical approaches and give several examples for each approach. As an aid in visualizing the defects as they may exist in native DNA we use the model shown in Fig. 1, in which the various lesions are designated as types I–IV (see Table I).

3.1 Physical Methods

3.1.1 Lowering of the Melting Temperature (T_m)

We define as the melting temperature (T_m) of DNA the midpoint of the thermal-induced transition from native to denatured DNA. It is easily followed by measuring the increase in absorbance at 260 nm. Damage from UV radiation results in loss of base stacking and hydrogen bonding, which in turn will lower the T_m. Marmur et al. (1961) first showed that the T_m of irradiated DNA decreased in proportion to the amount of UV radiation absorbed.

Setlow and Carrier (1963) found that reversal of pyrimidine dimers by short-wavelength radiation failed to produce an equivalent increase in the melting temperature. Shown in Fig. 2 are the melting profiles for DNA irradiated first at 280 nm to give 20% $\widehat{TT}_1$ and then at 239 nm—a wavelength at which photoreversal of $\widehat{TT}_1$ occurs—to give 3% $\widehat{TT}_1$. We note that the T_m is lowered by 14°C following 280 nm irradiation but increases by only ~4°C after photoreversal. It is concluded from this that photoproducts other than pyrimidine dimers contributed significantly to the lowering of the T_m.

Fluence response curves are presented in Fig. 3 for the lowering of the T_m following irradiation of DNA at various wavelengths. Of interest is the continuous decrease in the T_m even after a photosteady concentration of thymine dimers has been attained. At 254 nm, for example, a photosteady concentration of thymine dimers is reached between 25 and 50 Kergs/mm². This result again suggests that nondimer damage is contributing to the T_m lowering.

8. DENATURATION IN UV-IRRADIATED DNA 235

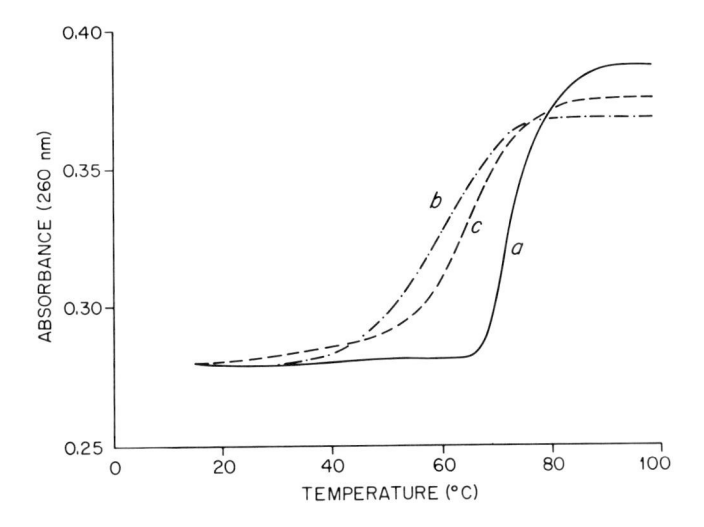

Fig. 2. Thermal melting curves for unirradiated and irradiated *Escherichia coli* DNA (before and after short wavelength reversal). The tabulation contains the corresponding dimer yields and ΔT_m valves. (Unpublished data of Rahn and Landry, 1969.)

Sample	λ Irradiation 280 nm	λ Irradiation 239 nm	%T as $\widehat{TT}$	ΔT_m (°C)
a	0	0	0	0
b	300,000	—	20	14
c	300,000	43,000	3	10

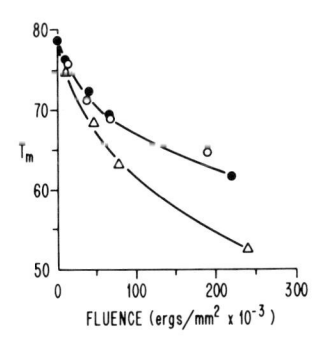

Fig. 3. Decrease in the T_m of calf thymus DNA as a function of the exposure upon irradiation at three different wavelengths. ●, 254 nm; △, 265 nm; ○, 280 nm. (Adapted from Zavil'gel'skii *et al.*, 1965.)

It is possible to observe UV-induced denaturation of DNA at 25°C by irradiating DNA in the presence of a denaturant, such as ethylene glycol or propylene glycol. The T_m of DNA in the presence of various amounts of different glycols is shown in Fig. 4a. In 90% ethylene glycol, for example, the T_m is ~35°C. Hence, as shown in Fig. 4b, irradiation (254 nm) with a dose sufficient to lower the T_m by about

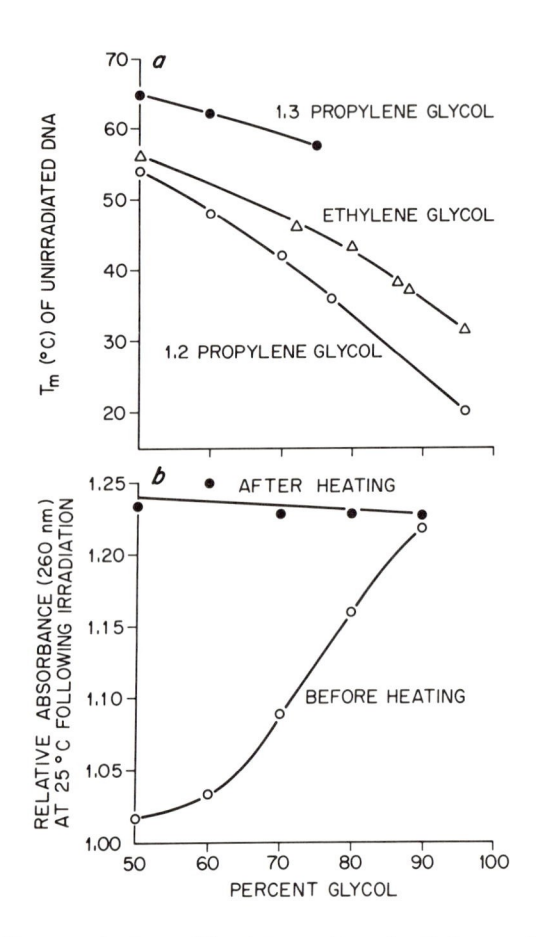

Fig. 4. (a) Decrease in the melting temperature of calf thymus DNA in solutions containing various percentages of the indicated glycol. The ionic strength was maintained constant at 0.01 M phosphate buffer, pH 7. (b) Increase in the relative absorbance of calf thymus DNA following irradiation (254 nm, 300 Kergs/mm²; in the presence of 50-90% ethylene glycol, 0.01 M phosphate buffer, pH 7. The absorbance was also measured after heating (100°C, 10 minutes) to completely denature the DNA. (Unpublished data of Rahn and Landry, 1969.)

17°C, results in complete denaturation at 25°C when 90% ethylene glycol is present; subsequent heating of the DNA gives no further increase in absorbance. Lesser amounts of glycol are not as effective in lowering the T_m, so that, as shown in Fig. 4b, progressively smaller changes in absorbance occur when lesser amounts of glycol are present. Since the absorbance changes are independent of whether the glycol is present during irradiation or is added afterward, it is concluded that no photochemical reaction involving the glycol occurs.

To determine the contribution that thymine dimers alone make to T_m lowering, acetophenone sensitization was used to obtain DNA containing nearly exclusively $\widehat{TT}_1$. As shown in Fig. 5, the change in T_m plotted as a function of $\widehat{TT}_1$, for acetophenone-sensitized DNA, is linear all the way up to the maximum yield of dimers attainable, ~35%. Irradiation with 239 nm to reverse the dimers, reverses the T_m lowering by an equivalent amount until a photosteady concentration of dimers is reached and additional photoproducts are made. From Fig. 5 it is estimated that 5% $\widehat{TT}_1$ decreases the T_m by ~1°C. From the 4° increase in T_m upon reversal of the dimer yield from 20% to 3% (Fig. 2), it is estimated that ~4% pyrimidine dimers made with UV radiation lowers the T_m by 1°C. This result is in close agreement with the acetophenone result.

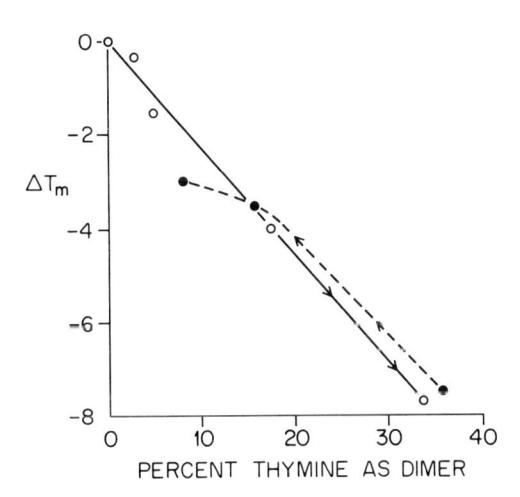

FIG. 5. Decrease in the melting temperature of acetophenone-sensitized DNA as a function of the thymine dimer content. Included in the figure is the increase in the T_m upon reversal of the dimers with 239 nm radiation. ○—○, Formation; ●---●, reversal. (Adapted from Rahn, 1972, and unpublished data of Rahn and Landry, 1969.)

3.1.2 Hydroxyapatite Chromatography

Native and denatured DNA are easily separated from one another on a hydroxyapatite column. As shown in Fig. 6, denatured DNA is eluted from the column at a lower concentration of phosphate than native DNA. Dasgupta and Mitra (1970) showed that irradiated DNA loses its native character and is eluted at a phosphate concentration intermediate between native and denatured DNA, depending upon the dose of radiation. Heavily irradiated DNA (1.44×10^6 ergs/mm^2) has an elution profile which barely overlaps that of either native or denatured DNA (see Fig. 6). This result demonstrates that irradiated DNA does not contain strands of DNA which are completely denatured.

Hydroxyapatite chromatography may be of some use in separating heavily irradiated DNA from native DNA, but the technique is limited because even a dose of 1.1×10^5 ergs/mm^2 does not render DNA sufficiently denatured as to allow for appreciable separation from unirradiated DNA.

3.1.3 Fluorescence of Bound Dye

Zavil'gel'skii et al. (1964) have followed the UV-induced denaturation in DNA by measuring changes in the fluorescence lifetime of bound acridine orange (AO). Acridine orange binds to native DNA as a monomer and to denatured DNA mainly as a dimer. Borisova et al. (1963) have shown that fluorescence of free AO changes from 2 to 5

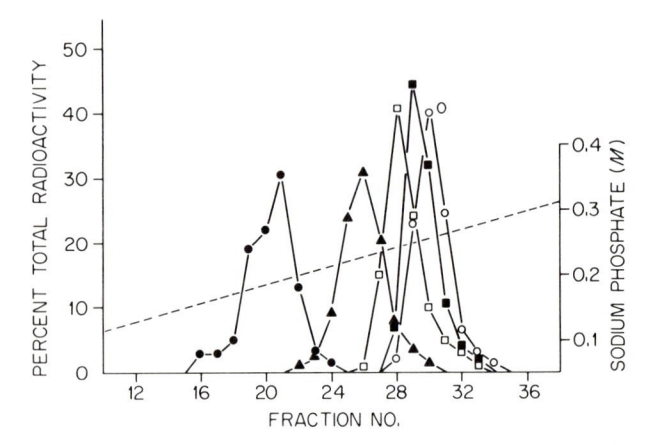

Fig. 6. Hydroxyapatite chromatography of native DNA irradiated with the indicated exposures at 254 nm. Thermally denatured DNA (●—●) is included for comparison. Elution of the DNA was achieved in a stepwise fashion with a gradient of sodium phosphate. (Adapted from Dasgupta and Mitra, 1970.) Irradiated—exposures in ergs/mm^2: ▲—▲, 1.44×10^6; □—□, 7.2×10^5; ■—■, 1.1×10^5; ○—○, none.

nsec when bound as a monomer and increases to 22 nsec when bound as a dimer. Since the monomer and dimer fluoresce with different wavelength maxima (see Table II), measurements of the fluorescence lifetime at different wavelengths provide a sensitive means of following the formation of denatured regions in irradiated DNA. Filters were used to select preferentially for emission from either the monomer region (530 nm) or the dimer region (640 nm), and the corresponding lifetimes were measured with a phase fluorometer. The measured lifetimes at a given wavelength is an average

$$\bar{\tau} = \sum_i \alpha_i \tau_i$$

where τ_i is the lifetime of the ith component and

$$\alpha_i = I_i / \left(\sum_i I_i \right)$$

is the fraction of the intensity due to the ith component (either free monomer, bound monomer, or bound dimer). Lifetimes are denoted by τ_g in the green region (530 nm) and τ_r in the red region (640 nm).

The relative intensities and lifetimes for native, denatured, and irradiated DNA complexed with AO are compared in Table II. The concentration of DNA is about 10-fold greater than that of AO, and

TABLE II

RELATIVE INTENSITY AND LIFETIME OF ACRIDINE ORANGE (AO) FLUORESCENCE IN THE PRESENCE OF VARIOUS STATES OF DNA[a,b]

Conditions	Relative intensity			Lifetime (nsec)		
	530 nm	640 nm	640:530	τ_g (530 nm)	τ_r (640 nm)	$\tau_g : \tau_r$
Free monomer	100	20	0.2	2.4	2.4	1
Native DNA (100% bound monomer)	460	40	0.1	5.7	5.7	1
Bound dimer	5	50	10	22	22	1
Denatured DNA (50% bound dimer, 50% free monomer)	50	35	0.7	3.6	19	0.19
Irradiated DNA	175	30	0.17	4.3	7	0.61

[a] From Zavil'gel'skii et al. (1964).
[b] The concentration of AO was 2×10^{-5} M and of DNA 60 µg/ml. The irradiated DNA received 80,000 ergs/mm² at 254 nm.

we presume that all the AO is bound to the native DNA. Hence τ_g and τ_r are both 5.7 nsec. However, in denatured DNA the binding of the dye is considerably weaker, and only 50% of the dye is bound, the rest remaining unbound. Consequently, the lifetime measured for AO with denatured DNA reflects contributions from both free monomer and bound dimer. Since the free monomer ($\tau = 2.4$ nsec) fluoresces 20 times stronger at 530 nm than the dimer ($\tau = 22$ nsec), τ_g for denatured DNA is 3.6 nsec. However, at 640 nm, the dimer fluoresces more strongly than the free monomer, and τ_r is 19 nsec. Hence, $\tau_g:\tau_r$ for denatured DNA is 0.19. Thus, the presence of denatured regions in DNA can be detected by the departure of $\tau_g:\tau_r$ from unity. As shown in Table II, $\tau_g:\tau_r$ for DNA irradiated with 80,000 ergs/mm^2 is 0.61. According to Zavil'gel'skii et al. (1964), changes in this ratio were detected with doses as low as 800–1000 ergs/mm^2.

Another way of following the formation of denatured regions is simply by measuring the increase in τ_r itself. Zavil'gel'skii et al. (1965) showed that when τ_r is plotted as a function of exposure, a maximum in τ_r occurs when an exposure of 0.5 to 1×10^6 ergs/mm^2 is reached. The initial rate of increase, up to 10 to 20×10^4 ergs/mm^2, however, is quite rapid and may reflect thymine dimer formation, which begins to saturate above this dose. Unfortunately, no attempts were made to quantitate these results in terms of defect concentration.

3.2 Chemical Methods

3.2.1 Reaction with Formaldehyde

Formaldehyde reacts with the amino groups of the free bases in DNA (Grossman, 1968). However, at 25°C the bases are hydrogen bonded and the reaction is severely retarded. At elevated temperatures, the "breathing" of the DNA helix is more pronounced; at 55°C, for example, reaction with 3% formaldehyde leads to complete denaturation within several hours. This denaturation can be followed by measuring the absorbance at 255 nm, a wavelength that reflects loss of base stacking due to unwinding but not to chemical changes of the chromophore resulting from formaldehyde binding.

Marmur et al. (1961) observed that the rate of reaction at 55°C increased after UV irradiation of the DNA. They interpreted this to mean that UV radiation induced denatured regions in the DNA that readily reacted with formaldehyde; these denatured regions then grew in size as the loss of hydrogen bonding and base stacking facilitated additional reaction with formaldehyde.

Shown in Fig. 7 is the increase in absorbance of DNA which has been irradiated at 280 nm followed by 239 nm and then treated with

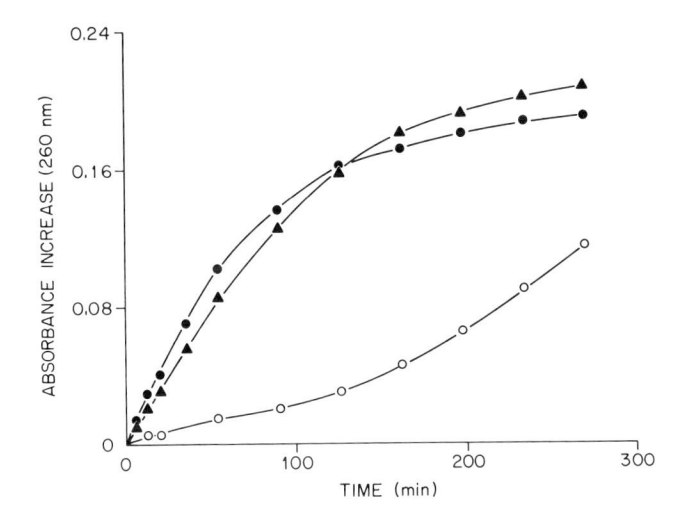

FIG. 7. Reaction of irradiated calf thymus DNA with formaldehyde in $10^{-3} M$ phosphate buffer. A dose of 400 Kergs/mm² at 280 nm was followed by a dose of 67 Kergs/mm² at 239 nm. Formaldehyde (3.7%) was added after irradiation, and the absorbance was measured at 47°C as a function of time. ○—○, No UV; ●—●, 280 nm; ▲—▲, 280 nm + 239 nm. (Unpublished data of Rahn and Landry, 1969.)

formaldehyde. We note a large increase in the rate of reaction following irradiation at 280 nm. Reversal of the cyclobutane dimers with 239 nm radiation results in only a slight diminution of the rate. This result again demonstrates that the UV-induced defects are only partially due to pyrimidine dimers. The rate of reaction of irradiated DNA with formaldehyde can thus be used as a means of quantitatively measuring defects in the DNA (see Section 4.1).

3.2.2 REACTION WITH CARBODIIMIDE

A water-soluble derivative of carbodiimide, N-cyclohexyl-N'-β-(4-methylmorpholinium)ethyl carbodiimide p-toluenesulfonate (CMEC) reacts specifically with the guanine and thymine residues of denatured, but not of native, DNA (Augusti-Tocco and Brown, 1965). For example, with nonhydrogen-bonded thymine this compound is thought to displace the N_3 proton to form

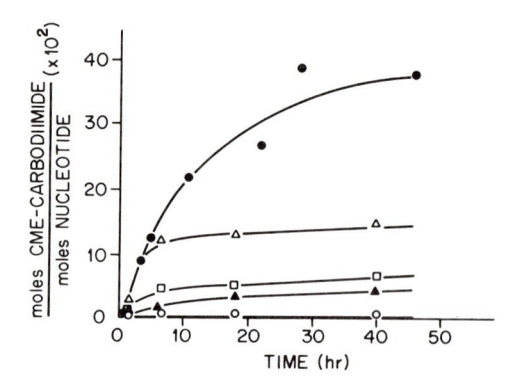

FIG. 8. The time course of reaction of CME-carbodiimide with the exposed bases in DNA following UV irradiation with the indicated fluences in ergs/mm² at 254 nm. ●—●, Denatured DNA; △—△, 1.5×10^6; ■—□, 5.0×10^5; ▲—▲, 2.5×10^5; ○—○, none. (Adapted from Drevich et al., 1966.)

By working with ¹⁴C-labeled CMEC, Drevich et al. (1966) measured the extent of binding as a function of time to DNA irradiated with various doses of UV. As shown in Fig. 8, the amount of CMEC bound after 5 hours of reaction is proportional to the radiation fluence. A dose of 1.5×10^6 ergs/mm² ultimately results in 15% of the nucleotides being able to combine with CMEC. By comparison, 40% of the bases in heat-denatured DNA (i.e., most of the G and T residues) reacted with CMEC.

The reaction of CMEC with irradiated DNA may take place in two steps as with formaldehyde, the second step being unwinding. However, judging from the plateaus in the reaction curves of Fig. 8, which suggest little additional denaturation, this unwinding step would have to be very slow. Perhaps the bulkiness of the CMEC prevents it from readily interacting with the helical portions of the DNA.

3.3 Biochemical Methods

3.3.1 Enzymatic Hydrolysis

An endonuclease isolated from *Neurospora crassa* is specific for single-stranded, but not for double-stranded, DNA. Dasgupta and Mitra (1970) measured the amount of acid-soluble material produced upon treatment of irradiated DNA with this enzyme. The results are shown in Table III for DNA isolated from two different sources and irradiated with several doses. There is almost a linear dependence on dose for the amount of acid-soluble material obtained upon hydrolysis. The value of 25–26% for a dose of 1.44×10^6 ergs/mm² is taken to

TABLE III
SUSCEPTIBILITY OF DNA FROM TWO DIFFERENT SOURCES TOWARD *Neurospora crassa* ENDONUCLEASE AFTER IRRADIATION WITH THE INDICATED FLUENCES AT 254 nm[a]

Source of DNA	Fluence (ergs/mm^2)	% of DNA susceptible to *N. crassa* nuclease
Escherichia coli B	3.6×10^5	8
	7.2×10^5	15
	1.44×10^6	26
	2.88×10^6	50
	5.76×10^6	65
T7 phage	7.2×10^5	14
	1.44×10^6	25

[a] Adapted from Dasgupta and Mitra (1970).

represent the percentage of DNA which is denatured by radiation. Dasgupta and Mitra pointed out that the results presented in Table III were obtained under experimental conditions sufficient for degrading 90% of an equal amount of heat-denatured DNA in 30 minutes. By either increasing the concentration of enzyme or increasing the incubation period, greater portions of the DNA were degraded. This result suggests that native regions of the irradiated DNA begin to unwind and become hydrolyzed.

It is of interest that a 15-minute incubation of irradiated DNA with *N. crassa* endonuclease made 85% of the thymine dimers acid soluble but released only 6% of the thymine. This result supports the idea that thymine dimers are intimately associated with the denatured regions. However, further treatment with the enzyme results in additional release of thymine, even when no thymine dimers remain.

3.3.2 COMPLEMENT FIXATION

It is possible to obtain antibodies specific for adenine. Using the complement fixation technique, Seaman *et al.* (1908) showed that irradiated DNA reacted considerably more with antiadenosine sera than did nonirradiated DNA (Fig. 9). This result demonstrates the creation of nonhydrogen-bonded adenosine residues upon UV irradiation of DNA. Such residues are most likely to be found opposite a thymine dimer site. It remains to be demonstrated that antisera for denatured DNA will also bind to irradiated DNA. Presumably, high doses will be necessary for this reaction to occur since the denatured regions must be sufficiently large for recognition to occur.

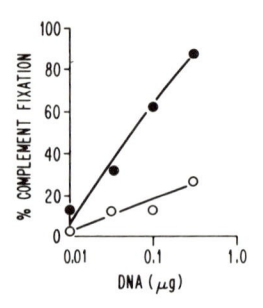

FIG. 9. Serological reaction of native *Proteus vulgaris* DNA with antiadenosine serum before and after irradiation at 270 nm. ●—●, +UV; ○—○, −UV. (Adapted from Seaman et al., 1968.)

4. Quantitative Detection of Defects

4.1 Kinetic Formaldehyde Method (KFM)

Trifonov et al. (1968) have developed a method for determining the concentration of denatured regions or defects in irradiated DNA by following its rate of reaction with formaldehyde. After irradiation, the DNA is kept at 55°C for several hours in the presence of ~2% formaldehyde, and the absorbance at 255 nm is monitored. The change in absorbance is expressed in terms of the degree of helicity θ, where

$$\theta = (OD_{max} - OD')/(OD_{max} - OD_{min})$$

In this expression, OD_{min} is the initial OD, OD' is the OD at time t of the reaction with formaldehyde, and OD_{max} is the final OD. The loss of helicity, $1 - \theta$, when plotted as a function of t, gives curves similar to those shown in Fig. 7.

For sufficiently large doses of irradiation, the relationship between θ and the time of reaction t according to Trifonov et al. (1968) is given by

$$-(\ln \theta / t) = 2cv + pvt$$

where c is the concentration of initial defects per base pair, v is the velocity of unwinding at a defect in base pairs per minute, and p is the rate constant for the creation of new defects. A plot of $-(\ln \theta / t)$ vs t (Fig. 10) results in a straight line with slope pv and extrapolated intercept at $t = 0$ of $2cv$. The initial rate of unwinding for a given molecule is defined as $I = 2cv$. As shown in Fig. 10 the intercept $I = 2cv$ increases with increasing amounts of radiation.

In order to determine c, it is necessary to know v for a given set of experimental conditions. At this point consider the model shown in Fig. 1 for DNA containing a variety of defects designated I–IV. Each

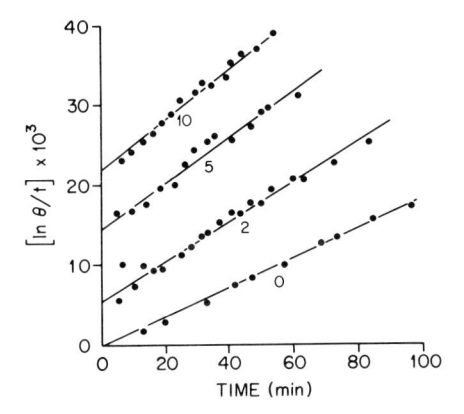

FIG. 10. Reaction rate of irradiated DNA from phage T2 (5 × 10⁷ daltons) at 54°C with 2.14% formaldehyde. The change in the degree of helicity θ is plotted as $-(\ln \theta/t)$ in order to obtain straight-line plots. The time of the irradiation in minutes is as designated in the figure. (Adapted from Trifonov et al., 1968.)

arrow points to an unwinding point, which is located at the end of a helical segment, and there are two unwinding points per defect. Defects are created either by double-strand breaks (I), a single-strand break (II), a cross-link denoted by (III), or a photochemical lesion such as a pyrimidine dimer or a cytosine hydrate (IV). At each one of these defects there are a certain number of nonhydrogen-bonded base pairs which are free to interact with formaldehyde. The basic assumption made by Trifonov et al. (1968) is that each unwinding point proceeds to unwind at a velocity v independent of the nature of the defect. Hence, it is possible to calibrate the method and determine v by measuring I for several DNA's differing only in size, i.e., containing only type I defects. If $\bar{n}$ is the mean number of base pairs per molecule, then for a molecule containing defects only at the ends, the concentration of defects per base pair is given by $c = 1/\bar{n}$. T₂ DNA was fragmented hydrodynamically, and the molecular weight was determined by either electron microscopy or sedimentation. Both methods gave similar results, little difference being noted between the mean-number and the mean-weight values of the molecular weight. From the molecular weight the values of $\bar{n}$ shown in Table IV were obtained using 325 as the average molecular weight of a nucleotide. The fragmented DNA was then allowed to react with formaldehyde, and the values of I shown in Table IV were obtained. In this way v could be determined for DNA varying widely in size, and, as indicated in Table IV, appears to be independent of the length of the DNA. The lower limit of detection is on the order of one defect per 10⁴ pairs of bases.

Using the value of v given in Table IV, Bannikov and Trifonov

TABLE IV
DETERMINATION OF THE VELOCITY OF UNWINDING (v) IN FORMALDEHYDE
FOR FRAGMENTED DNA OF VARIOUS SIZES[a]

MW	$\bar{n}$	$I \times 10^3$ (min^{-1})	v (min^{-1})
5×10^7	77,000	0.0 ± 0.3	—
2.5×10^6	3,800	1.4 ± 0.3	2.5 ± 0.7
2.2×10^5	340	15.7 ± 1.0	2.4 ± 0.3

[a] Adapted from Trifonov et al. (1968).

(1969) measured the concentration of single-strand breaks formed in DNA treated with pancreatic DNase. This enzyme makes nicks in a random fashion. The enzyme-treated DNA was treated with formaldehyde under the same conditions used to obtain v with fragmented DNA. The concentration of defects was then calculated from the intercept I and compared with the number of single-strand chain breaks as determined directly from the single-strand molecular weight obtained by sedimentation in an alkaline sucrose gradient. As shown in Table V, there is an excellent correlation between the two methods of measuring breaks, and the assumption is verified that the velocity v of unwinding in the presence of formaldehyde is the same for defects associated with both single- and double-strand defects.

The correlation between thymine dimers as determined chromatographically and UV-induced defects was examined by Shafranovskaya et al. (1972, 1973). These workers found, as shown in Fig. 11, that the concentration of defects as determined by the KF method (and calibrated with fragmented DNA) was about six times less than the concentration of thymine dimers when the concentration of thymine dimers was equal

TABLE V
COMPARISON OF THE CONCENTRATION OF DEFECTS WITH THE CONCENTRATION
OF SINGLE-STRAND BREAKS IN DNA TREATED WITH PANCREATIC DNASE[a,b]

Preparation	$\Delta I \times 10^3$ (min^{-1})	(Defects/base pair) $\times 10^3$	(Single-strand breaks/base pair) $\times 10^3$
1	3.0 ± 0.6	0.6 ± 0.1	0.4 ± 0.1
2	8.4 ± 0.6	1.7 ± 0.1	1.4 ± 0.4
3	20.1 ± 0.6	4.1 ± 0.1	3.6 ± 0.4

[a] Adapted from Bannikov and Trifonov (1969).
[b] Defects were measured by the kinetic formaldehyde method, and breaks were determined from sedimentation measurements.

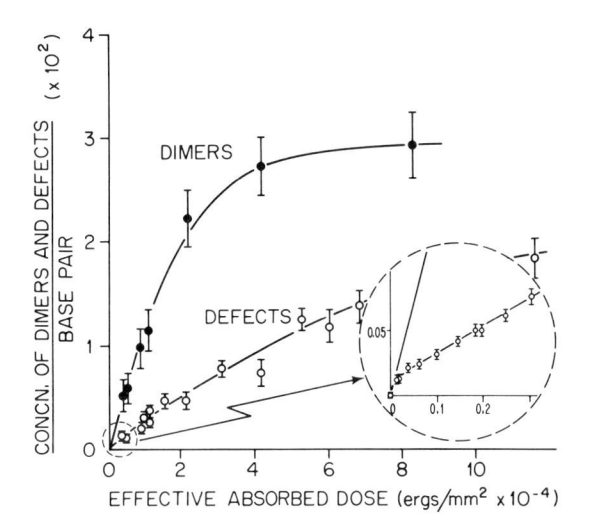

FIG. 11. Dose dependence of concentration of thymine dimers and concentration of secondary structure defects of irradiated (254 nm) T2 DNA. Inset contains the low dose data. (Adapted from Shafranovskaya et al., 1973.)

to or greater than 6 dimers/10^3 bases. The failure to observe at least one defect per dimer indicates that either the assumption of equal velocity of unwinding (v) for double-strand breaks and thymine dimers is wrong or else some other mechanism is occurring, such as long-range energy transfer resulting in preferential formation of thymine dimers at a single defect site. We shall return to this question of clustering of thymine dimers in Section 5.2.3.

4.2 Thermal Melting Analysis

Another method for quantitatively measuring defects was developed by Berestetskaya et al. (1970), who used the UV-induced broadening of the melting curve (see Fig. 2) to determine the concentration of defects generated during irradiation. Their method is based on the fact that the width of the melting curve ΔT broadens as the molecular weight decreases, i.e., as the number of ends increases. The width, ΔT, is determined by the relation (Lazurkin et al., 1970)

$$\Delta T = \frac{1}{|\partial\theta/\partial T|_{\max}}$$

θ being the degree of helicity. The temperature at which $|\partial\theta/\partial T|$ is a maximum is the melting temperature T_m. (For theoretical reasons, it turns out to be easier to relate changes in molecular weight to the width of the melting curve rather than the lowering of the T_m.) Furthermore, they assume, as Trifonov et al. (1968) had previously with the kinetic formaldehyde method, that the contribution to the T_m broadening by a defect, such as a thymine dimer, for example, is the same as the contribution from a double-strand chain break caused by fragmentation, i.e., both initiate two unwinding points in the starting DNA with comparable weakening of the surrounding helical portions. Hence, if ΔT represents the width of the melting curve for either a fragmented DNA or a DNA containing UV-induced defects, then according to Berestetskaya et al. (1970)

$$\delta(\Delta T)^2 = (\Delta T)^2 - (\Delta_0 T)^2$$
$$\sim 1/\bar{n}$$

where $\Delta_0 T$ corresponds to the melting width of the initial DNA and $\bar{n}$ is the mean number of base pairs either per molecule for fragmented DNA or between defects for UV'd DNA. Since $1/\bar{n} = c$ where c is the concentration of defects, then $\delta(\Delta T)^2 = Ac$ where A is a constant which is assumed to be the same for both types of defects. These workers showed that a plot of $\delta(\Delta T)^2$ vs $1/\bar{n}$ for fragmented DNA gives a straight line going through the origin in agreement with the theoretical prediction. From the value of A thus obtained, they calculate the number of defects in irradiated DNA assuming A is constant for both dimers and double-strand breaks. This method of determining defects is about one-tenth as sensitive as the kinetic formaldehyde method.

4.3 Comparison of the Kinetic Formaldehyde Method with the Thermal Melting Method

Berestetskaya et al. (1970) irradiated DNA and measured both the change in the breadth of the melting curve $\delta(\Delta T)$ and I, the rate of unwinding in formaldehyde. A plot of $\delta(\Delta T)^2$ vs I is shown in Fig. 12 and, as expected since both quantities are proportional to c (the concentration of defects), the curve is linear.

$\delta(\Delta T)^2$ and I for a single fragmented DNA sample was also determined and is plotted in Fig. 12. The fact that this point falls on the same curve as that obtained for UV-irradiated DNA indicates that

$$\frac{2c_{\text{uv}} \cdot v_{\text{uv}}}{2c_{\text{frag}} \cdot v_{\text{frag}}} = \frac{A_{\text{uv}} \cdot c_{\text{uv}}}{A_{\text{frag}} \cdot c_{\text{frag}}}$$

or

$$\frac{v_{\text{uv}}}{v_{\text{frag}}} = \frac{A_{\text{uv}}}{A_{\text{frag}}}$$

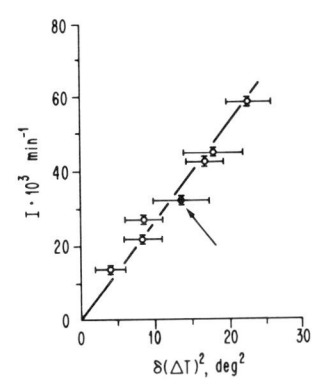

FIG. 12. Relationship between the I and $\delta(\Delta T)^2$ values for irradiated (254 nm) DNA (○) and fragmented DNA of 325,000 daltons (●). (Adapted from Berestetskaya et al., 1970.)

From this equivalency Berestetskaya et al. reasoned that

$$v_{uv} = v_{frag}$$

and

$$A_{uv} = A_{frag}$$

since different physical processes form the basis of these two methods of examining defects. This result was taken to represent experimental verification of the assumption that fragmented DNA and irradiated DNA contain equivalent defects with respect to thermodenaturation and reaction with formaldehyde.

5. Nature of Defect Region

5.1 Size of Defects

Shafranovskaya et al. (1972) estimated that a thymine dimer defect is 20–40 base pairs, as determined by kinetic measurements, although details of this measurement were not presented. On the other hand, a much smaller size for a dimer defect was obtained by Hayes et al. (1971). These workers synthesized a photodecamer $d(T_4\text{-}\widehat{TT}\text{-}T_4)$ and reacted it with poly(dA) to form a duplex polynucleotide. By comparing its melting temperature with duplexes made from poly(dA) and thymine oligonucleotides of varying sizes (Fig. 13), they estimated that ~4 of the hydrogen-bonded base pairs in the decamer were disrupted when a dimer was present. They interpreted this result to mean that base pairing on either side of the dimer was destroyed. Observations with molecular models agreed with these experimental results in that A·T

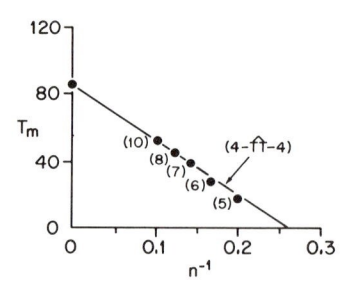

FIG. 13. T_m values of the oligodeoxythymidylate–polydeoxyadenylate interaction plotted against n^{-1} of dTn. The arrow points to the T_m of the dimer-containing complex. (Adapted from Hayes et al., 1971.)

base pairs on either side of a nonhydrogen-bonded region containing four base pairs formed hydrogen bonds easily with no distortion.

The size of the dimer defect as determined by Hayes et al. (1971)—4 base pairs—is very nearly the same size as that associated with a single-strand break as determined by Collyns et al. (1965), who measured chain scission and hydrogen-bond breakage on exposure of DNA to ionizing radiation. Chain breaks were determined from the amount of inorganic phosphate released following incubation of the irradiated DNA with phosphomonoesterase. Hydrogen-bond breaks were determined optically by using the equation of Applequist (1961) to relate hyperchromicity with the fraction of intact hydrogen bonds. The G value for chain breaks was 0.8, while for disruption of base pairing, G varied from 2.7 to 6.6, depending upon the concentration of the DNA used. Hence, 3–8 base pairs are lost per single break. The value of 3 may be closer to the real value, since Hagen (1967) found that 3 base pairs was the maximum distance 2 strand breaks could be apart, when located on opposite strands, and still give rise to a double-strand break.

5.2 Distribution of Defects

5.2.1 Base Content of Denatured Regions

Snake venom phosphodiesterase cannot split the phosphodiester bonds adjacent to nucleotides modified with CMEC (Drevich et al., 1967). Hence, irradiated DNA treated with CMEC contains regions that are resistant to enzymatic hydrolysis (also see below). Salganik et al. (1967) measured the distribution of thymine dimers between the mononucleotide fraction (A) and the oligonucleotide fraction (B) following enzymatic hydrolysis of irradiated DNA modified with CMEC. Using Sephadex chromatography, they found that when 9% of the thymine was converted into thymine dimers, 8% of the dimers were in fraction (B) and only 1% of the dimers were in fraction (A). The CMEC activity was found only in fraction (B). Hence, the nuclease-resistant fraction of the DNA contained the major portion of thymine dimer as well as the bound CMEC. Base composition analysis revealed that the content of A-T pairs in the oligonucleotide fraction was 1.8 times greater than that of G-C pairs while in the mononucleotide fraction it was 1.14 times greater. Since unirradiated DNA had $(A + T)/(G + C) = 1.3$, it was concluded that A-T rich regions are more sensitive to UV denaturation.

Unfortunately, Salganik et al. (1967) did not measure the enzymatic hydrolysis of irradiated DNA unmodified by treatment with CMEC. Since thymine dimers themselves provide blocks to enzymatic cleavage of the phosphodiester backbone (Setlow et al., 1964) it is difficult to judge what additional contribution bound CMEC makes toward inhibition of enzyme hydrolysis.

5.2.2 Distribution of Thymine Dimers in Pyrimidine Tracts

Evidence for the preferential formation of thymine dimers in long pyrimidine tracts has recently been obtained by Brunk (1973). He isolated pyrimidine runs up to 9 residues in length from irradiated DNA and measured the percentage of thymine converted to dimer in each of these runs. His results as shown in Table VI indicate that when 4.5% of the total thymine in the DNA was dimerized with 254 nm radiation, the probability of dimerization was 4- to 5-fold greater in a run of 9 pyrimidine residues than it was in a run of 2. Hence, defects due to thymine dimers are expected to be preferentially associated with the longer runs of thymine. To test the possibility that a greater percentage of the thymines were capable of being dimerized in the longer tracts, Brunk used acetophenone sensitization to make the maximum number of thymine dimers in each tract. He found, as shown in Table VI, that all the pyrimidine runs had ~70% of their thymine as dimer and concluded that the distribution of adjacent thymines is not responsible for the preferential dimerization of thymine in long pyrimidine runs.

5.2.3 Clustering of Thymine Dimers

In their study on the relationship between defects and thymine dimers, Shafranovskaya et al. (1972, 1973) maintained that low doses there was a 1:1 correlation between dimers and defects (see inset in Fig. 11) and that only at higher doses did this ratio become 6:1. They hypothesized, therefore, that once a thymine dimer is formed it makes a localized denatured region which acts as an energy trap. Transfer to

TABLE VI
Percentage of Thymine Dimerized in Runs of Pyrimidines Located in T. pyriformis DNA with an Adenine-Thymine Content of 0.78[a,b]

Peak	Percentage of thymine dimerized	
	8000 Ergs/mm² at 254 nm	Acetophenone sensitization at 313 nm
2	4.5 ± 0.4	75.4 ± 0.8
3	5.5 ± 0.3	64.4 ± 1.4
4	6.5 ± 0.8	69.4 ± 1.2
5	7.0 ± 0.7	67.8 ± 1.3
6	7.5 ± 0.6	67.6 ± 0.8
7	10.0 ± 0.3	71.5 ± 0.3
8	14.5 ± 1.1	70.6 ± 0.9
9	20.4 ± 1.4	70.9 ± 0.3
Total thymine	4.5 ± 0.7	52.8 ± 2.1

[a] Adapted from Brunk (1973).
[b] Dimers were formed either with 254 nm irradiation or upon extensive acetophenone sensitization.

this trap occurs leading to a preferential formation of dimers in this region without increasing the number of defects. They maintained that the increase in the ratio of dimers to defects from 1:1 to 6:1 occurred when there was only 1 defect per 10^3–10^4 nucleotide pairs. From this it was concluded that energy transfer occurs over several thousand base pairs.

However, the data obtained for defects in the low dose region are sparse, as indicated in Fig. 11, and no dimer yields were measured below a dose of 4000 ergs/mm² (they depended upon an extrapolation from high doses). Furthermore, the crucial measurements of defects are in the 1 defect/10^4 base pairs concentration range, which represents the lowest detectable level, and it is difficult to distinguish between the contribution from end effects and dimers at this level. Therefore, the low dose data upon which they base their conclusion that a 1:1 ratio exists between defects and dimers remains somewhat in question. Furthermore, their conclusion that energy transfer takes place over thousands of base pairs is not consistent with many of the existing data (for a review of energy transfer in polynucleotides see Guéron and Shulman, 1968).

As mentioned in Section 4.1, the basic assumption that underlies the kinetic formaldehyde method is that measurements of defects due to photoproducts, such as thymine dimers, can be made based on fragmented DNA as a calibration source. Experimental data in support of this assumption was presented in Section 4.3. However, an alternative interpretation of these data is possible. It might be that

$$A_{uv} = f \cdot A_{frag}$$

and

$$v_{uv} = f \cdot v_{frag}$$

where f is a constant. This situation would imply that type I and type IV defects (Fig. 1) differ quantitatively by a factor f in their ability to either interact with formaldehyde or broaden the T_m profile. Hence, the results of Shafranovskaya et al. (1972, 1973) in which six dimers were formed for every defect measured could be explained by assuming $f = 1/6$ and that dimers are one-sixth as effective as either double- or single-strand chain breaks in facilitating the reaction with formaldehyde. We rule out the possibility that the defect due to a dimer is smaller than that of a chain break and therefore has a smaller rate of unwinding. As noted in Section 5.1, both thymine dimers and single-strand breaks appear to disrupt the same number of hydrogen bonds, about 3–4 base pairs. It is felt, however, that a dimer defect should unwind more slowly than a chain break defect since the maintenance of both phosphodiester linkages in the vicinity of a dimer would restrict the unwinding of the strands to a greater extent than if one or both of the strands had a nick in it.

It is of interest that if dimers were preferentially formed in the defect regions, then dimerization should be favored at the ends of the DNA molecule, provided the DNA is on the order of 5×10^6 daltons or less in size. Evidence that dimers are in fact randomly distributed throughout the DNA is provided by the amount the molecular weight of irradiated DNA decreases upon treatment with UV-endonuclease. This enzyme makes a chain break at the site of a pyrimidine dimer (Carrier and Setlow, 1970). Upon UV irradiation of E. coli DNA, the number of thymine dimers and the inverse of the molecular weight (following endonuclease treatment) vary linearly with dose. Furthermore, a dose of 1600 ergs/mm² at 254 nm makes 0.83% pyrimidine dimer or on the average, 1 pyrimidine dimer every 0.32×10^6 daltons. In the same irradiated DNA, endonuclease treatment reduces the number average, single-strand molecular weight from 5×10^6 to 0.4×10^6 daltons. Hence, there is a chain break made by the enzyme every 0.43×10^6 daltons. Since the number of dimers and the number of chain breaks are, on the average, nearly the same number of nucleotides apart, and since the chain breaks are assumed to be randomly distributed, then the distribution of dimers should also be random.

6. Summary and Future Experiments

We have seen that, for acetophenone sensitization, there is a very clear correlation between thymine dimers and the lowering of the T_m. However, such a correlation has not been observed for UV irradiation. Clearly there are contributions made to the denaturation of DNA during UV irradiation which come from forms of damage other than pyrimidine dimers. As of yet, there is no clear indication of what these other forms of damage are.

Possible experiments in the future to ascertain the spectrum of damage responsible for the observed denaturation might include studies with model systems, such as poly[d(A-T)·d(A-T)] and poly[d(G-C)·d(G-C)]. Pyrimidine dimers could not form in these alternating systems, and contributions to the T_m lowering would most likely come from cross-linking reactions.

It would also be interesting to determine whether the formation of cross-links in DNA follows a sigmoid dose response curve as do the hydrates of cytosine, since partial denaturation would favor both these products.

It would be desirable to measure the rate of reaction of acetophenone-sensitized DNA with formaldehyde and compare the results obtained with the UV irradiation studies. The possibility exists that DNA which has been sensitized could serve as a suitable system for determining the velocity of unwinding (v) of a thymine dimer defect.

Acknowledgment

The author thanks R. B. Setlow for helpful comments on the manuscript.

References

Applequist, J. (1961). *J. Amer. Chem. Soc.* **83**, 3158.
Augusti-Tocco, G., and Brown, G. L. (1965). *Nature (London)* **206**, 683.
Bannikov, Yu. A., and Trifonov, E. N. (1969). *In* "Structure and Genetical Properties of Biopolymers," Vol. 2, p. 323. Kurchatov Inst. Atomic Energy, Moscow.
Berestetskaya, I. V., Kosaganov, Yu. N., Lazurkin, Yu. S., Trifonov, E. N., and Frank-Kamenetskii, M. D. (1970). *Mol. Biol.* (Russian) **4**, No. 1, 137.
Borisova, O. F., Koselev, L. L., and Tumerman, L. A. (1963). *Dokl. Akad. Nauk SSSR* **152**, 1001.
Brunk, C. F. (1973). *Nature (London) New Biol.* **241**, 74.
Carrier, W. L., and Setlow, R. B. (1970). *J. Bacteriol.* **102**, 178.
Collyns, B., Okada, S., Scholes, G., Weiss, J. J., and Wheeler, C. M. (1965). *Radiat. Res.* **25**, 526.
Dasgupta, R., and Mitra, S. (1970). *In* "Macromolecules in Storage and Transfer of Biological Information," p. 253. Dept. At. Energy, Bombay, India.

Drevich, V. F., Salganik, R. I., Knorre, D. G., and Mal'ygin, E.G.(1966).*Biochim. Biophys. Acta* **123**, 207.
Drevich, V. F., Knorre, D. G., Mal'ygin, E. G., and Salganik, R. I. (1967). *Mol. Biol.* **1**, 249.
Grossman, L. (1968). *In* "Methods in Enzymology," Vol. 12, (L. Grossman and K. Moldave, eds.), Pt. B, p. 467. Academic Press, New York.
Guéron, M., and Shulman, R. G. (1968). *Annu. Rev. Biochem.* **37**, 571.
Hagen, U. (1967). *Biochim. Biophys. Acta* **134**, 45.
Hayes, F. N., Williams, D. L., Ratliff, R. L., Varghese, A. J., and Rupert, C. S. (1971). *J. Amer. Chem. Soc.* **93**, 4940.
Lazurkin, Yu. S., Frank-Kamenetskii, M. D., and Trifonov, E. N. (1970). *Biopolymers* **9**, 1253.
Marmur, J., Anderson, W. F., Matthews, L., Berns, K., Gajewska, E., and Doty, P. (1961). *J. Cell. Comp. Physiol.* **58**, Suppl. 1, 33.
Rahn, R. O. (1972). *In* "Concepts in Radiation Cell Biology" (G. L. Whitson, ed.), p. 1. Academic Press, New York.
Rahn, R. O., and Landry, L. C. (1971). *Biochim. Biophys. Acta* **247**, 197.
Rahn, R. O., Landry, L. C., and Carrier, W. L. (1973). *Photochem. Photobiol.* (in press).
Salganik, R. I., Drevich, V. F., and Vasyunina, E. A. (1967). *J. Mol. Biol.* **30**, 219.
Seaman, E., Levine, L., and Van Vunakis, H. (1968). *In* "Nucleic Acids in Immunology" (O. J. Plescia and W. Braum, eds.), p. 157. Springer-Verlag, Berlin and New York.
Setlow, R. B., and Carrier, W. L. (1963). *Photochem. Photobiol.* **2**, 49.
Setlow, R. B., Carrier, W. L., and Bollum, F. J. (1964). *Biochim. Biophys. Acta* **91**, 446.
Shafranovskaya, N. N., Trifonov, E. N., Lazurkin, Yu. S., and Frank-Kamenetskii, M. D. (1972). *JETP Lett.* (in Russian) **15**, No. 7, 404.
Shafranovskaya, N. N., Trifonov, E. N., Lazurkin, Yu. S., and Frank-Kamenetskii, M. D. (1973). *Nature (London) New Biol.* **241**, 58.
Trifonov, E. N., Shafranovskaya, N. N., Frank-Kamenetskii, M. D., and Lazurkin, Yu. S. (1968). *Mol. Biol.* **2**, 887.
Varghese, A. J. (1972). *In* "Photophysiology" (A. C. Giese, ed.), Vol. 7, p. 207. Academic Press, New York.
Zavil'gel'skii, G. B., Borisova, O. F., Minchenkova, L. E., and Minyat, E. E. (1964). *Biokhimiya* **29**, 508.
Zavil'gel'skii, G. B., Minchenkova, L. E., Minyat, E. E., and Savich, A. P. (1965). *Biokhimiya* **30**, 652.

AUTHOR INDEX

Numbers in italics refer to the pages on which the complete references are listed.

A

Aagaard, J., 15, *59*
Akazawa, T., 150, *156*
Allen, C. F., 14, *61*, 109, *110*
Allen, F. L., 91, *93*
Allen, M. B., 158, *179*
Amelunxen, F., 104, *112*
Amesz, J., 98, 99, *111*
Andersen, K. S., 102, *110*
Anderson, J. M., 14, 15, 17, 20, 21, 22, 23, 40, 41, 43, 45, *59*, *60*, *63*, 81, *93*, *94*, *95*, 102, 104, *110*
Anderson, W. F., 234, 240, *255*
Andrews, T. J., 32, *61*, 152, *153*
Applequist, J., 250, *254*
Armstrong, J. J., 79, *96*
Arnon, D. I., 70, 74, 78, 81, 82, 85, 86, *93*, *94*, *95*, *96*, 161, 172, 173, 176, 177, *179*, *180*, *183*
Arntzen, C. J., 78, 81, *94*, 104, 105, 107, 108, 109, *110*
Atkinson, D. E., 162, *179*
Aubert, J.-P., 164, *180*
Augusti-Tocco, G., 241, *254*
Avron, M., 18, *59*, 88, *96*, 119, 124, 126, 127, 128, 133, 134, 139, 145, *153*, *154*, *156*
Azi, T., 109, *111*
Azzi, J. R., 87, *94*, 99, *110*

B

Baccarini-Melandri, A., 140, *153*
Bach, A. N., 168, *179*
Bachofen, R., 173, *179*
Bagg, J., 118, *153*
Bahr, J. T., 150, *154*
Bailey, J. L., 76, *94*
Bain, J. M., 102, *110*
Baldry, C. W., 29, *59*
Baltscheffsky, H., 158, *179*
Baltscheffsky, M., 158, *179*
Bamberger, E. S., 89, *94*
Bannikov, Yu. A., 245, 246, *254*

Bannister, T. T., 90, *96*, 142, *153*
Barber, J., 131, 132, 134, *153*
Barker, H. A., 159, *182*
Bartholomew, B., 47, *62*
Bassham, J. A., 4, 22, 36, *59*, 149, 151, *153*, *154*
Baszynski, T., 81, *94*
Bates, R. M., 115, *153*
Bayliss, N. S., 169, *179*
Bazzaz, M., 22, *59*
Beevers, H., 29, *61*
Ben-Amotz, A., 18, *59*
Bendall, D. S., 79, 85, *94*
Bendall, F., 98, *111*
Benemann, J. R., 158, 169, 174, *179*
Ben-Hayyim, G., 145, *153*
Bennett, R., 161, 177, *179*
Benzinger, T., 135, *153*
Berestetskaya, I. V., 247, 248, 249, *254*
Bergersen, F. J., 158, 165, 170, *179*, *180*
Berns, K., 234, 240, *255*
Berry, J. A., 22, 28, 29, 32, 47, 53, 57, 58, *59*, *60*, *61*, *62*, 102, *111*
Bertsch, W., 75, 87, *94*, *95*, 99, 101, 109, *110*, *111*
Biedermann, M., 166, *180*
Biggins, D. R., 170, *180*
Biggs, S., 14, *61*
Billings, W. D., 47, 48, *59*, *62*
Binet, A., 148, *153*
Bishop, D. G., 102, *110*
Bishop, N. I., 66, 67, 68, 71, 73, 76, 77, 78, 79, 82, 86, 87, 88, 89, *94*, *95*, *96*, 101, *110*
Björkman, O., 10, 12, 14, 15, 16, 17, 19, 20, 21, 23, 24, 28, 29, 31, 32, 33, 34, 38, 39, 40, 41, 43, 44, 45, 47, 48, 49, 50, 52, 53, 55, 56, 57, 58, *59*, *60*, *61*, *62*
Bjorn, L. O., 215, *228*
Black, C. C., Jr., 15, 22, 24, 26, 29, *60*, *61*, *62*, 102, *110*
Blinks, L., 98, *110*

257

Block, E., 210, 211, *228*
Boardman, N. K., 14, 15, 17, 20, 21, 22, 23, 40, 41, 43, 45, *59, 60, 61, 63,* 81, *93, 94, 95,* 102, 104, 106, *110*
Boehme, H., 93, *94*
Bollum, F. J., 251, *255*
Bolton, J. R., 101, *112*
Bonaventura, C., 89, *94,* 142, *153*
Borisova, O. F., 235, 238, 239, 240, *254, 255*
Bose, S. K., 161, *180*
Bourque, D. P., 47, 48, *59*
Bowes, G., 23, 32, *60, 62*
Boyer, J. S., 21, *61*
Boynton, J., 28, 53, 57, 58, *60*
Bradbeer, C., 169, 175, *182*
Brand, J., 81, *94*
Brand, M. J. D., 116, 121, *153*
Branton, D., 105, *111*
Braun, B. Z., 143, *155*
Bregoff, H. M., 165, *181*
Briantais, J.-M., 101, 105, 109, *111*
Brill, W. J., 168, 175, *180, 182, 183*
Brodie, A. F., 210, 211, *229*
Brown, C. M., 163, *182, 183*
Brown, G. L., 241, *254*
Brown, J. S., 99, 100, *111*
Brunk, C. F., 251, 252, *254*
Buchanan, B. B., 161, 172, 173, 178, *180, 182, 183*
Buchbinder, L., 210, *228*
Buchwald, H. E., 130, *156*
Buck, R. P., 115, 116, *153*
Bucke, C., 29, *59, 60*
Bulen, W. A., 160, 166, 170, 174, 175, *180, 181, 183*
Bulley, N., 17, *60*
Burchard, R. P., 210, 211, *228*
Burns, R. C., 158, 160, 169, 170, 171, 174, 176, *180, 181*
Burr, G. O., 24, *61*
Burris, R. H., 158, 159, 160, 162, 163, 164, 165, 169, 170, 174, 175, 176, 177, *180, 181, 182, 183*
Burton, K., 135, *153*
Butler, L. G., 149, *156*
Butler, W. L., 67, 77, 85, 87, 88, 89, 90, 92, *94, 96,* 213, 214, 216, 217, 218, *219, 220, 221,* 222, 223, 224, 225, 226, 227, *229*

C

Cammack, R., 172, 173, *181*
Campbell, F. W., 192, 193, *207*
Canvin, D. T., 32, *62*
Carmeli, C., 136, 141, *153*
Carnahan, C., 161, *182*
Carnahan, J. E., 168, 172, 173, *180*
Carrier, W. L., 232, 234, 251, 253, *254, 255*
Castle, J. E., 168, 172, 173, *180*
Chabot, B. F., 47, 48, *59*
Chain, R. K., 70, 74, 78, 81, 82, *93, 94, 96*
Chance, B., 91, *94,* 131, 134, 135, *156,* 220, *228*
Chaney, T. H., 93, *95*
Chartier, P., 36, *60*
Chen, T., 29, *60*
Cheniae, G. M., 92, *94*
Chollet, P., 36, *60*
Christian, G. D., 117, *155*
Clayton, R. K., 13, *60,* 66, *96,* 131, *154*
Clements, F. E., 57, *61*
Cobb, W. A., 193, 195, *207*
Cogdell, R. J., 136, *153*
Cohen, W. S., 137, *153*
Cohen-Bazire, G., 166, *180*
Collins, V. G., 159, *180*
Collyns, B., 250, *254*
Cook, J. R., 211, 212, *228, 229*
Coombs, J., 24, 29, *59, 60*
Coste, F., 19, *62*
Cramer, W. A., 86, 93, *94,* 177, *180*
Crane, F. L., 81, *94,* 105, *110, 112*
Craston, A., 141, *155*
Crofts, A. R., 124, 125, 126, 127, 131, 132, 133, 134, 136, 137, 138, 141, 143, 150, *153, 154, 156,* 167, *180*
Cusanovich, M. A., 173, *180*

D

Daesch, G., 165, *180*
Dalton, H., 158, 165, *166,* 169, 170, 171, *180*
Daniel, J. W., 212, 213, *228*
Dasgupta, R., 238, 242, 243, *254*
Datko, E. A., 75, *95*
Davidson, J. B., 87, *94,* 99, *110*
Davis, L. C., 168, *180, 183*
Deamer, D. W., 126, 127, *153*

AUTHOR INDEX

Decker, K., 161, 162, *181*, *183*
De Faya, J., 212, 213, *229*
Degani, H., 133, *153*, *156*
de Kouchkovsky, Y., 19, *62*
Deters, D., 140, *155*
Detroy, R. W., 170, *180*
D'Eustachio, A. J., 174, *180*, *181*
DeVoe, R. G., 195, *208*
Dewolfe-Slade, D., 214, *229*
Dilley, R. A., 14, *61*, 78, 81, *94*, 104, 105, 107, 108, 109, *110*, *112*, 124, 130, 132, 138, 141, 148, 150, *153*, *154*, *155*, *156*
Dixon, R. A., 179, *180*
Dixon, R. O. D., 160, *180*
Döring, G., 108, *111*
Dolzmann, P., 68, *94*
Donze, M., 102, *111*
Doty, P., 234, 240, *255*
Downes, R. W., 31, 55, *60*
Downton, J., 22, 26, 29, *59*, *60*, *63*
Downton, W. J. S., 102, *111*
Drevich, V. F., 242, 250, 251, *255*
Drews, G., 158, 166, *180*, *182*
Drozd, J., 166, *180*
Dudas, I., 210, *228*
Dugger, W. M., 212, *229*
Duranton, J., 148, *153*
Duval, D., 148, *153*
Duysens, L. N. M., 98, 99, *111*, 129, 142, 143, *153*
Dworkin, M., 210, 211, *228*

E

Eagles, C. F., 44, *60*
Eckfeldt, E. L., 115, *153*
Edge, H., 104, *111*
Edmondson, D. E., 173, *180*
Edsall, P. C., 214, 215, *229*
Edwards, G., 22, 26, 29, *60*, *62*
Egle, K., 5, 31, *60*, *61*
Ehrenberg, M., 212, 213, 214, *228*, *229*
Eisenberg, M. A., 161, *180*
El-Badry, A. M., 149, *153*
Elmerich, C., 164, *180*
El-Sharkawy, M., 52, 54, *60*
Elstner, E., 80, *96*
Emerson, R., 16, *60*, 98, *111*, 145, *154*
Emrich, H. M., 130, *154*
Engel, E. K., 89, *95*
Engel, H., 210, 211, *229*

Entsch, B., 18, *62*
Epel, B., 211, 212, 213, 215, 216, 218, *229*
Epel, B. L., 67, 88, 89, 90, 92, *94*, *96*, 213, 214, 216, 217, 218, *219*, *220*, *221*, 222, 223, 224, 225, 226, 227, *229*
Erixon, K., 67, 85, 88, 89, 90, *94*, *96*
Ernster, L., 167, *181*
Ettlinger, G., 193, 195, *207*
Evans, E. H., 136, *153*
Evans, H. J., 173, *181*
Evans, M. C. W., 136, 137, 138, 140, *156*, 169, 170, 172, 173, 177, 178, *180*, *181*, *182*, *183*

F

Fan, H. N., 86, *94*
Farron, F., 141, *155*
Fedorov, M. V., 159, *181*
Fiat, R., 137, *154*
Fisher, R. J., 167, *182*
Fiver, D. W., 215, *229*
Fleet, B., 117, *154*
Fleischman, D. E., 131, *154*
Fock, H., 5, 31, *60*, *61*
Fork, D. C., 99, 102, 110, *111*
Forrester, M. L., 31, *61*
Forti, G., 83, *94*
Foust, G. P., 173, *182*
Franck, J., 91, *93*
Frank-Kamenetskii, M. D., 244, 245, 246, 247, 248, 249, 251, 253, *254*, *255*
French, C. S., 10, 22, *61*, 98, 99, 102, *111*, 123, *154*
Frenkel, A. W., 158, 164, 165, 167, 176, *181*, *182*
Frey-Wyssling, A., 212, 213, 222, 223, 224, *229*
Fuhrman, J. S., 136, 137, 138, 139, 140, *155*
Fuller, R. C., 161, 177, *179*

G

Gaensslen, R. E., 124, 127, 133, 136, 141, *154*
Gaffron, H., 14, *62*
Gajewska, E., 234, 240, *255*
Galmiche, J. M., 137, *154*
Garewal, H. S., 86, *94*
Gates, D., 46, *61*

AUTHOR INDEX

Gauhl, E., 23, 24, 29, 31, 44, 47, 50, 52, 53, 57, *59*, *60*, *61*
Gee, R., 76, 77, *94*
Gest, H., 140, *153*, 158, 159, 160, 161, 162, 165, 178, *180*, *181*, *182*
Gibbs, M., 31, 32, *61*, 69, *95*
Giese, A. C., 214, *229*
Girault, G., 137, *154*
Givan, A. L., 67, 75, 76, 77, *94*
Godfrey, P. J., 47, 48, *59*
Goedheer, J. C., 99, 100, *111*
Goldberg, E., 214, *229*
Goldstrohm, D. D., 212, 213, *229*
Goldsworthy, A., 31, 33, *61*
Good, N. E., 104, *111*, 133, 138, 139, 144, 145, *154*
Good, P., 109, *110*
Goodchild, D. J., 12, 14, 17, 20, 21, 23, 40, 41, 43, 45, *60*, *61*, 104, *112*
Goodenough, U. W., 66, 67, 69, 79, *94*, *95*, *96*
Gorman, D. S., 67, 80, 81, 83, *94*, *95*
Gottschalk, W., 71, 72, 73, *95*
Govindjee, 22, *59*, 142, 143, *154*, *155*
Graham, J., 15, *62*
Grahl, H., 14, *63*
Granick, S., 66, *95*
Grau, F. H., 159, *181*
Gray, C. T., 162, *181*
Greenblatt, C. L., 89, *95*
Gregory, R. P. F., 75, *95*, 101, 109, *111*
Griffith, M., 66, *96*
Gromet-Elhanan, Z., 126, *154*
Gross, E., 148, *154*
Gross, R. G., 212, *229*
Grossman, L., 240, *254*
Grünhagen, H. H., 130, *154*
Grunwald, T., 124, 126, 128, 133, 134, 139, *156*
Guerin, B., 212, 213, *229*
Guéron, M., 252, *255*
Gurney, E., 179, *183*

H

Hadfield, K. L., 174, 175, *181*
Hageman, R. H., 23, 32, *60*, *63*
Hagen, U., 250, *255*
Hall, A. E., 23, *61*
Hall, D. O., 104, *111*, *112*, 172, 173, *181*
Hall, H. M., 57, *61*

Halldal, P., 11, *61*
Halliday, A. M., 195, 206, *208*
Hamilton, P. B., 159, *181*
Hamilton, W. D., 175, *182*
Haque, M. Z., 178, *181*
Hardy, R. W. F., 158, 159, 170, 171, 174, 175, *180*, *181*
Harm, H., 210, 211, *229*
Harris, B., 29, *62*
Harrison, A. T., 38, 47, 55, 56, *60*, *62*
Hart, R. W., 123, *154*
Hartt, C. E., 24, *61*
Hatch, M. D., 24, 28, 31, 52, *61*
Hauska, G. A., 147, *154*
Haveman, J., 102, *111*
Hayaishi, O., 171, *182*
Hayes, F. N., 249, 250, *255*
Heath, R. L., 127, *154*
Heber, U. W., 71, 72, 73, *95*, 102, *111*
Hedén, C.-G., 179, *181*
Heldt, H. W., 149, *154*
Helgager, J. A., 21, *62*
Hems, R., 135, *153*
Heron, J. R., 207, *208*
Herscovici, A., 139, *156*
Hesketh, J. D., 31, 52, 54, *60*, *61*
Hiesey, W. H., 23, 47, 48, 50, *60*
Highkin, H., 14, *61*
Hilgenberg, W., 5, *60*, *61*
Hill, R., 79, 81, *94*, *95*, 98, *111*
Hiller, R. G., 81, 93, *94*, *95*
Hind, G., 79, 80, 86, *95*, 114, 115, 117, 127, 139, 150, 151, *154*
Hinkson, J., 160, 174, *180*
Hoch, G. E., 73, 74, 78, *95*, *96*, 145, 147, *156*, 160, *181*
Hollaender, A., 210, 211, *229*
Holmgren, P., 39, 44, *60*, *61*
Holsten, R. D., 170, 171, *180*
Homann, P. H., 68, 92, *95*
Huang, A., 29, *61*
Hudock, G. A., 68, *95*
Hughes, D. E., 166, *181*
Huzisige, H., 109, *111*

I

Iyama, J., 52, *62*
Izawa, S., 104, *111*, 115, 117, 127, 133, 137, 138, 139, 144, 145, *154*

AUTHOR INDEX

J

Jackson, E. K., 175, *181*
Jackson, J. B., 131, 132, 133, 134, 136, *153*, *154*
Jackson, W., 31, *61*
Jacobi, G., 103, *111*
Jagendorf, A. T., 114, 120, 123, 124, 128, 137, 138, 140, *153*, *154*, *155*, *156*
Jagger, J., 210, 211, 228, *229*
Jarvis, M. S., 39, *61*
Jarvis, P. G., 39, *61*
Jeffreys, D. A., 195, *208*
Jeng, D., 170, *181*
Jensen, R. G., 149, 150, 151, *154*
Jermolieva, Z. V., 168, *179*
Johannson, G., 116, *154*
Johnson, H. B., 29, 51, *62*
Johnson, H. S., 24, *61*
Joliot, A., 91, *95*
Joliot, P., 91, *95*
Jones, B. K., 159, *181*
Jones, L. W., *61*, 88, *95*
Jones, T. L., 175, *182*
Junge, W., 130, 135, *154*, *156*
Jungermann, K., 162, *183*

K

Kahn, A., 104, *110*
Kahn, J. S., 127, *154*
Kalina, M., 104, *111*
Kalininskaya, T. A., 159, *181*
Kamen, M. D., 159, 160, 165, *181*
Kamp, B. M., 98, *111*
Kanai, R., 24, 26, *60*, *61*
Kandler, O., 82, *96*
Karlish, S. J. D., 138, *154*
Karpilov, Y. S., 59, *61*
Kashket, E. R., 210, 211, *229*
Kawaguchi, K., 178, *181*
Ke, B., 93, *95*, 130, *155*
Keck, R. W., 14, 21, *61*
Keister, D. L., 132, *156*, 162, 177, *181*
Kelly, M., 167, 170, 172, *180*, *181*
Kennedy, I. R., 167, *181*
Khalifa, M. M., 23, *63*
Kikuti, T., 109, *111*
Kindergan, M., 89, *94*
Kirk, J. T. O., 66, *95*, 105, *111*
Kirk, M. R., 149, *153*
Kirshteine, B. E., 168, *182*

Kitzinger, C., 135, *153*
Klein, R. M., 214, 215, *229*
Klein, S., 107, 108, 109, *112*
Knaff, D. B., 70, 74, 78, 82, 85, *94*, *95*
Knorre, D. G., 242, *255*
Kobayashi, M., 178, *181*, *182*
Koch, B., 169, 173, *181*
Kok, B., *61*, 73, 75, 88, *95*
Kondrat'eva, E. N., 158, *181*
Kortschak, H. P., 24, 29, *61*
Koryta, J., 115, *154*
Kosaganov, Yu. N., 247, 248, 249, *254*
Koselev, L. L., 238, *254*
Kowallik, W., 212, *229*
Kraan, G. P. B., 131, *153*
Kraayenhof, R., 126, 135, *154*
Krauss, R. W., 211, 212, 213, 215, 216, 218, *229*
Krebs, W. M., 116, *155*
Kretovich, V. L., 168, *182*
Kreutz, W., 76, *94*
Krogmann, D. W., 81, *94*, 138, *153*
Krotkov, G., 31, *61*, *63*
Krull, I., 115, 116, *153*
Kugler, G. C., 117, 118, *155*
Kuneida, R., 144, 145, *155*
Kurup, C. K., 210, *229*

L

Laetsch, W. M., 28, 29, *61*, 102, *111*
Lal, S., 117, *155*
Landry, L. C., 232, 235, 236, 237, 241, *255*
Larsen, H., 210, *228*
Lasada, M., 177, *179*
Latzko, E., 69, *95*
Lazurkin, Yu. S., 244, 245, 246, 247, 248, 249, 251, 253, *254*, *255*
LeComte, J. R., 160, 166, 170, 174, 176, *180*
Lee, C.-P., 167, *181*
Lee, S. B., 29, *60*, 160, *182*
Lehmann, H., 103, *111*
Lemon, E. R., 52, *62*
Levine, L., 243, 244, *255*
Levine, R. P., 66, 67, 68, 69, 71, 75, 76, 77, 79, 80, 81, 82, 83, 84, 86, 87, 92, *94*, *95*, *96*, 101, *111*, 140, *156*
Lewis, C. M., 16, *60*
Lewis, S. M., 159, *182*

Lichtenthaler, H., 19, *61*
Lie, T. A., 158, *182*
Lilly, V. G., 212, 213, *229*
Lin, D. C., 149, 151, *155*
Lindstrom, E. S., 159, *182*
Lipmann, F., 141, *155*
Little, H. M., 160, *181*
Long, S., 29, *60*
Lorimer, G. H., 32, *61,* 152, *153*
Lozier, R., 89, 90, *96*
Ludlow, M., 12, 17, 38, 39, 40, *60*
Ludwig, L. J., 32, *62*
Luzzana, M., 121, *155*
L'vov, N. P., 168, *182*
Lyman, H., 66, *96*
Lyubimor, V. I., 168, *182*

M

McCarty, R. E., 124, 127, 128, 132, 133, 136, 137, 138, 139, 140, 141, 143, *154, 155*
McCormick, A. V., 150, 151, *156*
McCree, K., 16, *62*
Mackay, D. M., 193, 195, *208*
McLeod, G. C., 99, *111*
McNary, J. E., 174, *182*
McNaughton, S. J., 21, *62*
McSwain, B. D., 70, 74, 78, 81, 82, *93, 94, 96*
Maffei, L., 192, 193, *207*
Malkin, S., 88, *95*
Mal'ygin, E. G., 242, 250, *255*
Mansfield, T. A., 35, *62*
Mantai, K. E., 88, 89, 90, *95*
Marcus, L., 167, *182*
Marmur, J., 234, 240, *255*
Martin, I. F., 92, *94*
Marx, R., 166, *180*
Massey, V., 173, *182*
Mathews, M. M., 210, *229*
Matile, P., 212, 213, 215, 222, 223, 224, *229*
Matsubara, H., 172, 173, *182*
Matthews, L., 234, 240, *255*
Mayhew, S. G., 173, *182*
Mayne, B. C., 15, 22, *60, 62,* 102, 107, 108, 109, *110, 112*
Medina, E., 23, *62*
Meers, J. L., 163, *182, 183*

Meidner, H., 35, 52, *62*
Mel, H. C., 119, *155*
Menke, W., 101, *112*
Michael, W. F., 195, 206, *208*
Miller, E. C., 159, *180*
Miller, R. E., 163, *182*
Milner, B. A., 207, *208*
Minchenkova, L. E., 235, 238, 239, 240, *255*
Minyat, E. E., 235, 238, 239, 240, *255*
Mitchell, P., 114, 123, 130, 141, *155*, 167, *182*
Mitra, S., 238, 242, 243, *254*
Mohanty, P., 143, *155*
Moll, B., 68, *95*
Mooney, H. A., 38, 47, 48, 55, 56, *60, 62*
Moore, E. W., 118, *155*
Morrow, P., 17, 40, *60*
Mortenson, L. E., 158, 161, 165, 167, 168, 169, 170, 171, 172, 173, 174, *180, 181, 182*
Morton, H. B., 193, 195, *197*
Mosolova, I. M., 215, *229*
Moss, D. N., 52, *62*
Mousseau, M., 19, *62*
Moustafa, E., 167, 171, 172, *182*
Mower, H. F., 168, 172, 173, *180*
Mühlethaler, K., 104, *111*
Muhle, M., 136, 138, *156*
Mulder, E. G., 158, *182*
Muller-Neugluck, M., 210, 211, *229*
Munday, J. C., Jr., 142, *154*
Munson, T. O., 165, 176, 177, *182*
Murakami, S., 144, 145, 146, 147, *155*
Murata, N., 104, 110, *111,* 123, 142, 143, 144, 145, 146, *154, 155*
Murata, Y., 52, *62*
Musgrave, R. B., 52, *61, 62*
Mutze, B., 210, 211, *229*
Myers, D. K., 214, *229*
Myers, J., 15, *62,* 98, 99, *111,* 142, 153, *155*

N

Nagatani, H., 163, *182*
Naim, V., 137, 138, *155*
Nakatani, H. Y., 86, *95,* 150, 151, *154*
Nakayama, N., 150, *156*
Nakazawa, T., 171, *182*
Nakos, G., 171, *182*

AUTHOR INDEX

Neales, T. F., 23, *62*
Nelson, C. D., 17, 31, *60, 61, 63*
Nelson, H., 137, 138, 140, *155*
Nelson, N., 137, 138, 140, *155*
Neumann, J., 67, 84, *96,* 104, *110,* 124, 128, 130, 137, 138, 140, *153, 154, 155, 156*
Newton, J. W., 163, *183*
Nilsson, J., 215, *228*
Ninnemann, H., 214, 215, 222, 223, 224, 225, 226, 227, *229*
Nobel, P. S., 119, 144, 147, 149, 151, *155*
Nobs, M. A., 23, 28, 47, 48, 50, 52, 53, 57, 58, *60, 62*
Norberg, K., 116, *154*
Norman, C., 214, *229*
Nozaki, M., 171, 177, *179, 182*

O

Oelze, J., 158, *182*
Ogawa, T., 106, 107, *112*
Ogren, W. L., 23, 32, *60, 62, 63*
Ohad, I., 68, 81, *95, 96*
Ohki, R., 144, 145, *155*
Ohrloff, C., 161, *181*
Okada, S., 250, *254*
Okayama, S., 89, 90, *94, 96*
Okuda, A., 178, *182*
Olson, J. M., 79, *95*
Olson, R. A., 89, *95*
Olson, R. S., 214, *229*
Oosting, H., 203, *208*
Oppenhein, J., 167, *182*
Orme-Johnson, W. H., 168, 175, *180, 182*
Ormerod, J. G., 160, 161, 165, *181, 182*
Ormerod, K. S., 160, 161, 165, *181, 182*
Osmond, C. B., 22, 24, 29, 30, 31, 32, 33, 34, 52, 58, 59, *61, 62, 63*
Ottenheym, H. C. J., 149, *153*

P

Packer, L., 144, 145, 146, 147, *155*
Padilla, G. M., 211, 212, *229*
Palade, G. E., 68, *95*
Papageorgiou, G., 142, 143, *154, 155*
Parejko, R. A., 165, 170, *180, 182*
Park, R. B., 74, 89, *94, 96,* 99, 104, 109, *110, 111, 112,* 144, 147, *155*
Parker, C. A., 165, *182*

Parlange, J. Y., 35, *62*
Parshall, G. W., 175, *181*
Paszewski, A., 69, *95*
Pearcy, R. W., 28, 38, 47, 48, 49, 53, 55, 56, 57, 58, *60, 62*
Pengra, R. M., 165, *182, 183*
Perley, G. A., 115, *153*
Perlmutter, A., 214, *229*
Perrella, M., 121, *155*
Peters, G. A., 78, 81, *94,* 107, 108, 109, *110*
Petrack, B., 141, *155*
Pfennig, N., 158, 159, *182*
Phelps, E. B., 210, *228*
Picaud, M., 101, 105, *111*
Piemeisel, L. M., 55, *62*
Piette, L. H., 87, *95*
Pine, M. J., 159, *182*
Pinsky, M. J., 159, *182*
Pittman, P. R., 140, *155*
Poincelot, R., 29, *62*
Polya, G. M., 120, *155*
Porterfield, I. D., 214, *229*
Portis, A. R., Jr., 128, *155*
Postgate, J. R., 158, 165, 166, 167, 169, 170, 172, 175, 179, *180, 181, 182*
Powls, R., 67, *96*
Pratt, D. C., 164, 165, 167, *182*
Pratt, L. H., 76, 77, *96*
Pressman, B. C., 114, 119, *155*
Prince, R. C., 126, 127, *153*
Prino, B., 214, *229*
Punnett, T., 104, *111,* 147, *155*
Pyliotis, N., 12, 14, 17, 20, 21, 23, 40, 41, 43, 45, *60, 61*

R

Racker, E., 140, 141, 143, *155*
Radunz, A., 101, *112*
Rahn, R. O., 232, 235, 236, 237, 241, *255*
Randles, H., 78, *96*
Randles, J., 145, 147, *156*
Rao, K. K., 173, *182*
Raps, S., 75, *95,* 101, 109, *111*
Ratliff, R. L., 249, 250, *255*
Raveed, D., 106, 107, *112*
Raven, J. A., 23, 36, *62*
Rechnitz, G. A., 116, 117, 118, 121, *153, 154, 155, 156*
Reeves, S. G., 104, *111, 112*

AUTHOR INDEX

Regan, D., 187, 189, 191, 192, 193, 194, 195, 196, 198, 200, 201, 203, 206, 207, *208*
Reich, R., 130, *156*
Reinwald, E., 130, *155*
Remy, R., 109, *112*
Renger, G., 108, *111*
Rice, G., 142, *153*
Richards, R. L., 172, *181*
Richards, W., 193, 201, *208*
Rigopoulos, N., 161, 177, *179*
Rilling, H. C., 210, *229*
Rippel-Baldes, A., 210, *229*
Ripps, H., 195, *208*
Rosing, J., 135, *155*
Ross, J. W., 118, 119, *155*
Rossi-Bernardi, L., 121, *155*
Rottenberg, H., 124, 126, 128, 133, 134, 139, *156*
Rounds, D. E., 214, *229*
Rüppel, H., 130, *156*
Rumberg, B., 124, 128, 129, 130, 135, 136, 138, *154, 155, 156*
Rupert, C. S., 249, 250, *255*
Rupprecht, E., 161, 162, *181, 183*
Rurainski, H. J., 74, 78, *96,* 145, 147, *156*
Russell, G. K., 66, *96*
Russell, S., 169, *181*
Ryrie, I. J., 140, *156*

S

Salganik, R. I., 242, 250, 251, *255*
Saltman, P., 76, 77, *94*
Sane, P. V., 74, *96,* 99, 104, *111, 112,* 144, 147, *154, 155*
San Pietro, A., 73, *96,* 101, 106, *112,* 132, 133, 140, 148, *153, 154, 156*
Santamaria, L., 214, *229*
Sasaki, R. M., 172, 173, *182*
Sato, V. L., 67, 84, *96,* 140, *156*
Sauer, F., 149, *154*
Sauer, K., 99, *112,* 145, 146, 147, *156*
Savich, A. P., 240, *255*
Scarisbrick, R., 81, *95*
Schaub, H., *60*
Schick, H.-J., 164, 165, 167, 177, *182*
Schiereck, P., 102, *111*
Schiff, J. A., 66, *96*
Schliephake, W., 130, *156*
Schmid, G. H., 14, *62,* 68, *95,* 101, *112*

Schmidt, S., 130, *156*
Schmidt-Mende, P., 129, *156*
Schneider, K. C., 169, 175, *182*
Scholes, G., 250, *254*
Schon, G. H., 210, 211, *229*
Schröder, H., 135, 136, 138, *154, 156*
Schröder, J., 166, *180*
Schroener, B., 220, *228*
Schuldiner, S., 81, *96,* 126, 127, 128, *156*
Schwartz, M., 136, 138, *156*
Schwelitz, F. D., 105, *112*
Scutt, P. B., 165, *182*
Seaman, E., 243, 244, *255*
Selman, B. R., 90, *96*
Senger, H., 68, 78, *94, 96*
Setlow, R. B., 234, 251, 253, *254, 255*
Shafranovskaya, N. N., 244, 245, 246, 247, 248, 249, 251, 253, *255*
Shah, V. K., 168, 175, *180, 182, 183*
Shanmugan, K. T., 161, 172, 173, *183*
Shantz, H. L., 55, *62*
Shavit, N., 88, *96,* 132, 133, 138, 139, *153, 156*
Shaw, E. R., 78, 81, *94,* 101, 106, 107, 108, 109, *110, 112*
Sheppy, F., 141, *155*
Sheridan, R., 15, *62*
Shibata, K., 123, *156*
Shimizu, M., 163, *182*
Shropshire, F., 47, *62*
Shulman, R. G., 252, *255*
Shumway, L. K., 104, *112*
Siekevitz, P., 68, *95*
Siggel, U., 124, 128, 129, 130, *155, 156*
Silverstein, R., 170, *183*
Simmons, S., 149, *156*
Sinclair, J., 146, *156*
Singh, J., 86, *94*
Singh, R. N., 169, 175, *182*
Sisakyan, N. M., 215, *229*
Sisler, F. D., 159, *183*
Sistrom, W. R., 15, *59,* 66, *96,* 166, *180,* 210, *229*
Skerra, B., 129, *156*
Skulachev, V. P., 133, 134, *156*
Slack, C. R., 24, 29, *61, 62*
Slater, E. C., 135, *155*
Slatyer, R. O., 24, 31, 52, 54, 55, *61, 62*
Slonimski, P. P., 212, 213, *229*

AUTHOR INDEX

Smillie, R. M., 18, *62*, 67, 87, *95, 96*, 102, *110*
Smith, R. V., 169, 170, 172, 173, 177, *180, 181, 183*
Sofrova, D., 85, *94*
Solon, A. G., 210, 211, *228*
Soloway, M., 210, *228*
Sorger, G. J., 166, *183*
Sorokin, C., 211, 212, *229*
Spekreijse, H., 190, 191, 193, 195, 196, 198, 203, 204, 205, 206, *208*
Spencer, D., 22, *59*
Sperling, H. G., 196, *208*
Springer-Lederer, H., 149, *153*
Stadtman, E. R., 163, *182*
Stafford, R. S., 210, *229*
Stanier, R. Y., 66, *96*, 158, 166, *180, 183*
Steinback, K. E., 99, *111*
Stepanian, M. P., 168, *179*
Stewart, W. D. P., 158, *183*
Stocking, C. R., 104, *111, 112*
Stokes, D. M., 150, 151, *156*
Strandberg, G. W., 165, 168, *183*
Streicher, S., 179, *183*
Strichartz, G. R., 129, 131, 134, 135, *156*
Stubbe, W., 101, *112*
Sugahara, K., 143, *155*
Sugiyama, T., 150, *156*
Sulkowski, E., 212, 213, *229*
Sun, A. S. K., 145, 146, 147, *156*
Surzycki, S. J., 79, *96*
Sutton, B., 30, *62*
Sutton, S., 207, *208*
Suzzuki, Y., 215, *228*
Swart-Fuchtbauer, H., 210, *229*
Sweers, H. E., 99, *111*
Szarek, S. R., 29, 51, *63*

T

Takahashi, E., 178, *181*
Takamiya, A., 144, 145, 146, *155*
Talens, A., 142, *153*
Tanner, W., 82, *96*
Tashiro, H., 144, 146, *155*
Telfer, A., 136, 137, 138, 140, *156*, 170, 172, 173, *181, 182*
Tempest, D. W., 163, *182, 183*
Thauer, R. K., 161, 162, *181, 183*
Thompson, H. I., 117, *156*
Thore, A., 132, *156*, 158, *179*

Thornber, J. P., 76, *96*
Thorne, S. W., 14, 15, 17, 20, 21, 22, 23, 40, 41, 43, 45, *60, 63*, 104, *110*
Tiezen, L. L., 21, *62*
Tilney-Basset, R. A. E., 66, *95*, 105, *111*
Ting, I. P., 29, 51, 58, 59, *62*
Togasaki, R. K., 69, *95, 96*
Togawa, K., 177, *179*
Tolbert, N. E., 31, 32, 33, *61, 63*, 152, *153*
Tove, S. R., 159, *182*
Trebst, A., 73, 80, *96*
Treffry, T. E., 104, *110*
Tregunna, E. B., 17, 22, 29, 31, *59, 60, 63*, 102, *111*
Treharne, K. J., 23, 44, *60, 62, 63*
Trifonov, E. N., 244, 245, 246, 247, 248, 249, 251, 253, *254, 255*
Trosper, T., 109, *110*
Troughton, J. H., 31, 58, *61, 63*
Tso, M.-Y., 175, *182*
Tsuchiya, Y., 136, 137, 138, 139, 140, *155*
Tsujimoto, H. Y., 70, 74, 78, 81, 82, 91, *93, 94, 96*
Tumerman, L. A., 238, *254*
Turner, G. L., 170, *180*
Tyszkiewicz, E., 137, *154*

U

Umbreit, W. W., 160, *183*
Uribe, E., 123, 137, *154*
Usiyama, H., 109, *111*

V

Valentine, R. C., 158, 161, 163, 169, 172, 173, 174, *179, 182, 183*
Vandecasteele, J.-P., 170, *183*
Van Der Tweel, L. H., 190, 191, 193, 195, 196, 203, 204, *208*
Van Vunakis, H., 243, 244, *255*
Varghese, A. J., 232, 249, 250, *255*
Vasyunina, E. A., 250, 251, *255*
Vater, J., 129, *156*
Vater, S., 108, *111*
Vaughan, H. G., Jr., 195, *208*
Verduyn Lunel, H. F. E., 191, 193, 203, *208*
Vernon, L. P., 93, *95*, 101, 106, 107, 108, 109, *112*, 150, *153*, 158, *183*
Vinen, R., 118, *153*

Volfin, P., 148, *153*
Volk, R. J., 31, *61*
von Stedingk, L.-V., 132, 136, *154*
von Wettstein, D., 68, *96*
Voorn, G., 107, 108, *112*
Vredenberg, W. J., 152, *156*

W

Wada, K., 86, *96*
Wagenknecht, A. C., 163, *183*
Waggoner, P., 35, *62*
Walker, D. A., 124, 150, 151, 152, *156*
Wall, J. S., 163, *183*
Walz, D., 119, *156*
Wang, L. C., 169, 175, *182*
Warden, J. T., 101, *112*
Wareing, P. F., 23, *62, 63*
Wasserman, A. R., 86, *94*
Weaver, E. C., 67, 76, 77, 87, *94, 96*, 101, *112*
Weaver, H., 101, *112*
Weier, T. E., 104, *112*
Weikard, J., 129, *156*
Weiss, J. J., 250, *254*
Wells, P. H., 214, *229*
Wessels, J. S. C., 80, *96*, 107, 108, 109, *112*
West, M., 47, *62*
Wheeler, C. M., 250, *254*
White, F. G., 107, 108, 109, *112*
Wiessner, W., 104, *112*
Wild, A., 14, *63*
Williams, D. L., 249, 250, *255*
Williams, G. J., 21, *63*
Wilson, P. W., 159, 160, 163, 164, 165, 167, 168, 169, 170, 175, *180, 181, 182, 183*
Wimpenny, J. W. T., 166, *181*

Winter, H. C., 172, 176, 177, *183*
Wise, W. C., 210, *229*
Witt, H. T., 108, *111*, 124, 129, 130, 134, *154, 156*
Witz, D. F., 170, *180*
Wojcieska, U. B., 23, *63*
Wolf, F. T., 31, *63*
Wolff, C., 130, *156*
Wolff, E. G., 215, *229*
Wong, J., 67, 87, 88, 89, *94, 95, 96*
Wong, P. P., 165, *183*
Woo, K. C., 22, *59, 63*, 102, *110*
Woolhouse, H. W., 23, *63*
Wraight, C., 123, *154*
Wraight, C. A., 123, 143, *154, 156*
Wright, L. J., 210, *229*

Y

Yamaguchi, M., 178, *182*
Yamamoto, Y., 109, *111*
Yamano, T., 171, *182*
Yamashita, T., 88, *96*
Yamazaki, R. K., 31, *63*
Yates, M. G., 166, *183*
Yike, N. J., 162, 177, *181*
Yoch, D. C., 165, 172, 173, 177, *183*
Yocum, C. F., 73, *96*
Yonetani, T., 220, 221, 222, *229*

Z

Zanetti, G., 83, *94*
Zavil'gel'skii, G. B., 235, 238, 239, 240, *255*
Zelitch, I., 31, *63*
Zickler, H. O., 14, *63*
ZoBell, C. E., 159, *183*
Zuidema, Th., 196, *208*

SUBJECT INDEX

A

Absorptance, spectral leaf, 10
Absorption spectrum, photosystems I and II, 99, 100
Alocasia, 12, 17, 18
Alocasia macrorrhiza, 38, 43
Amaranthus retroflexus, 26
Anabaena, 102
Anabaena cylindrica, 18, 170
Anacystis nidulans, 18
Atriplex, 17, 18, 26, 32, 39, 43, 49, 57, 58
Atriplex hastata, 54, 55
Atriplex patula, 27
Atriplex patula ssp. *hasta*, 17, 39, 40, 52, 53
Atriplex patula ssp. *spicata*, 49
Atriplex rosea, 27, 52, 53
Atriplex spongiosa, 26, 54, 55
Azoferredoxin, 170
Azotobacter, 159, 162, 166, 171

B

Bacteriochlorophyll, 11, 15
Benson-Calvin cycle, mutations affecting, 68–70
Blue-green algae, heterocyst, 102

C

C-550, 85, 86, 88, 89, 90, 92
Calvin-Benson pathway, 7, 22–23, 28
Cation flux, chloroplast activity, 113–156
C_4 Dicarboxylic acid pathway, 15, 24–29, 30, 32, 102
  adaptation and evolution, 51–59
Chlamydomonas reinhardi, photosynthetic studies, 66–92, 140
Chlorella, 14, 15
Chlorella pyrenoidosa, 16, 212
Chlorobium, 163
Chlorophyll, 10
  absorbance changes, 129–131
Chlorophyll a, 106, 108, 109
  fluorescence, 99–100, 142
  immunology, 101
Chlorophyll b, 106, 108, 109
Chloroplast,
  cation flux, role of, 113–156
  fractionation procedures, 106–109
  membrane potential, 129–135
  photosystem localization, 103–105
Chloropseudomonas ethylicum, 172, 177–178
Chromatic aberration, 198–200, 201
Chromatium, 163, 171, 173, 177
Chromatophore,
  quenching, 126
  hydrogen accumulation, 132
  coupling factor, 140
Clostridium, 163
Clostridium pasteurianum, 170
Cordyline rubra, 38, 43
Crassulacean acid metabolism, 7, 25, 29–31, 35
  adaptation and evolution, 51–59
Cytochrome-553, 81–84
Cytochrome a_3,
  photosensitivity, 219–222, 224–225
Cytochrome b-559, 86, 87, 89, 90, 91, 92, 93
Cytochrome b-563, 79
Cytochrome f, 81–84
Cytochrome oxidase, 220, 224, 227–228

D

Dactylis glomerata, 44
DNA denaturation, 231–255
Dunaliella parva, 19

E

Electron flow, 138–139
  fluorescence lowering, 143
Electron paramagnetic resonance spectroscopy, 101
Electron transport, 100–101
Energy transduction,
  ATP formation, 135–138
  electrochemical gradient, 124–135
Euglena, 104, 105, 127
Euglena gracilis, 212
Euphorbia, 57
Evoked potential recording, 188–192
Evoked potential, visual perception and, 192–201

F

Fagus sylvatica, 19
Ferredoxin, 72–73, 172–173
Flavodoxin, 172–173
Fluorescence, 99–100
  DNA denatured, 238–240
  enhancement, 145–148
  lowering, 143
Froelichia gracilis, 29

G

Genetics,
  photosynthesis analysis, 65–96
Growth, light inhibition, 209–229

H

Hill reaction, 19, 21, 75
Hydrogen, photoevolution, 159
Hydrogenase, nitrogenase relationship, 159–162
Hydrogen ion, electrochemical gradient, 124–135

I

Ion flux measurement, 114–123

K

Klebsiella pneumoniae, 159, 163, 165, 170

L

Light artifacts, 120
Light intensity, photosynthetic adaptation, 16–18, 38–45

M

Magnesium, 144
  fluorescence enhancement, 145–147
  photosynthesis control, 148–152
Manganese, 92
Membrane potential, 129–135
Mimulus cardinalis, 46
Molybdoferredoxin, 170, 175

N

Nitella, 152
Nitrogen fixation,
  photosynthetic, 157–183
Nostoc, 159

O

Oenothera, 68, 70
Ostreobium reinecky, 11
Oxygen,
  nitrogen fixation, 165–167
  photosensitivity and, 222, 225–226
  photosynthetic inhibition, 31–34

P

P700, 14, 15, 20, 73–79, 100, 106, 147
Panicum maximum, 26
Perilla, 23
Phargmites communis, 47
Photoinhibition, growth and respiration, 209–229
Photorespiration, 31–34, 152
Photophosphorylation, 77, 84–85, 108, 130, 133
  electrochemical gradient, 135–139
Photosynthesis,
  C_4 dicarboxylic acid pathway, 15, 24–29, 30, 32, 102
  crassulacean acid metabolism, 7, 25, 29–31, 35, 51–59
  higher plant comparative, 1–63
  magnesium, control by, 148–152
  mutational studies, 65–96
  quantum yield, 15–18
Photosynthetic adaptation, 37–59
  light climates, 38–45
  temperature, 45–50
Photosynthetic bacteria, nitrogen fixation, 157–183
Photosynthetic efficiency, 37, 39
Photosynthetic electron transport, 18–22, 41, 70–84
Photosynthetic unit size, 12–15
Photosystem I, 6, 14, 18, 20
  reaction center, 73–79
  separation of, 97–112
Photosystem II, 6, 14, 18, 20, 22, 75, 77
  mutations affecting, 85–93
  separation of, 97–112
Phycocyanin, 11
Phycoerythrin, 11
Phytoflavin, 18
Plastocyanin, 79–81
Plastoquinone, 87

Prototheca zopfii, photoinhibition of, 211, 213–222

Q

Quantum yield, photosynthesis, 15–18

R

Respiration, light inhibition, 209–229
Retina, spike generation, 203–206
Rhodopseudomonas capsulata, 140
Rhodopseudomonas spheroides, 15, 131
Rhodospirillum rubrum, 136, 159–161, 165, 170, 175–177

S

Saccharomyce cerevisiae, photoinhibition, 222–226
Scenedesmus obliquus, photosynthetic studies, 66–93
Solanum dulcamara, 44
Solidago, 44
Sorghum, 15
Stomate,
 conductance, 39
 diffusion paths, 34–36

T

Temperature, photosynthetic adaptation, 45–50
Teucrium scorodonia, 19
Thylakoid, 105
 cation flux, 114, 125, 149
 salt requirement, 148
Thymine dimers, 251–253
Tidestromia oblongifolia, 55, 56
Transpiration, 46

U

Ultraviolet irradiation, 89
 DNA denaturation, 231–255

V

Vicia faba, 71, 72
Visual cortex, 206–207
Visual information processing, 185–208

W

Water, photolysis, 91–93

Z

Z scheme, 98, 99